CALIFORNIA NATURAL HISTORY GUIDES

**PESTS OF THE NATIVE
CALIFORNIA CONIFERS**

California Natural History Guides

Phyllis M. Faber and Bruce M. Pavlik, General Editors

PESTS
of the Native California Conifers

David L. Wood
Thomas W. Koerber
Robert F. Scharpf
Andrew J. Storer

UNIVERSITY OF CALIFORNIA PRESS
Berkeley   Los Angeles   London

The section entitled "Damage by Larger Animals" draws on information in Hygnstrom, Timm, and Larson, editors, Prevention and Control of Wildlife Damage (Lincoln: University of Nebraska Cooperative Extension, 2000).

California Natural History Guide Series No. 70

University of California Press
Berkeley and Los Angeles, California

University of California Press, Ltd.
London, England

© 2003 by the Regents of the University of California

Library of Congress Cataloging-in-Publication Data

Pests of the native California conifers / David L. Wood ... [et al.].
    p.   cm. – (California natural history guides; 70)
   Includes bibliographical references and index.
   ISBN 0–520–23327-1 (hardcover : alk. paper) — ISBN 0–520–23329-8 (pbk. : alk. paper)
   1. Conifers—California—Diseases and pests—Identification.   I. Wood, David L. II. Series.

SB608.C7 P46 2003
571.9′25′09794—dc21   2002029141

Manufactured in China

10   09   08   07   06   05   04   03
10 9 8 7 6 5 4 3 2 1

The paper used in this publication is both acid-free and totally chlorine-free (TCF). It meets the minimum requirements of ANSI/NISO Z39.48–1992 (R 1997) (*Permanence of Paper*). ♾

The publisher gratefully acknowledges the generous
contribution to this book provided by

the Moore Family Foundation
Richard and Rhoda Goldman Fund
and
the General Endowment Fund of the
University of California Press Associates

CONTENTS

Acknowledgments	ix
Introduction	xi

Diagnosing Pest-damaged Trees: Seven Basic Steps — 1

Damage by Insects — 3

- To Reproductive Structures — 4
- To Foliage and Shoots — 15
- To Branch Tips and Terminals — 47
- To Stems and Larger Branches — 58

Biotic Diseases — 97

- Needle Diseases — 98
- Canker Diseases — 106
- Mistletoes — 118
- Rust Diseases — 126
- Heart Rots — 137
- Root Diseases — 146

Abiotic Diseases — 157

- Air Pollution — 158
- Drought — 160
- Flooding — 162
- Frost — 163
- Heat — 164
- Herbicides — 165

Salt	166
Winter Burn and Winter Drying	169

Damage by Larger Animals — 171

American Black Bear (*Ursus americanus*)	172
Deer and Elk (*Odocoileus* and *Cervus* spp.)	173
Jackrabbits and Hares (*Lepus* spp.)	175
Livestock	177
Mountain Beaver (*Aplodontia rufa*)	178
Pocket Gophers (*Thomomys* spp.)	180
Porcupine (*Erethizon dorsatum*)	182
Woodrats (*Neotoma* spp.)	183

Guide to Damage by Symptom Location — 186

On Reproductive Structures	186
On Smaller Branches or Treetops	186
On Stems, Larger Branches, or Both	188
Overall Tree Decline	189

Guide to Damage by Host Species — 192

Douglas-fir	192
Grand Fir	193
Red Fir	194
White Fir	194
Jeffrey Pine	195
Lodgepole Pine	196
Monterey Pine	198
Ponderosa Pine	199
Sugar Pine	200
Other Pines	201
Incense-cedar	202
Port Orford–cedar	203

Glossary — 205
General References and Resources — 215
Illustration Credits — 217
Index of Agents — 219
General Index — 223

ACKNOWLEDGMENTS

The authors wish to express their appreciation for the considerable support of this project from many colleagues. We thank David H. Adams, Jack P. Marshall, Donald R. Owen, and Steven M. Jones of the California Department of Forestry and Fire Protection; Gregory A. Giusti and David M. Rizzo of the University of California, Davis; John T. Kliejunas, Nancy G. Rappaport, and David E. Schultz of the USDA Forest Service; and Robert C. Heald and Reginald H. Barrett of the University of California, Berkeley, for their helpful reviews of the manuscript.

We also thank everyone who contributed photographic slides to this work. Additional slides were provided by the authors. Some of these were from collections such as the University of California Forest Entomology Collection, the University of California Cooperative Extension Collection, and the Pacific Southwest Station, USDA Forest Service Collection.

We thank the USDA Forest Service, the California Department of Forestry and Fire Protection, and the University of California, Berkeley, for financial support, and Donald C. Perkins and Brian R. Barrett of the California Department of Forestry and Fire Protection and John A. Neisess of the USDA Forest Service for providing the initial impetus for this project.

INTRODUCTION

This book is intended to be a guide to the identification of damaging agents or "pests" commonly encountered in the coniferous forests of California. Therefore, we have focused our attention on those native conifers most likely to be encountered by resource managers and the public. We have not emphasized many of the tree species in the high-elevation forests of the Sierra Nevada or in the Coast Ranges, although many pests of lower and mid-elevation Sierra Nevada forests are found damaging these species. We define a pest as an organism that has an adverse effect on some resource management objective. Insects, pathogens, and vertebrate animals may reduce the quantity or quality of a wood product; they may degrade wildlife habitat; they may increase fire hazard or create other safety hazards to humans or structures in the forest; and they may impact the aesthetic values of the resources. We hesitate to characterize humans as forest pests, although they frequently damage or degrade forest resources. Air pollution and salt injury are examples of such human-caused damage to forests in California. We have included these agents under "Abiotic Diseases."

Many thousands of organisms live in the forest. Most can be considered beneficial organisms, as they play a role in recycling nutrients from the longest lived organisms in the earth's ecosystems. Only a very small fraction of them ever come to our attention as pests. Even those that are considered pests are held in check most of the time by natural regulating agents, including parasites, predators, and disease-causing microorganisms. We have included in this book only organisms with the potential to cause significant or very obvious damage. Hence, users of this guide may encounter minor pest problems that have not been described here.

Minor native pests not described in this book have the potential to increase in significance with time. This may occur because of changes in climate, management practices, economics, or other factors. In addition, new pests may be introduced into California. In both situations it is difficult to predict pest impacts on coniferous forest ecosystems. In 2001, the pathogen *Phytophthora ramorum,* a previously undescribed species, was identified as the cause of a canker disease of oaks referred to as "sudden oak death." This pathogen has killed large numbers of coast live oaks and tanoaks and has caused a wide range of leaf, shoot, and stem symptoms on a large number of other nonconiferous trees and shrubs. Recently, coast redwood, *Sequoia sempervirens,* and Douglas-fir, *Pseudotsuga menziesii,* were confirmed as hosts for this new pathogen on the central coast of California. Symptoms of infection of coast redwood included discolored needles and cankers in small branches throughout the crowns of infected understory trees. To date, the importance of *P. ramorum* in causing branch dieback and death of mature redwoods is not known. The pathogen is apparently able to kill sprouts of redwood and may cause loss of vigor in saplings. In Douglas-fir, symptoms on saplings include cankers on small branches that result in the wilting of new shoots, dieback of branches, and loss of needles from branch tips. Symptoms on overstory Douglas-fir have not been observed, and the importance of *P. ramorum* infection in Douglas-fir is also unknown. As more is learned about *P. ramorum* biology and its impacts on coniferous forests, we will be in a better position to predict its significance in these forests. Up-to-date information about *P. ramorum* and the diseases it causes can be found at www.suddenoakdeath.org.

Information about seedling pests and other pests not included in this guide can be found in the general references. The target audience for this guide includes foresters, arborists, and all land and resource managers who have forested areas in their purview. This guide will also be useful to the general public; hikers, campers, residents, and others can use this guide to make a preliminary diagnosis of the cause of damage, whether dramatic or subtle. Also, this guide should be useful in college-level courses that consider California's coniferous forest ecosystems.

For each damaging agent we describe typical symptoms and illustrate important diagnostic features with color photographs. We briefly describe the life cycles and habits of these organisms, together with site conditions and geographic locations in which damage is likely to occur.

At the end of the discussion of each pest we list references that provide (1) further information about the life history and management of that pest, and (2) references to the recent scientific literature for that pest. In some cases, these references are many years old, indicating that little new information about a particular agent has emerged. A short list of general references and a glossary of pest-related terms are also presented. An index of agents and listings of agents by tree species and parts of trees affected are provided.

Some information is provided about treatment and management strategies for reducing pest-caused losses; however, the technology and policies governing pest management practices are continually changing and are often complex. Therefore, we do not attempt to provide specific recommendations concerning, for example, the application of available pesticides. A short list of sources of up-to-date pest management advice is provided. One or more of these sources should be consulted in connection with any pest management activity.

David L. Wood
Thomas W. Koerber
Robert F. Scharpf
Andrew J. Storer
October 2002

DIAGNOSING PEST-DAMAGED TREES: SEVEN BASIC STEPS

Diagnosis of tree health problems involves identifying the cause from the symptoms (expressions of damage to the tree), signs (evidence of the cause), and patterns of occurrence. Each of these provides clues that can be useful in making a diagnosis. The following sequence of investigative steps may assist in diagnosing a tree health problem.

1. *Locate the damage.* Determine the part of the tree that is actually affected. Note, for example, whether the impact is on one-year-old needles, scattered branches, or the entire tree crown; is the damage limited to the lower or upper part of the crown, or one side of the crown?
2. *Identify the species.* Note what species are affected and whether some individuals are less affected than others. Check the condition of adjacent trees.
3. *Observe the pattern of occurrence.* Is the problem more severe in some areas than others, and are there differences between these areas? Are the problems limited to a particular environmental zone or related to a particular cultural activity? For example, some diseases are more prevalent along stream courses or lake shores.
4. *Look for obvious causes.* These include damage by large or small animals, frost, lightning, injuries, fire, and so on.
5. *Look for the presence of fungi, insects, or parasites, such as mistletoes.* Try to determine whether the organisms found are the main cause of the problem or just secondary. For example, insects will frequently infest trees that are weakened by disease. Trees can be affected by insects and pathogens over a long period of time before they succumb to one of these biological agents.

6. *Examine the roots.* If the whole tree is dead or exhibits symptoms (e.g., yellowing of needles), and nothing is found above the ground to indicate a cause, expose the roots and root crown for examination. Also, a small patch of bark can be removed to observe the condition of the phloem and sapwood.
7. *Determine when the problem was first noticed in the area.* Inquire about cultural practices in the area, such as the use of herbicides, fertilizers, irrigation, road salting, and so on. Also determine whether there have been recent, unusually severe weather conditions.

With some or all of the above information, it should be possible to make a preliminary determination of the cause of the health problem of a tree. If this is not possible, there are a number of people who can be contacted for help. These experts include the local County Agriculture Commissioner's office, the University of California Cooperative Extension, the California Department of Forestry and Fire Protection, and the USDA Forest Service's Forest Pest Management office.

Be prepared to list your observations, and try to provide photographs and samples for observation and analysis.

DAMAGE BY INSECTS

Damage by Insects

WESTERN CONIFER SEED BUG *Leptoglossus occidentalis*
Pls. 1–3

HOSTS: Common on Douglas-fir and on knobcone pine, Monterey pine, ponderosa pine, sugar pine, and western white pine. Seeds of other species of conifers may be fed upon in late summer and fall.

DISTRIBUTION: Throughout California, including non-forested regions.

SYMPTOMS AND SIGNS: Seeds are damaged or destroyed by the feeding activity of this insect. The character of the damage changes as the season progresses. When feeding occurs before the seed coats harden, the insects completely remove the seed contents and the seed coats collapse. Flattened brown seed coats adhere tightly to the cone scales and remain in the cones at maturity. Later in the season, after the seed coats harden, all or part of the contents may be removed. These seeds retain the size and shape of normal seeds and fall out of the cones at maturity. However, these seeds are empty or partially consumed, or contain a shrunken, spongy endosperm (pl. 1)(pl. 2).

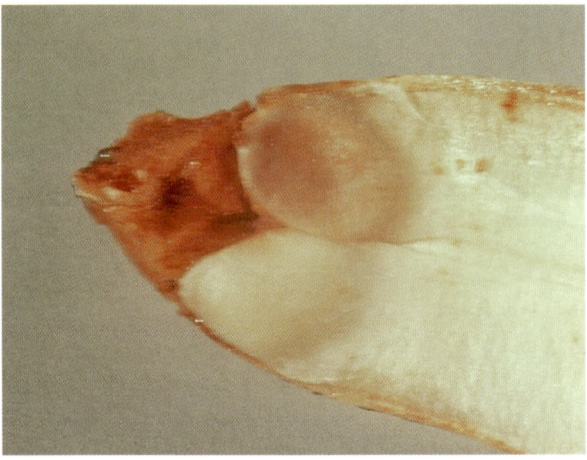

Plate 1. Immature ponderosa pine seeds. The collapsed upper seed indicates damage by *Leptoglossus occidentalis,* the Western Conifer Seed Bug; the lower seed is normal.

Plate 2. Douglas-fir seed damaged by *Leptoglossus occidentalis*.

LIFE CYCLE: Adults leave their hibernation sites about mid-May and feed on new shoots and the staminate cones of pines. When the seeds begin to develop, the insects feed on seeds within the growing cones. The adults are about 1.3 to 1.9 cm (.5 to .75 in.) long and .6 cm (.25 in.) wide, with long legs and antennae. The body is reddish brown to dark brown, with alternating yellow and dark brown spots along the sides of the abdomen. The wings are held flat against the dorsum. There is a distinctive white, zigzagging line across the middle of the wings. A thin, tubular beak, over half as long as the insect, is attached to the front of the head and held along the underside between the bases of the legs. The eggs are about 3 mm (about .12 in.) long, light brown, and shaped like half-cylinders. They are firmly glued end to end in a row of four to 12 on a needle of the host tree from late May to July. The eggs hatch in about 10 days. The nymphs have a brown head and thorax and a bulbous, orange abdomen. Both nymphs and adults feed by inserting the long, tubular mouthparts through the cone scales into the seeds developing within the cones. Enzymes produced by the salivary glands are injected into the seeds to dissolve the contents, which are then sucked out. The nymphs transform into winged adults in August and September. In September and October, the new adults disperse from the host trees and seek sheltered locations to spend the winter. At this time they often enter buildings, and are sometimes found in the Sacramento Valley far from the forests in which they developed (pl. 3).

SIGNIFICANCE: This insect can destroy substantial quantities of seed. Losses are difficult to estimate because there are other causes of empty seed. Reductions in seed yield as high as 40 percent in Douglas-fir and 25 percent in western white pine have been attributed to this insect.

Damage by Insects

Plate 3. *Leptoglossus occidentalis*, the Western Conifer Seed Bug: Wide, flat tibia of hind leg *(arrow)* is distinctive.

SIMILAR PESTS: *Leptoglossus zonatus* feeds on juniper berries. The body of this species is narrower than that of *L. occidentalis* and it has a faint yellow band rather than a white, zigzagging line across the wings. Several other *Leptoglossus* species are pests of fruit and nut crops in agricultural regions.

MANAGEMENT OPTIONS: Both contact and systemic insecticides have been used to protect pine seed crops in the Southeast. In seed orchards removal or burning of brush piles, lumber stacks, and so on, where the adults find shelter for the winter, may reduce survival. Hand-pollinated cones may be grown inside nylon or plastic mesh bags to exclude the insects.

For More Information

Blatt, S. E., and J. H. Borden. 1999. Physical characteristics as potential host selection cues for *Leptoglossus occidentalis* (Heteroptera: Coreidae). *Environmental Entomology* 28:246–54.

Blatt, S. E., J. H. Borden, H. D. Pierce, R. Gries, and G. Gries. 1998. Alarm pheromone system of the Western Conifer Seed Bug, *Leptoglossus occidentalis*. *Journal of Chemical Ecology* 24:1013–31.

Koerber, T. W. 1963. *Leptoglossus occidentalis* (Hemiptera: Coreidae): A newly discovered pest of coniferous seed. *Annals of the Entomological Society of America* 56:229–34.

Schowalter, T. D., and J. M. Sexton. 1990. Effect of *Leptoglossus occidentalis* (Heteroptera: Coreidae) on seed development of Douglas-fir at different times during the growing season in western Oregon. *Journal of Economic Entomology* 83:1485–86.

Damage by Insects

DOUGLAS-FIR CONE MOTH *Barbara colfaxiana*
Pls. 4, 5

HOST: Douglas-fir.

DISTRIBUTION: Sierra Nevada and Coast Ranges. It is more common in the Sierras and inner Coast Ranges than on the coast.

SYMPTOMS AND SIGNS: Infested cones have irregular tunnels packed with pitch-soaked, brown frass pellets. Heavily damaged cones may be partly killed and distorted into a curled shape. In late summer and through the winter, one or more pitch-covered cocoons may be present in the center of the cone (pl. 4)(pl. 5).

LIFE CYCLE: Emergence of the adult moths from old cones is synchronized with the opening of the cone buds. Adult moths have a wing span of 1.3 to 1.9 cm (.5 to .75 in.). The fore wings have gray, brown, and silver irregular bands. The hind wings are gray. Green, flattened, scalelike eggs are laid singly on the bracts of new cones. The eggs hatch in 10 days to two weeks and the new

Above: Plate 4. Entry point of *Barbara colfaxiana*, the Douglas-fir Cone Moth, in a young Douglas-fir cone. *(arrow)* Frass is produced by tunneling larva.

Left: Plate 5. *Barbara colfaxiana* feeding damage in a Douglas-fir cone.

larvae immediately tunnel into the cone tissues. The entry point is marked by a drop of pitch and a few light-brown frass pellets. The larvae feed on scale tissues and seeds. Young larvae are white with a black head. Mature larvae are 1.3 cm (about .5 in.) long with a white body and brown head. The larvae reach maturity in late June or early July. They construct a tough cocoon of brownish silk coated with pitch near the central axis of the cone. The insect remains in the pupal stage from July until the next spring. A variable portion of the population delays transformation to the adult stage until the second or third spring after completing larval development.

SIGNIFICANCE: The Douglas-fir Cone Moth is one of the most common and destructive pests of Douglas-fir seed crops. A single larva will destroy about two-thirds of the seeds in a cone. Heavy infestations regularly reduce Douglas-fir seed yields in seed orchards and reduce natural seed fall on cut-over or burned areas.

SIMILAR PESTS: Other *Barbara* species with identical life cycles and habits are found in the cones of true firs. *B. ulteriorana*, which closely resembles *B. colfaxiana*, is also found in Douglas-fir cones. At least two species of *Dioryctria* feed in Douglas-fir cones. The larvae of these moths are brown or reddish and produce dry frass pellets held together with webbing. The *Dioryctria* species feeding period is mid to late summer.

MANAGEMENT OPTIONS: The extent of damage to the seed crop varies greatly from year to year. Abundant cone crops typically suffer little damage, whereas nearly all the seed in a light cone crop will be destroyed. The feeding of this insect is completed by early July, leaving two months to evaluate the cone crop and, if necessary and possible, to secure seed supplies from lightly damaged cone crops to be stored for future use. Systemic insecticide sprays have been used to protect cone crops in seed orchards. Systemic insecticide implants may be used to protect cone crops on selected trees.

For More Information

Hedlin, A. F. 1960. On the life history of the Douglas-fir Cone Moth, *Barbara colfaxiana* (Kft.) (Lepidoptera: Olethreutidae) and one of its parasites, *Glypta evetriae* Cush. (Hymenoptera: Ichneumonidae). *The Canadian Entomologist* 92:826–34.

Sahota, T. S., A. Ibaraki, and S. H. Farris. 1985. Pharate adult diapause of *Barbara colfaxiana* (Kft.): Differentiation of 1- and 2-year dormancy. *The Canadian Entomologist* 117:873–76.

Damage by Insects

PINE SEEDWORM   *Cydia miscitata*
Pl. 6

HOST: Ponderosa pine.

DISTRIBUTION: Throughout the California range of ponderosa pine.

SYMPTOMS AND SIGNS: Seeds of ponderosa pine are mined out and packed with fine, granular, brown frass. The damaged seeds adhere to the cone scales and remain in the cone at maturity. White larvae are present in the seeds during the summer and, later, in the pith core of mature cones. Infested cones show no external damage symptoms (pl. 6).

LIFE CYCLE: The insects overwinter in the larval stage, occupying a tunnel in the pith core of a cone. Pupation occurs in early spring and is followed by transformation to the adult stage when the second-year cones begin elongation. The adult moths are dark gray with silver cross bands on the front wings. The females deposit reddish brown eggs among the scales on a cone stem. The eggs hatch in about two weeks and the new larvae immediately tunnel into the cone between the scales and mine inward to reach the seeds. Larvae feed on seed contents throughout the summer, leaving the seeds filled with frass. On reaching maturity in late summer, the larvae burrow from the last seed destroyed into the core of the cone to spend the winter.

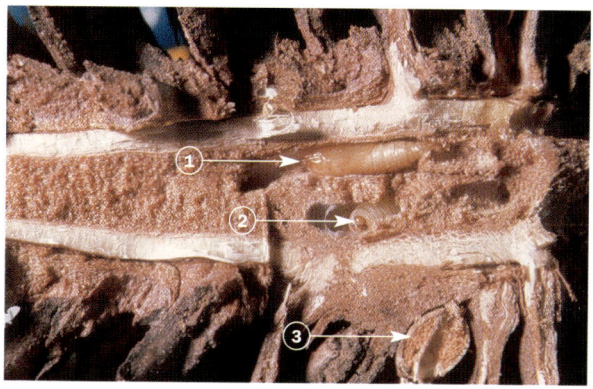

Plate 6. *Cydia miscitata*, the Pine Seedworm, in a ponderosa pine cone: (1) Pupa, (2) larva, and (3) damaged seed.

SIGNIFICANCE: The larvae of this species consume an average of five seeds each. Population levels are usually in the range of one to five larvae per cone and result in reductions of 20 percent or less of the potential seed yield. On rare occasions, seed losses have reached nearly 50 percent.

SIMILAR PESTS: Two very similar species infest pine cones in California. *C. piperana* feeds on seeds of ponderosa pine and Jeffrey pine. The life cycle and habits are identical, except that the eggs are placed on the cone scales rather than on the stem. The mature larvae and adults are slightly larger than those of *C. miscitata*. *C. injectiva* feeds on the seeds of Jeffrey pine and spends the winter in the pupal stage rather than the larval stage.

MANAGEMENT OPTIONS: Insecticide applications have been used to control similar species in the southeastern United States. Collecting and burning cones containing overwintering larvae might reduce infestation in the following year in isolated seed orchards.

For More Information

Hedlin, A.F. 1967. The Pine Seedworm, *Laspeyresia piperana* (Lepidoptera: Olethreutidae), in cones of ponderosa pine. *The Canadian Entomologist* 99:264–67.

CONE BEETLES *Conophthorus* spp.

Pls. 7–10

HOSTS: Most pines in California.

DISTRIBUTION: Throughout the forested regions of California.

SYMPTOMS AND SIGNS: Immature cones die and turn reddish brown. Depending on the time of infestation, cone development may be aborted, resulting in a small cone that does not develop further. Cones killed by these beetles never open, unlike cones killed by many other insects, and the cones of sugar pine and western white pine drop to the ground throughout the growing season. A pitch tube marks the entrance to a tunnel in the stem or base of an infested cone (pl. 7)(pl. 8).

LIFE CYCLE: Adult beetles leave overwintering sites in cones or twigs in late spring. Adults are cylindrical black beetles about 3 mm (.12 in.) long. Female beetles tunnel into the base or stem of a cone and excavate a flat, spiral tunnel, severing the water-conductive tissues of the cone. A pitch tube forms around the entry point. The female then excavates a straight gallery along the

Damage by Insects

Above: Plate 7. Ponderosa pine cone whorl. The reddish brown cone to the right has been killed by cone beetles (a *Conophthorus* spp.); the other cones are normal.

Left: Plate 8. Cone beetle entry tunnel and pitch tube in the stem of a western white pine cone.

cone axis, from the base toward the tip of the cone, and deposits ovoid white eggs along the sides of the gallery. On completing the egg gallery in the first cone, the female beetle moves to another cone. The eggs hatch in about a week to produce white, C-shaped legless larvae. The larvae feed on seeds and scales, reaching maturity in four to six weeks. Mature larvae transform immediately to the pupal stage, which lasts 10 to 14 days. Cones of sugar pine and western white pine drop to the ground about four weeks after a beetle attack, so that the larval and the pupal stages are completed mostly in dead cones lying on the ground. The new adult beetles remain in the dead cones for the rest of the summer. Some of the adults leave the cones in fall and tunnel into twigs of the host tree, making galleries up to 7.5 cm (3 in.) long where they spend the

Damage by Insects

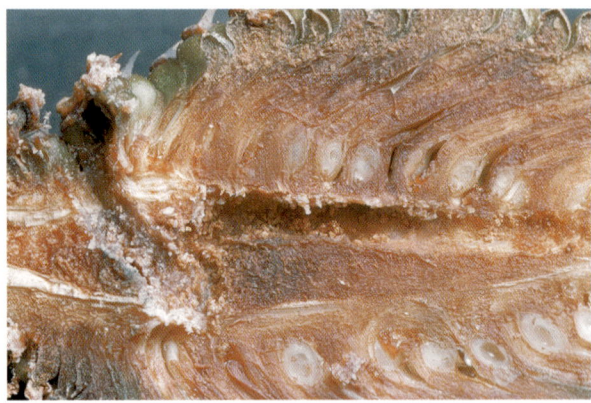

Plate 9. Cone beetle entrance tunnel and egg gallery in a ponderosa pine cone.

Plate 10. Cone beetle larvae and pupae in a western white pine cone.

winter. Adult beetles in ponderosa pine cones remain in the cones through the winter. Adults of the Monterey Pine Cone Beetle, *C. radiatae*, emerge from some Monterey pine cones in early fall and attack nearby developing first-year cones. In the absence of a cone crop beetles may survive the summer by mining in the twigs of host trees (pl. 9)(pl. 10).

SIGNIFICANCE: Cone beetles are the most serious pest of cone crops of sugar pines and western white pines in California. A loss of 75 percent, or more, of the cone crop of these host species has been frequently observed. The Monterey Pine Cone Beetle also acts as vector for the pitch canker fungus, *Fusarium circinatum*, carrying it to first-year conelets and second-year cones of Monterey pine.

SIMILAR PESTS: The larvae of deathwatch beetles (Anobiidae: *Ernobius* spp.), which typically feed on dead wood, are frequently found feeding in pine cones killed by cone beetles and other insect species. They are distinguished from *Conophthorus* spp. larvae by having three pairs of legs on the thoracic segments.

MANAGEMENT OPTIONS: Cone crops in seed orchards can be protected by insecticidal sprays, but it can take up to three applications to control cone beetles. Clearing the larger vegetation from under and around sugar pines and western white pines exposes fallen cones to direct sunlight, resulting in internal temperatures that kill most of the developing larvae. Burning the fallen cones in which beetles overwinter may reduce infestation of the next cone crop. Hand-pollinated cones in tree-breeding programs have been grown inside nylon or plastic mesh bags to exclude attacking beetles.

For More Information

Bedard, W. D. 1966. High temperature mortality of the Sugar Pine Cone Beetle, *Conophthorus lambertianae* Hopkins (Coleoptera: Scolytidae). *The Canadian Entomologist* 98:52–57.

Furniss, M. M. 1997. *Conophthorus ponderosae* (Coleoptera: Scolytidae) infesting lodgepole pine cones in Idaho. *Environmental Entomology* 26:855–58.

Hoover, K., D. L. Wood, A. J. Storer, J. W. Fox, and W. E. Bros. 1996. Transmission of the pitch canker fungus, *Fusarium subglutinans* f. sp. *pini* to Monterey pine, *Pinus radiata*, by cone- and twig-infesting beetles. *The Canadian Entomologist* 128:981–94.

Rappaport, N. G., M. Haverty, P. J. Shea, and R. E. Sandquist. 1994. Efficacy of esfenvalerate for control of insects harmful to seed production in disease-resistant western white pines. *The Canadian Entomologist* 126:1–5.

DOUGLAS-FIR SEED CHALCID *Megastigmus*
Pls. 11, 12 *spermotrophus*

HOSTS: Bigcone Douglas-fir and Douglas-fir.

DISTRIBUTION: Throughout the range of Douglas-fir.

SYMPTOMS AND SIGNS: Each infested Douglas-fir seed contains one white, legless larva. Adults make circular emergence holes upon leaving infested seeds (pl. 11)(pl. 12).

Damage by Insects

Right: Plate 11. Larva of *Megastigmus spermotrophus*, the Douglas-fir Seed Chalcid.

Below: Plate 12. *Megastigmus spermotrophus* emergence holes in Douglas-fir seeds: Note that there is one hole per seed.

LIFE CYCLE: Seed chalcid larvae spend the winter inside seeds. Transformation to the pupal stage occurs in early spring, followed by emergence of the adults after the new Douglas-fir cones turn downward into the pendant position on the twigs. The adults are yellow, ant-sized wasps with clear wings. The female chalcid has an ovipositor, exceeding the length of her body, that is used to insert eggs through the cone scales into developing seeds. The eggs produce white, legless larvae that consume the contents of the developing seeds. Each larva completes its development within a single seed, and there is only one larva in each infested seed. Some larvae live two to three years in their seeds before emerging.

SIGNIFICANCE: Reduction of Douglas-fir seed yield due to seed chalcid infestation is usually less than 15 percent. Several other insects commonly found in Douglas-fir cones often cause much more damage. It is very difficult to distinguish infested from normal seed because there is no external sign of larval infestation. Furthermore, normal seed-cleaning operations do not completely separate infested from uninfested seed. As a result, this pest has been widely distributed in seed shipments and is established in most areas where Douglas-fir has been planted.

SIMILAR PESTS: Other species of seed chalcids with habits and life cycles very similar to those of *M. spermotrophus* infest the seed crops of true firs, western hemlock, and ponderosa pine.

MANAGEMENT OPTIONS: Systemic insecticide sprays have been used to protect Douglas-fir seed crops in seed orchards. Insecticide implants may be used to reduce seed chalcid damage on selected individual trees.

For More Information

Blatt, S. E., and J. H. Borden. 1998. Interactions between the Douglas-fir Seed Chalcid, *Megastigmus spermotrophus* (Hymenoptera: Torymidae), and the Western Conifer Seed Bug, *Leptoglossus occidentalis* (Hemiptera: Coreidae). *The Canadian Entomologist* 130:775–82.

Niwa, C., and D. L. Overhulser. 1992. Oviposition and development of *Megastigmus spermotrophus* (Hymenoptera: Torymidae) in unfertilized Douglas-fir seed. *Journal of Economic Entomology* 85:2323–28.

Schowalter, T. D., M. I. Haverty, and T. W. Koerber. 1985. Cone and seed insects in Douglas-fir, *Pseudotsuga menziesii* (Mirb.) Franco, seed orchards in the western United States: Distribution and relative impact. *The Canadian Entomologist* 117:1223–30.

LODGEPOLE NEEDLEMINER *Coleotechnites milleri*
Pls. 13–16

HOST: Lodgepole pine.

DISTRIBUTION: Sierra Nevada lodgepole pine forests above 2,400 m (about 7,900 ft). Outbreaks have been limited to Yosemite National Park and Kings Canyon National Park.

SYMPTOMS AND SIGNS: Needles of lodgepole pine are eaten from the inside, leaving only the epidermis. Damaged needles are light yellow at first, later becoming rusty brown. Heavily defoliated lodgepole stands appear rusty brown from a distance (pl. 13) (pl. 14)(pl. 15)(pl. 16).

LIFE CYCLE: The Lodgepole Needleminer requires two years to complete development. Female moths lay eggs in groups of two to

Damage by Insects

Above: Plate 13. *Coleotechnites milleri*, the Lodgepole Needleminer: (1) Exit hole; (2) larva in lodgepole pine foliage.

Right: Plate 14. Mine produced by *Coleotechnites milleri*, first instar.

Below: Plate 15. Larva of *Coleotechnites milleri* inside a lodgepole pine needle.

Damage by Insects

Plate 16. Lodgepole pines defoliated by *Coleotechnites milleri*.

20 on lodgepole pine foliage in July and early August of odd-numbered years. The space inside previously mined needles and the base of needle pairs are favored locations for oviposition. After hatching in mid to late August the larvae bore into the curved side of a needle and mine toward the tip. The mine appears as an irregular yellow line on the needle surface. The larvae spend winter inside the needle. Feeding resumes in spring, with the larvae hollowing out the needle and tunneling toward the base. In midsummer the larvae leave the first needle and bore into another, entering the needle near the base on the flat side. Two or more needles are eaten during the remainder of the summer. The larvae again remain in a mined needle through the winter and resume feeding in spring. Each larva eats one or two additional needles before reaching maturity in late May or early June. The larvae are about 8 mm (.3 in.) long and yellow to reddish orange at maturity. Before pupation each larva cuts an exit hole near the tip on the flat side of the needle. The jet-black pupae are found inside the mined needles. The adults emerge in July to early August.

SIGNIFICANCE: The Lodgepole Needleminer has repeatedly defoliated many thousands of acres of lodgepole pine forest, mainly in the northern half of Yosemite National Park. Healthy lodgepole pines will tolerate three complete defoliations; because the insect has a two-year life cycle, this requires six years. Tree mortality may result from defoliation alone, but more often results from Mountain Pine Beetle, *Dendroctonus ponderosae,* attack on the weakened trees. In the dry, cold climate of the High Sierra Nevada dead trees decay slowly. Thousands of hectares (acres) of standing dead trees, termed "ghost forests," remain from outbreaks as far back as 1960. Large numbers of dead trees in a heavily used recreation area pose a serious hazard to facilities and users. Removal of dead trees in heavily used areas and clearing fallen trees from trails represent a continuing expense in areas affected by Lodgepole Needleminer outbreaks.

SIMILAR PESTS: A needleminer with identical habits and life cycle, but that flies in even-numbered years, is found around Sentinel Meadow in the Inyo National Forest, east of Yosemite. A *Coleotechnites* species with a one-year life cycle has caused extensive defoliation of Jeffrey pine in the San Bernardino Mountains. Another *Coleotechnites* species with a one-year life cycle has defoliated lodgepole pine in southern Oregon and probably occurs in northeastern California.

MANAGEMENT OPTIONS: Lodgepole pines will survive years of intense defoliation by needleminers; this, combined with the fact that needleminer outbreaks are usually terminated by unfavorable weather conditions, suggests caution in the use of pesticides against this insect. Insecticidal sprays directed against the adults during the flight period or against the maturing larvae are known to temporarily reduce population densities. A systemic insecticide implant will control needleminer infestations of individual trees.

For More Information

Koerber, T. W., and G. R. Struble. 1971. *Lodgepole Needleminer.* Forest Pest Leaflet 22. Washington, D.C.: USDA Forest Service.

Mason, R. R., and H. G. Paul. 1999. Long-term dynamics of Lodgepole Needle Miner populations in central Oregon. *Forest Science* 45 (1):15–25.

Struble, G. R. 1972. *Biology, ecology and control of the Lodgepole Needleminer.* Technical Bulletin 1458. Washington, D.C.: USDA Forest Service.

Damage by Insects

PINE NEEDLE SHEATHMINER *Zelleria haimbachi*
Pls. 17–20

HOSTS: Most commonly recorded hosts: Jeffrey pine, lodgepole pine, Monterey pine, and ponderosa pine.

DISTRIBUTION: Throughout the pine forests of California.

SYMPTOMS AND SIGNS: Partly elongated needles of pines die and drop off. Thin, tubular silk webs may be seen between the needle bases. An irregular pattern of missing needles on otherwise normal shoots remains as long as the affected shoots retain some needles (pl. 17)(pl. 18)(pl. 19)(pl. 20).

LIFE CYCLE: Adult moths lay eggs singly on foliage during midsummers. Larvae bore directly from the eggs into the underlying needle. The larvae feed within the needles for the rest of summer and remain in the mined needles through winter. Each mine appears as a thin yellow line along the edge of the infested needle. The following spring, when the new needles start elongating, each larvae exits the mined needle and feeds on the base of a new needle by boring through the needle sheath and eating the part

Plate 17. *Zelleria haimbachi*, the Pine Needle Sheathminer: Webbing and pupae on a ponderosa pine shoot.

Plate 18. Ponderosa pine shoot showing needles killed by *Zelleria haimbachi*.

Damage by Insects

Plate 19. Surviving needles remain after those killed by *Zelleria haimbachi* drop off.

Plate 20. Ponderosa pine defoliated by *Zelleria haimbachi*.

of the needle within the sheath. Each larva feeds on six to 10 needle fascicles during the next three to four weeks. The larvae are tan with two lengthwise, dull orange stripes. Each larva constructs a thin, tubular web of white silk between the needle bases. At maturity, the larvae transform to the pupal stage within the silk webbing. Further development to the adult stage proceeds immediately, requiring about 10 days.

SIGNIFICANCE: Severe defoliation of pine plantations has been recorded on several occasions, possibly resulting in reduced growth rates. Tree mortality following defoliation by this insect has not been observed. Numerous parasitic and predatory insects have been recorded to feed on the larvae and pupae, usually terminating outbreaks within two to three years.

SIMILAR PESTS: Numerous other species of small caterpillars feed on pine foliage. However, the webbing between the needle bases and feeding restricted to the needle base within the needle sheath are distinctive characteristics of the Pine Needle Sheathminer.

MANAGEMENT OPTIONS: Damage by this insect has usually not been severe enough to justify management intervention. Control efforts have been restricted to highly valued trees in arboreta, seed orchards, and Christmas tree plantations. Insecticidal sprays

Damage by Insects

applied at the start of the needle sheath feeding phase of the life cycle and systemic insecticide implants inserted during the needle mining phase of the life cycle have prevented defoliation.

For More Information
Stevens, R. E. 1959. *Biology and control of the Pine Needle Sheathminer, Zelleria haimbachi Busck (Lepidoptera: Yponomeutidae)*. Pacific Southwest Forest and Range Experiment Station Technical Paper 30. Berkeley: USDA Forest Service.

———. 1971. *Pine Needle Sheathminer*. Forest Pest Leaflet 65. Washington, D.C.: USDA Forest Service.

GIANT CONIFER APHID   *Cinara ponderosae*
Pl. 21

HOST: Ponderosa pine.
DISTRIBUTION: Throughout the range of ponderosa pine.
SYMPTOMS AND SIGNS: Dense colonies of tan and yellow insects appear on new shoots of pine trees. The adult aphids are about 2.5 mm (.1 in.) long and may be winged or wingless. The insects secrete a sweet liquid called honeydew that may support the growth of a black sooty mold. The colonies are often tended by ants that collect the honeydew and protect the aphids from insect predators. Heavy infestations may cause yellowing of the foliage and reduction of shoot growth (pl. 21).

LIFE CYCLE: The winter is spent in the egg stage on needles and bark of the host tree. The eggs hatch in spring, producing female aphids that give birth asexually to living young. These offspring are also all females that, in turn, give birth asexually to living

Plate 21. *Cinara ponderosae*, Giant Conifer Aphids, on ponderosa pine.

females. Thus many generations, and potentially large numbers of aphids, can be produced rapidly. As the season progresses winged females are produced that disperse to other trees. When the new shoots begin to harden a final generation of sexual males and females is produced. After mating, the females deposit eggs on the bark and foliage of the trees. The eggs remain dormant until the following spring.

SIGNIFICANCE: Aphid infestations have been reported to cause growth reductions of young seedlings. However, the population densities usually seen have negligible impact on trees. Parasites, predators, and adverse weather conditions usually prevent the development of damaging populations.

SIMILAR PESTS: *C. curvipes*, a black aphid, feeds on true firs, and *C. occidentalis* feeds on white fir. *Essigella californica*, a slender, light green species, feeds on the foliage of Monterey pine and ponderosa pine.

MANAGEMENT OPTIONS: Normally, natural control agents reduce aphid populations before they cause significant damage. Contact insecticides are effective in controlling aphids on ornamental or other highly valued trees.

For More Information

Brown, L. R., and C. O. Eads. 1967. *Insects affecting ornamental conifers in southern California*. Bulletin 834. Berkeley: University of California Agricultural Experiment Station.

Johnson, N. E. 1965. Reduced growth associated with infestations of Douglas-fir seedlings by *Cinara* species (Homoptera: Aphididae). *The Canadian Entomologist* 97:113–19.

Tilles, D. A., and D. L. Wood. 1982. The influence of Carpenter Ant (*Camponotus modoc*) (Hymenoptera: Formicidae) attendance on the development and survival of aphids (*Cinara* spp.) (Homoptera: Aphididae) in a giant sequoia forest. *The Canadian Entomologist* 114:1133–42.

BALSAM TWIG APHID *Mindarus abietinus*
Pls. 22, 23

HOST: White fir.

DISTRIBUTION: Not known at this time.

SYMPTOMS AND SIGNS: New shoots of trees are infested by greenish yellow aphids covered with a powdery white wax. New needles and twigs are twisted and stunted. Needles drop from heavily infested shoots in late summer (pl. 22)(pl. 23).

Damage by Insects

Plate 22. Mature white fir shoots damaged by *Mindarus abietinus*, the Balsam Twig Aphid.

Plate 23. New white fir twig showing *Mindarus abietinus* damage.

LIFE CYCLE: This aphid spends the winter in the egg stage on the bark and foliage of host trees. The eggs hatch in spring, producing female aphids that reproduce asexually. Successive generations of female aphids in both wingless and winged forms are produced through the summer. The winged forms disperse to new host trees and continue reproducing asexually. In late summer sexual male and female aphids are produced and mate. The females deposit eggs that remain dormant until the next spring.

SIGNIFICANCE: These insects feed by sucking fluid from the needles and new shoots of their host plants. As many as 100 aphids have been found on a single seedling. The removal of fluids and nutrients, possibly combined with effects of toxic saliva, cause stunting and defoliation of the infested plants. Infestations in nurseries have resulted in the loss of a significant portion of the white fir seedling crop. Seedlings carrying aphid eggs would be expected to have poorer growth and reduced survival after transplantation.

SIMILAR PEST: *Cinara curvipes* is a much larger, black aphid feeding on white fir. It does not produce powdery white wax or cause twig twisting.

Damage by Insects

MANAGEMENT OPTIONS: Infestations of seedlings in nursery beds have been controlled by applications of insecticidal soap and introductions of a predatory green lacewing, *Chrysoperla rufilabris*.

For More Information
Saunders, J. L. 1969. Occurrence and control of the Balsam Twig Aphid on *Abies grandis* and *A. concolor*. *Journal of Economic Entomology* 62:1106–9.

Voegtlin, D. 1995. Notes on the *Mindarus* spp. (Homoptera: Aphididae) of North America with descriptions of two new species. *Proceedings of the Entomological Society of Washington* 97:178–96.

COOLEY SPRUCE GALL ADELGID — *Adelges cooleyi*
Pls. 24–26

HOSTS: Douglas-fir and Sitka spruce.

DISTRIBUTION: Coast Ranges and northern Sierra Nevada.

SYMPTOMS AND SIGNS: On Douglas-fir, pinhead-sized dark brown insects covered with fluffy white wax are found on current-year needles and cones in spring and summer months. In late summer, infested needles become twisted, turn yellow, and drop off. On spruce, distinctive galls resembling cones are formed. Current-year galls are 1.2 to 7.5 cm (.5 to 3 in.) long, green with pink and purple markings, with the insects living inside cavities at the bases of needlelike projections. After the first year the galls turn brown and may remain on the tree for several years. Galls are produced on spruce only in areas where Douglas-fir occurs (pl. 24)(pl. 25)(pl. 26).

Plate 24. *Adelges cooleyi*, Cooley Spruce Gall Adelgids, on Douglas-fir foliage. Note that the foliage is blackened by sooty mold growing on honeydew produced by the adelgids.

Damage by Insects

Left: Plate 25. *Adelges cooleyi* on Douglas-fir cones.

Below: Plate 26. Gall produced by *Adelges cooleyi* on an Engelmann spruce.

LIFE CYCLE: The Cooley Spruce Gall Adelgid has a complex life cycle that may involve six generations over a two-year period on two hosts. Each generation has egg, nymph, and adult stages and although a cycle on spruce and Douglas-fir takes two years, all generations are present in any one year. On Douglas-fir the sistens generation overwinters on the needles. This generation matures as parthenogenic females and produces eggs in spring. Some of the eggs yield a wingless progrediens generation, members of which crawl to the nearest new foliage to feed. The rest of the eggs produce a winged sexupara generation, which reaches maturity in about a month and then flies to spruce trees. On spruce, the winged insects lay eggs that produce the sexualis generation, which through sexual reproduction gives rise to an overwintering fundatrix generation. These mature as parthenogenic females in spring and lay eggs that give rise to the gall-forming gallicola migrans generation. As the gall containing the gallicola migrans generation opens in midsummer, the winged parthenogenic females emerge and migrate to Douglas-fir, where eggs are laid that develop into the sistens generation. Where spruce is absent or rare, two or more parthenogenic generations per year are produced on Douglas-fir needles.

SIGNIFICANCE: Heavy infestations may cause minor growth reductions, particularly of young Douglas-fir trees growing on poor sites. The insects produce a sticky secretion (honeydew) that supports the growth of a black sooty mold. Honeydew coating the cones of infested trees interferes with the extraction and processing of seed by clogging machinery and causing seed and debris to stick together.

SIMILAR PESTS: The Balsam Woolly Aphid, *A. piceae*, an introduced European pest, is found in Oregon and Washington, where it is a serious pest of true fir species. The Balsam Woolly Aphid feeds on the stem, branches, and twigs of the host tree rather than the foliage. Twig infestations cause gall-like deformations of the twig tips. Stem infestations cause formation of abnormal, dense, red sapwood. Heavily infested trees are progressively weakened and eventually die. The Hemlock Woolly Adelgid, *A. tsugae*, feeds on twigs and branches of western hemlock.

MANAGEMENT OPTIONS: Usually, natural control agents reduce adelgid populations before significant damage occurs. Insecticide applications may sometimes be justified for highly valued ornamental trees.

For More Information

Cumming, M. E. P. 1959. The biology of *Adelges cooleyi* (Gill.) (Homoptera: Phylloxeridae). *The Canadian Entomologist* 91:601–17.

BLACK PINELEAF SCALE   *Nuculaspis californica*
Pls. 27, 28

HOSTS: All pine species and Douglas-fir.

DISTRIBUTION: Throughout the forested regions of California.

SYMPTOMS AND SIGNS: Foliage is sparse, and needles are shorter than normal and blotched with yellowish, necrotic areas. Gray to black, oval-shaped scale insects about 2 mm (about .1 in.) long appear on the needles (pl. 27)(pl. 28).

LIFE CYCLE: Adult scales overwinter on the foliage. In early summer winged males emerge and mate with the legless, immobile females. Each female produces a mass of yellowish eggs that remain under her protective shell. The eggs hatch within a few weeks to produce the nymphal stage, called crawlers, which look like miniature aphids. The crawlers disperse to new feeding locations within two to three weeks. The crawlers insert their mouthparts into a needle, secrete their characteristic grayish black covering, and remain immobile for the rest of the life cycle. In the northern part of California, *N. californica* produces one generation per year. In southern California there are two generations, and possibly a partial third, per year.

SIGNIFICANCE: Heavy infestations reduce both radial and terminal growth rates. If sizable populations of scale persist for several years, trees gradually weaken and become susceptible to bark beetle infestation. Normally, parasitic wasps, predatory beetles, and unfavorable weather conditions keep the population density of this insect well below

Plate 27. Extensive needle loss on ponderosa pine caused by *Nuculaspis californica*, Black Pineleaf Scale, infestation.

Damage by Insects

Plate 28. *Nuculaspis californica*, scales.

damaging levels. However, certain environmental conditions, such as dusty conditions along roads and around construction sites, as well as insecticides applied repeatedly to control mosquitoes or agricultural pests, interfere with the predators and parasites of this scale, allowing its population to increase.

SIMILAR PESTS: The Pine Needle Scale, *Chionaspis pinifoliae*, causes the same type of damage but its scales are white rather than gray to black. Infestations by aphids or spider mites and ozone injury also produce yellow blotches on foliage.

MANAGEMENT OPTIONS: The use of road oil or water to reduce dust will reduce scale populations by favoring parasite and predator activity. Contact insecticides during the crawler stage have been applied to protect highly valued trees. Trunk-implanted systemic insecticide treatments may also be used.

For More Information

Edmunds, G. F., Jr. 1973. Ecology of Black-pine Leaf Scale (Homoptera: Diaspididae). *Environmental Entomology* 2:765–77.

Ferrell, G. F. 1986. *Black Pineleaf Scale*. Forest Insect and Disease Leaflet 91. Washington, D.C.: USDA Forest Service.

PINE NEEDLE SCALE   *Chionaspis pinifoliae*
Pl. 29

HOSTS: All Douglas-fir and pine species.

DISTRIBUTION: Throughout the forested regions of California and on ornamental trees in urban areas.

SYMPTOMS AND SIGNS: Foliage turns a blotchy yellow and needles are shorter than normal. White, elongated oval scales about 3 mm (.1 in.) long appear on the needles (pl. 29).

LIFE CYCLE: The winter is spent in the egg stage under dead female scales on pine needles. The eggs hatch in spring, producing nymphs, called crawlers. The crawlers look like miniature aphids. They disperse actively and may also be wind transported. Within a few weeks the crawlers find new feeding sites, insert their mouthparts into needles, and molt to an immobile stage that secretes the characteristic white scale covering. The scales reach maturity in midsummer. Winged males emerge from under their protective scales to mate with the immobile females. Eggs are produced immediately and another generation is completed by the end of summer.

SIGNIFICANCE: Heavy infestations weaken trees, and young trees may be directly killed by scale infestations. Older trees weaken over a period of two or three years and may become susceptible to bark beetle colonization. Sustained infestations cause reductions in both radial and terminal growth. Ladybird beetles and parasitic wasps usually keep scale populations below damaging levels. Dust, however, interferes with predator and

Plate 29. *Chionaspis pinifoliae*, the Pine Needle Scale.

parasite activity, and hence trees along dusty roads are often heavily infested.

SIMILAR PESTS: The Black Pineleaf Scale, *Nuculaspis californica*, causes similar damage but the scales are gray or black rather than white. Ozone injury and feeding by spider mites and aphids also produce yellow blotches on foliage.

MANAGEMENT OPTIONS: Dust control will prevent damaging infestations along roadsides. In one instance, discontinuing insecticidal fogging to control adult mosquito populations was promptly followed by predator population increase and reduction of the scale population. Highly valued ornamentals and Christmas trees may be sprayed with insecticides. Insecticides are most effective when applied during the crawler stage. Trunk-implanted systemic insecticides may also be used.

For More Information

Burden, D. J., and E. R. Hart. 1993. Parasitoids associated with *Chionaspis pinifoliae* and *Chionaspis heterophyllae* (Homoptera: Diaspididae) in North America. *Journal of the Kansas Entomological Society* 66:383–91.

Cumming, M. E. P. 1953. Notes on the life history and seasonal development of the Pine Needle Scale, *Phenacaspis pinifoliae* (Fitch) (Diaspididae: Homoptera). *The Canadian Entomologist* 85:347–52.

Luck, R. F., and D. L. Dahlsten. 1975. Natural decline of a Pine Needle Scale (*Chionaspis pinifoliae* [Fitch]), outbreak at South Lake Tahoe, California, following cessation of mosquito-control with malathion. *Ecology* 56:893–904.

GOUTY PITCH MIDGE *Cecidomyia piniinopsis*
Pls. 30–32

HOST: Ponderosa pine.

DISTRIBUTION: Throughout the range of ponderosa pine in California.

SYMPTOMS AND SIGNS: Tufts of needles die in late summer. The dead needles are yellow at first, changing to reddish brown. In heavy infestations all the needles on individual twigs die. Severely infested twigs become twisted. Small orange or red larvae are found in pitch pockets in the cortical tissue of the twigs (pl. 30) (pl. 31).

LIFE CYCLE: Adult midges are minute flies resembling mosquitoes. They fly in early summer, at which time the females lay minute red eggs on the new twig growth of ponderosa pine. On hatching, the larvae bore into the twigs. The larvae form small cavities in the

Damage by Insects

Plate 30. Twigs damaged by *Cecidomyia piniinopsis*, the Gouty Pitch Midge: Susceptible tree on the left, resistant tree on the right.

Plate 31. *Cecidomyia piniinopsis* larva and pitch pocket in ponderosa pine twig.

growing twig tissues. The surface of the shoot will show a swelling over each cavity. The larvae live in the pitch-filled cavities until they reach maturity. In early spring the larvae tunnel to the surface of the shoot and crawl out onto the needles, where they spin cocoons about the size and shape of rice grains. Transformation to the adult stage requires two to three weeks. The adults emerge by neatly removing a cap from one end of the cocoon. There is a single generation per year (pl. 32).

TO FOLIAGE AND SHOOTS

Damage by Insects

Plate 32. *Cecidomyia piniinopsis* cocoons on ponderosa pine needles.

SIGNIFICANCE: Heavy infestations by the Gouty Pitch Midge cause branches to die and reduce the growth rate of affected trees. When terminal shoots are killed, a crook or fork in the tree bole may result. Often, severely infested plantation trees have been stressed by the heavy growth of competing vegetation. In these instances growth is severely retarded and tree mortality may result.

SIMILAR PESTS: Ponderosa Pine Tip Moth, *Rhyacionia zozana*, and twig beetles, *Pityophthorus* spp., kill young shoots of ponderosa pine and herbicide exposure produces twisted twigs with dead foliage. Damage by the Gouty Pitch Midge can be distinguished by the presence of small, pitch-filled cavities scattered along the length of the affected shoots. In shoots that are not killed the pitch-filled pockets remain embedded in the wood and can be used to date and estimate the severity of past infestations.

MANAGEMENT OPTIONS: Vigorous, rapidly growing trees tolerate infestation without shoot deformation or mortality. Elimination of competing vegetation and proper matching of planting site and seed source will result in plantations of vigorous trees that tolerate Gouty Pitch Midge infestations without serious damage. Ponderosa pines with new shoots that exhibit smooth, dry or waxy surfaces are less susceptible to infestation than pines with shoots that have sticky surfaces. With enough advance planning it should be possible to produce planting stock with smooth, dry or waxy shoots for use in locations with a history of Gouty Pitch Midge damage.

For More Information

Austin, A., J. S. Yuill, and K.G. Brecheen. 1945. Use of shoot characters in selecting ponderosa pines resistant to resin midge. *Ecology* 26:288–96.

Eaton, C. B., and J. S. Yuill. 1971. *Gouty Pitch Midge*. Forest Pest Leaflet 46. Washington, D.C.: USDA Forest Service.

Damage by Insects

PANDORA MOTH *Coloradia pandora*
Pls. 33–36

HOSTS: Jeffrey pine and ponderosa pine.

DISTRIBUTION: Eastern Sierra Nevada and mountains of northeastern California.

SYMPTOMS AND SIGNS: Ponderosa pine and Jeffrey pine become defoliated. Needles are completely consumed except for stubs about 6 mm (.25 in.) long at the base. Buds remain intact (pl. 33).

LIFE CYCLE: Adults appear in late June and July. They are large moths, having a wing span of 7 to 10 cm (2.75 to 4 in.). The fore wings are dark gray, and the hind wings are pale pink with a black spot in the center and a gray cross-band. Females attach globular, yellowish brown eggs in clusters to needles and branches of host trees. In August, black larvae covered with short, stiff spines hatch from the eggs. The larvae feed in groups on the new foliage until late fall. They remain in groups among the needle fascicles through the winter. The larvae resume feeding in spring, reaching maturity in late June or early July. Mature larvae are 6 to 7.5 cm (2.4 to 3 in.) long with stout, branching spines. They are brown to yellow green with a light stripe down the back. Mature larvae crawl down the trees and burrow into the soil to a depth of 2.5 to 15 cm (1 to 6 in.), where they transform to the pupal stage. Pupae are dark brown and about 2.5 cm (1 in.) long. The insects remain in the pupal stage through the next winter, emerging as adults the following June or July. However, some pupae remain in diapause for up to five years (pl. 34)(pl. 35)(pl. 36).

Plate 33. Defoliation of young Jeffrey pine by *Coloradia pandora*, the Pandora Moth.

Damage by Insects

Above: Plate 34. *Coloradia pandora* adult.

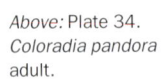

Right: Plate 35. *Coloradia pandora* eggs.

Below: Plate 36. *Coloradia pandora* larva on a Jeffrey pine shoot: Note that the needle bases remain uneaten.

34 TO FOLIAGE AND SHOOTS

SIGNIFICANCE: The Pandora Moth periodically occurs in epidemic populations that defoliate host trees over many thousands of hectares. Defoliation decreases the radial growth of trees and may weaken them, making them vulnerable to bark beetle infestation. In California, Pandora Moth outbreaks have usually been of short duration without causing significant tree mortality. The feeding period of mature larvae is completed shortly before the new needles appear, so the new foliage escapes damage. As the life cycle requires two years, infested trees may partly recover their foliage complement and nutrient reserves. The larvae are susceptible to a polyhedrosis virus that spreads rapidly in dense populations, causing nearly 100 percent mortality. Native Americans captured and ground the larvae as a source of food.

SIMILAR PESTS: Various other caterpillars defoliate pine trees; however, the large size and distinctive appearance of Pandora Moth larvae make them easily recognizable. In the absence of larvae the needle stubs left on the twigs indicate feeding by Pandora Moth larvae.

MANAGEMENT OPTIONS: Insecticides have not been used to reduce damage by this moth; Pandora Moth outbreaks are typically suppressed by natural control agents, such as polyhedrosis virus, before extensive tree death occurs.

For More Information

Gerson, E. A., and R. G. Kelsey. 1997. Attraction and direct mortality of Pandora Moths, *Coloradia pandora* (Lepidoptera: Saturniidae), by nocturnal fire. *Forest Ecology and Management* 98:71–75.

Miller, K. K., and M. R. Wagner. 1989. Effect of Pandora Moth (Lepidoptera: Saturniidae) defoliation on growth of ponderosa pine in Arizona. *Journal of Economic Entomology* 82:1682–86.

Ross, D. W. 1996. Phenology of Pandora Moth (Lepidoptera: Saturniidae) adult emergence and egg eclosion in central Oregon. *Pan-Pacific Entomologist* 72:1–4.

SILVERSPOTTED TIGER MOTH   *Lophocampa argentata*
Pls. 37–39

HOSTS: Principally Douglas-fir, but also grand fir, lodgepole pine, Monterey pine, Sitka spruce, western hemlock, and western redcedar.

DISTRIBUTION: Coast Ranges of central and northern California.

SYMPTOMS AND SIGNS: Large, irregular webs of white silk containing dead needles, frass pellets, and hairy brown caterpillars

Damage by Insects

appear, usually in the upper crown of the tree. The branches near the web are completely defoliated (pl. 37)(pl. 38)(pl. 39).

LIFE CYCLE: Adults emerge from their cocoons in late July and August. These moths have a wing span of 3.5 to 5 cm (1.4 to 2 in.).

Top: Plate 37. *Lophocampa argentata*, Silverspotted Tiger Moth, colonies in webs on Sargent cypress.

Right: Plate 38. *Lophocampa argentata* larva.

Below: Plate 39. *Lophocampa argentata* larvae in feeding web.

The fore wings and thorax are reddish brown with numerous silver spots, and the hind wings are yellowish white. Females deposit masses of green eggs on the foliage and twigs of host trees. They hatch in about three weeks. New larvae are about 3 mm (.12 in.) long and covered with brown hairs. The young larvae live colonially in a silken web among the branches of the host tree and feed on the nearby foliage. As the larvae grow they enlarge the web, which may ultimately measure as much as 60 by 90 cm (24 by 35 in.). Larvae spend the winter in their web and resume feeding in spring. As the larvae near maturity they crawl out from the web and feed individually. Mature larvae are about 3.5 cm (1.4 in.) long. They have long, brown hairs along the sides and dense tufts of shorter, black hair along the back. Two tufts of longer yellow and brown hairs project forward from behind the head. The larvae mature by late June and spin oval cocoons on the host tree or in the litter on the ground. The cocoons are brown and composed of silk and larval body hairs.

SIGNIFICANCE: Extensive defoliation of forest trees by Silverspotted Tiger Moth larvae is rare in California. The insect is potentially a pest in Christmas tree plantations. Naturally occurring parasites and predators usually prevent the development of damaging populations. Larval hairs are poisonous and can cause a rash on some people.

SIMILAR PESTS: The larvae of the Douglas-fir Tussock Moth, *Orgyia pseudotsugata*, are hairy and similar in size to those of the Silverspotted Tiger Moth. However, they do not live in a web and the hair tufts on the backs of larvae are white rather than black. The larvae of the Fall Webworm, *Hyphantria cunea*, are hairy and live in colonies in a web, but they live on hardwoods rather than conifers.

MANAGEMENT OPTIONS: Normally, natural control agents eliminate Silverspotted Tiger Moth populations before serious damage occurs. On occasion, individual colonies on highly valued trees are sprayed with an insecticide.

For More Information

Shaw, D. C. 1998. Distribution of larval colonies of *Lophocampa argentata* Packard, the Silverspotted Tiger Moth (Lepidoptera: Arctiidae), in an old growth Douglas-fir, *Pseudotsuga menziesii*, western hemlock, *Tsuga heterophylla*, forest canopy, Cascade Mountains, Washington State. *Canadian Field Naturalist* 112:250–53.

Silver, G. T. 1958. Studies on the Silverspotted Tiger Moth, *Halisidota argentata* Packard (Lepidoptera: Arctiidae) in British Columbia. *The Canadian Entomologist* 90:65–80.

Damage by Insects

DOUGLAS-FIR TUSSOCK MOTH *Orgyia pseudotsugata*
Pls. 40–44
HOSTS: Douglas-fir and White fir.
DISTRIBUTION: Throughout the forested regions of California.
SYMPTOMS AND SIGNS: In early summer new foliage is partly eaten and turns reddish brown. Later, both new and old foliage is completely consumed. Trees are defoliated from the top down and from the outside of the crown inward to the trunk. Distinctive tussock moth larvae are present in mid to late summer. Cocoons covered with gray-brown hair and characteristic egg masses are present through the fall and winter (pl. 40)(pl. 41)(pl. 42)(pl. 43).
LIFE CYCLE: Adult moths emerge from cocoons in late summer. Males have a wing span of a little over 2.5 cm (1 in.). The fore wings are dark gray-brown and the hind wings are a lighter brown. Unlike the males, which typically fly in late afternoon, females have rudimentary wings and do not fly. They are about 2 cm (.75 in.) long and densely clothed in dark gray-brown

Plate 40. White fir defoliated by *Orgyia pseudotsugata*, the Douglas-fir Tussock Moth.

Damage by Insects

Plate 41. White fir stand defoliated by *Orgyia pseudotsugata*.

Plate 42. *Orgyia pseudotsugata* larva.

Plate 43. *Orgyia pseudotsugata* cocoon and egg mass.

scales. After emergence, females mate and deposit up to 300 spherical, white eggs in a mass on top of her cocoon. Eggs are enclosed in a mass of stiff, whitish, foamy material mixed with scales from the female's abdomen. Egg hatch the following spring is synchronized with the opening of foliage buds on the host trees. New larvae are about 6 mm (.25 in.) long and covered with dark gray hairs. Larvae disperse from the egg masses by climbing upward and then dropping on a long silk thread. The thread, and the hairs on their bodies, permit the larvae to be carried long distances on the wind. Young larvae feed initially on the underside of new needles. The damaged needles die and turn reddish brown. As the larvae grow larger, they consume both new and old

Damage by Insects

Plate 44. Webbing left on white fir by dispersing first instar *Orgyia pseudotsugata* larvae.

needles. When large populations are present trees may be completely defoliated. Mature larvae are 2.5 to 3.8 cm (1 to 1.5 in.) long and either cream colored or brown. Their sides are covered with tufts of long, brown hair. Two long tufts of black hair extend forward from behind the head and a similar, single tuft extends backward like a tail from the last segment. The dense tufts of short white hairs on the first four abdominal segments are a key character of tussock moths. Mature larvae spin gray-brown cocoons covered with larval hairs on foliage, tree branches, and in bark crevices. During heavy infestations large masses of cocoons may accumulate in sheltered locations, on the bark of the trunk, and along the underside of larger branches (pl. 44).

SIGNIFICANCE: The Douglas-fir Tussock Moth has defoliated many thousands of hectares, causing extensive tree mortality. Top kill and radial increment reductions occur regularly. The body hairs of the Douglas-fir Tussock Moth larvae contain a toxic substance that causes extreme skin irritation in sensitive individuals and may cause respiratory problems when inhaled. Consequently, dense larval populations may become a medical hazard in recreation areas. Outbreaks usually collapse after three to four years as the result of numerous species of insect parasites and predators, bird predation on egg masses during the winter, and a highly virulent nucleopolyhedrosis virus that causes devastating epidemics among the dense larval populations.

SIMILAR PESTS: Other species of tussock moth include both adults and larvae similar in appearance to the Douglas-fir Tussock Moth. Hardwood trees are usually the preferred hosts of these similar species. The larvae of the Silverspotted Tiger Moth,

Lophocampa argentata, are hairy and similar in size to tussock moth larvae. However, they live in colonies in a silken web and have tufts of black rather than white hair on the back.

MANAGEMENT OPTIONS: White fir stands growing on warm, dry sites such as ridge tops are more susceptible to Douglas-fir Tussock Moth damage than stands on cooler, moister sites. Often these stands have developed as a result of long-term fire suppression and selective removal of pines. Conversion of these sites to pine stands will reduce the risk of damage by Douglas-fir Tussock Moth. Systematic surveys of Douglas-fir Tussock Moth populations, using pheromone-baited traps, can provide advance warning of rising moth populations and allow the timely organization of large-scale control programs. Aerially applied contact insecticides have been used to suppress populations. Recently, aerial applications of *Bacillus thuringiensis* have been used. Individual, highly valued trees can be protected from defoliation by using systemic insecticide implants.

For More Information

Brookes, M. H., R. W. Stark, and R. W. Campbell, eds. 1978. *The Douglas-fir Tussock Moth: A synthesis.* Technical Bulletin 1585. Washington, D.C.: USDA Forest Service.

Dahlsten, D. L., D. L. Rowney, W. A. Copper, and J. M. Wenz. 1992. Comparison of artificial pupation shelters and other monitoring methods for endemic populations of Douglas-fir Tussock Moth, *Orgyia pseudotsugata* (McDunnough) (Lepidoptera: Lymantriidae). *The Canadian Entomologist* 124:359–69.

Mason, R. R., B. E. Wickman, and H. G. Paul. 1997. Radial growth response of Douglas-fir and grand fir to larval densities of the Douglas-fir Tussock Moth and the Western Spruce Budworm. *Forest Science* 43:194–205.

SPRUCE BUDWORMS *Choristoneura* spp.

Pls. 45–47

HOSTS: Bigcone Douglas-fir, Douglas-fir, grand fir, red fir, and white fir.

DISTRIBUTION: Four closely related species are each found in various parts of the forested regions of California. *Choristoneura carnana californica* is found in the Sierra Nevada and North Coast Ranges, where it feeds on Douglas-fir. *C. carnana* is found in the mountains of southern California, where it feeds on bigcone

Damage by Insects

Douglas-fir, and along the west slopes of the Sierra Nevada, where it feeds on Douglas-fir. The Modoc Budworm, *C. retiniana* (= *C. viridis*), feeds on true firs, principally white fir, from northeastern California south through the Sierra Nevada to the mountains of southern California. The Western Spruce Budworm, *C. occidentalis,* occurs in the northernmost counties of California and feeds on grand fir, white fir, and Douglas-fir.

SYMPTOMS AND SIGNS: Buds, new foliage, and cones of host trees are eaten by budworm larvae in spring and early summer. In heavy infestations old needles may also be consumed. Usually, upper tree crowns are more severely defoliated than the lower branches. By July the larvae have matured and begin spinning irregular white silk webbing among the needles. Full-grown larvae are light brown with rows of light yellow spots and a darker brown head, except for the larvae of *C. retiniana*, which are green (pl. 45)(pl. 46).

Above: Plate 45. Defoliation of Douglas-fir by spruce budworm, a *Choristoneura* species.

Right: Plate 46. *Choristoneura* spp. larva on a Douglas-fir twig.

Damage by Insects

Plate 47. *Choristoneura* pupae on a white fir twig.

LIFE CYCLE: Flat, scalelike, light green eggs are deposited in masses on the underside of needles. Eggs hatch in about 10 days, producing light green larvae with brown heads. First-stage larvae spin silken shelters (hibernacula) under bark scales and in other hidden locations, where they remain until the following spring. When the buds start to swell the larvae leave their hibernacula and feed in the expanding buds. They continue to eat new foliage until reaching maturity in late June or early July. Mature larvae are about 2.5 cm (1 in.) long. They transform to the pupal stage in the webbing among the needles. Pupae are brown except for those of the Modoc Budworm, which are green. The moths emerge in late July and early August. Adults are extremely variable in color and pattern. The fore wings are usually marked with rectangular patches in various shades of brown, gray, and black and have a prominent white mark midway along the leading edge (pl. 47).

SIGNIFICANCE: Spruce budworms can cause severe defoliation resulting in reduction of radial increment, top kill, and tree mortality. The Western Spruce Budworm has been responsible for extensive, long-term defoliation throughout the western United States and Canada. In northeastern California outbreaks of the Modoc Budworm have been recorded. One outbreak of *C. carnana californica* in northwestern California has been recorded. Low populations of budworms that would not cause defoliation or affect tree growth can severely reduce seed production, because of preferential feeding on the cones.

SIMILAR PESTS: Two other budworm species, *C. subretiniana* and *C. lambertiana*, with habits similar to those of spruce budworms are found feeding on lodgepole pine and sugar pine, respectively, in California. The Spruce Coneworm, *Dioryctria reniculelloides*, feeds on Douglas-fir and white fir foliage, produces webbing like that of spruce budworms, and is often found mixed with spruce budworm larvae. The larvae of the Spruce Coneworm may be distinguished by their longitudinally striped pattern.

MANAGEMENT OPTIONS: Fire control and past timber harvest practices have created dense, evenly aged stands with high proportions of white fir and Douglas-fir: conditions that favor spruce budworm outbreaks. Forest management practices that create mixed species stands with reduced white fir and Douglas-fir components and density control to maintain tree vigor are believed to reduce susceptibility to spruce budworm damage. Large-scale aerial applications of insecticides or *Bacillus thuringiensis* have been used to control spruce budworm outbreaks. High-volume hydraulic insecticide treatments have been applied to protect Douglas-fir seed crops. Trunk implants of systemic insecticides have been used to protect individual superior trees in defoliated areas and to prevent seed crop damage.

For More Information

Brookes, M. H., R. W. Campbell, J. J. Colbert, R. G. Mitchell, and R. W. Stark (Technical coordinators). 1987. *Western Spruce Budworm.* Technical Bulletin No 1694. Washington, D.C.: USDA Forest Service.

Fellin, D. G., and J. E. Dewey. 1986. *Western Spruce Budworm.* Forest Insect and Disease Leaflet 53. Washington, D.C.: USDA Forest Service.

Mason, R. R., B. E. Wickman, and H. G. Paul. 1997. Radial growth response of Douglas-fir and grand fir to larval densities of the Douglas-fir Tussock Moth and the Western Spruce Budworm. *Forest Science* 43:194–205.

DEVASTATING GRASSHOPPER   *Melanoplus devastator*
Pls. 48–50

HOSTS: Almost any green vegetation, although forbs and grass are much preferred to conifers. Most frequently damaged conifers are plantation-grown Douglas-fir and ponderosa pine.

DISTRIBUTION: Throughout California. Conifer plantations below 1,200 m (about 4,000 ft) have been most frequently impacted.

SYMPTOMS AND SIGNS: Trees are defoliated from the ground up. Defoliation is likely to be more severe around the edges of a plantation. In severe infestations the bark and buds are eaten, reducing young trees to bare, wooden sticks (pl. 48)(pl. 49).

Damage by Insects

Plate 48. Ground-upward defoliation of ponderosa pine seedling by *Melanoplus devastator*, the Devastating Grasshopper. Note the dry grass surrounding the seedling: grasshoppers begin feeding on pines when their preferred food, green grass and forbs, is no longer available.

Plate 49. Young ponderosa pine defoliated by *Melanoplus devastator*.

LIFE CYCLE: Young grasshoppers hatch in early spring from eggs buried in the soil. Immature grasshoppers, called nymphs, are miniature wingless versions of the adults. They feed on grasses and forbs and develop to the adult stage by late June. Adults are about 2.5 cm (1 in.) long, amber to brown, with dark spots on the fore wings. The fan-shaped hind wings are transparent. The hind femur is marked with three dark cross-bands. The hind tibia is blue

TO FOLIAGE AND SHOOTS

Damage by Insects

Plate 50. Adult *Melanoplus devastator* on a ponderosa pine bud.

at the base, becoming amber toward the tarsus. Adult grasshoppers continue to feed on grasses and forbs until the usual summer drought causes the vegetation to dry up. Often, the only green vegetation remaining is the planted conifers, and the grasshoppers prefer green conifer foliage to dry grass. A dense grasshopper population moving to young trees will sometimes eat all the foliage in a few days and then eat the buds and bark. At this point starvation is likely to eliminate most of the grasshopper population. Females deposit eggs in the soil in late summer and fall. The eggs are cemented together in masses of 20 to 200 and are encased in a rigid foam protective pod. The egg pods are buried 2 to 3 cm (.75 to 1.25 in.) deep in the soil and remain there through the winter (pl. 50).

SIGNIFICANCE: Grasshoppers are potentially a serious threat to conifer plantations, which can be destroyed very quickly. Trees less than five years of age in low-elevation plantations are especially vulnerable.

SIMILAR PESTS: At least nine species of grasshoppers potentially destructive to conifer plantations may be found in California. In addition to *M. devastator*, three other species in the genera *Melanoplus*, *Camnula*, and *Oedaleonotus* have been reported to cause damage to trees. All are similar in appearance, life cycle, and habits.

MANAGEMENT OPTIONS: Large grasshopper populations do not develop in the absence of abundant grasses, forbs, or both. Control of competing vegetation in plantations can reduce the probability of grasshopper population buildup and also maximize the growth rate of the trees; thus they remain in the vulnerable size range for only a short time. The types of plants needed to sustain grasshopper populations will not persist after crown closure occurs. Insecticide applications have been used to control grasshopper populations and a bait containing a disease organism, *Nosema locustae*, is available. Low-elevation plantations with grass or forb cover should be checked for grasshoppers in early spring.

For More Information
Thompson, R. M., and G. M. Buxton. 1964. *An index of the Acridoidea (Orthoptera) of California, with selected references.* Occasional Paper 5. Sacramento: Bureau of Entomology, California Department of Agriculture.

PONDEROSA PINE TIP MOTH   *Rhyacionia zozana*
Pls. 51–53

HOSTS: All pine species.

DISTRIBUTION: Throughout the pine-growing regions of California. Most commonly found from 600 to 1,500 m (about 2,000 to 5,000 ft).

SYMPTOMS AND SIGNS: Foliage on terminal portions of current-year shoots fades to yellow and finally brown. The damaged portions of twigs are dry, brittle, and riddled with larval feeding tunnels (pl. 51).

LIFE CYCLE: The insect spends the winter in the pupal stage in a tough cocoon firmly attached to the root collar of a tree at, or just below, the ground line. The moths emerge in late spring, when new pine needles are breaking through the needle sheaths. Eggs are deposited in small groups between the needle bases on new pine shoots. The flattened, scalelike eggs are greenish yellow at first and then turn orange. When the eggs hatch the new larvae spin small, silken tents between the shoot and needle bases. Under the shelter of the tent they begin to feed on the base of the needle bundle and on the succulent tissue of the new shoot. As they grow larger they tunnel into the shoot, eating all tissues except the outer bark. Several larvae, which are reddish orange, typically infest each shoot. Immature larvae have a dark brown or

Damage by Insects

Plate 51. Ponderosa pine terminal shoot killed by *Rhyacionia zozana*, the Ponderosa Pine Tip Moth.

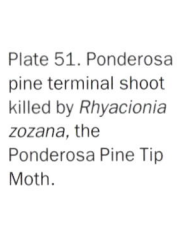

black head capsule, whereas older larvae have a tan head capsule. Larvae reach maturity in midsummer, and most of them leave the mined-out shoot to crawl down the trunk to the root collar, where they spin cocoons. A few larvae may spin cocoons in bark crevices or in mined shoots. By late summer the larvae transform to the pupal stage (pl. 52) (pl. 53).

SIGNIFICANCE: Feeding by tip moth larvae kills new shoots. If the terminal shoot of a tree is killed the tree will develop one or more new leaders, sometimes resulting in a crook or fork in the stem. Repeated killing of terminal shoots retards height growth. If competing vegetation is present, retarded height growth of the trees may allow the competing species to overtop the trees, thereby greatly reducing their growth rate. After trees reach 1.5 to 1.8 m (5 to 6 ft) in height they are no longer susceptible to attack by this insect.

SIMILAR PESTS: The Nantucket Pine Tip Moth, *R. frustrana*, has similar habits and appearance. This species pupates within the mined shoots and multiple generations occur each year. At present this species causes the most damage in southern California,

Damage by Insects

Above: Plate 52. *Rhyacionia zozana* cocoons at the base of a young tree.

Left: Plate 53. *Rhyacionia zozana* feeding site on a ponderosa pine terminal shoot: Silk tent sheltering larva *(arrow)*. Note the stage of needle growth.

where it is a severe pest of Monterey pine Christmas trees. The Western Pine Shoot Borer, *Eucosma sonomana,* feeds within the growing shoots of several pine species, with feeding confined to the pith core. Larvae of the Lodgepole Terminal Weevil, *Pissodes terminalis,* feed within terminal shoots of lodgepole pine. White, legless larvae and pupae may remain within the damaged shoots throughout the summer and winter.

MANAGEMENT OPTIONS: Good site preparation and control of competing vegetation will result in rapid height growth and minimize the number of years the trees are subject to damage. Systemic insecticides applied at the time the eggs are laid have successfully protected highly valued trees from infestation but have not been considered cost-effective for pine plantations.

For More Information

Niwa, C. G. 1988. Parasites and predators associated with the Ponderosa Pine Tip Moth, *Rhyacionia zozana* (Kearfott) (Lepidoptera: Tortricidae), in California and Oregon. *The Canadian Entomologist* 120:881–86.

Damage by Insects

Niwa, C. G., and G. E. Daterman. 1989. Pheromone mating disruption of *Rhyacionia zozana* (Lepidoptera: Tortricidae)—influence on the associated parasite complex. *Environmental Entomology* 18:570–74.

Stevens, R. E. 1971. *Ponderosa Pine Tip Moth*. Forest Pest Leaflet 103. Washington, D.C.: USDA Forest Service.

WESTERN PINE SHOOT BORER *Eucosma sonomana*
Pls. 54–56

HOSTS: Bishop pine, Jeffrey pine, knobcone pine, lodgepole pine, and ponderosa pine.

DISTRIBUTION: Common east of the Sierra Nevada crest, in northeastern California, and in the inner Coast Ranges. Rare on the western slopes of the Sierra Nevada.

SYMPTOMS AND SIGNS: Terminal shoot growth is stunted. Needles on the upper part of the shoot are shorter and closer together than those on the basal part of the shoot. The terminal shoot may be shorter than the lateral shoots of the top whorl. Infested lateral shoots are also stunted and have shortened needles. Lateral shoots are often killed, whereas terminal shoots are rarely killed. The pith core of the infested shoot is eaten by the larva, leaving a mine tightly packed with pitch-impregnated frass. An exit hole may be seen near the lower end of the mine. Infested lodgepole pine shoots do not show the stunted growth and shortened needles found in other species (pl. 54)(pl. 55)(pl. 56).

LIFE CYCLE: The insect spends winter in the pupal stage in a tightly woven cocoon under the duff layer covering the soil. Adults emerge in early spring as soon as the ground is free

Plate 54. Jeffrey pine terminal shoot infested by *Eucosma sonomana*, the Western Pine Shoot Borer. Note that the terminal shoot is shorter and has shorter needles than the adjacent lateral shoots.

Damage by Insects

Plate 55. Ponderosa pine terminal shoot killed by *Eucosma sonomana*. Note that a lateral shoot has turned up to become the new terminal shoot.

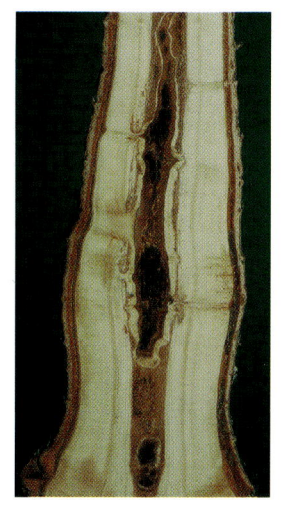

Plate 56. *Eucosma sonomana* larval mine in a Jeffrey pine stem. Note the swelling of the stem.

from snow. Female moths deposit eggs under the bud scales on terminal shoots and upper whorl lateral shoots. Larvae, which are brownish to grayish white with a tan head capsule, emerge from the eggs and bore directly into the tip of the growing shoot. Larvae mine downward in the pith core of the growing shoot for 22 to 30 cm (9 to 12 in.). Feeding is confined to the pith core of the shoot, leaving the surrounding xylem layer intact. Usually, only one larva is found per shoot. Larvae reach maturity in about eight weeks, cut an exit hole in the side of the shoot near the lower end of the mine, and drop to the ground. There they spin cocoons under the duff layer, transform to the pupal stage, and remain dormant until the following spring.

SIGNIFICANCE: Infestation by the Western Pine Shoot Borer reduces height and volume growth rates of infested stands. Growth reductions have been estimated in the range of 25 to 30 percent. Infested internodes are shorter than normal, so that lumber from infested trees will have increased numbers of knots. If the terminal shoot is infested in successive years, it may lose dominance.

When a terminal shoot is killed, one or more lateral shoots may replace the terminal, resulting in a crook or fork in the tree bole. Trees become susceptible to infestation when they reach .9 to 1.2 m (3 to 4 ft) in height and are most likely to suffer significant growth reductions and deformations when they are 1.2 to 3.0 m (4 to 10 ft) tall. Larger trees continue to be infested but tolerate infestation with less growth reduction.

SIMILAR PESTS: Pine tip moths, *Rhyacionia* spp., feed within growing shoots. All parts of the shoot are eaten and the infested shoot is killed. The larvae of Lodgepole Terminal Weevil, *Pissodes terminalis*, feed within terminal shoots. The infested shoots are killed, and white, legless larvae or pupae may remain within the shoots through fall and winter.

MANAGEMENT OPTIONS: Experimental applications of synthetic sex pheromones have been successfully used to disrupt mating of this insect, resulting in reduced infestations of ponderosa pine and Jeffrey pine plantations. Large-scale pheromone applications have not been attempted. Stand density control and control of competing vegetation may help increase the height growth rate and reduce the number of years trees remain in the most vulnerable size range.

For More Information

Prueitt, S. C., and D. W. Ross. 1998. Effects of environment and host genetics on *Eucosma sonomana* (Lepidoptera: Tortricidae) infestation levels. *Environmental Entomology* 27:1469–72.

Sower, L. L., and R. G. Mitchell. 1994. Mating disruption of Western Pine Shoot Borer (Lepidoptera: Olethreutidae) in lodgepole pine. *Journal of Economic Entomology* 87:144–47.

Stoszek, K. J. 1973. Damage to ponderosa pine plantations by the Western Pine Shoot Borer. *Journal of Forestry* 71:701–5.

PINE REPRODUCTION WEEVIL   *Cylindrocopturus eatoni*
Pls. 57–60

HOSTS: Gray pine, Jeffrey pine, ponderosa pine, and sugar pine.

DISTRIBUTION: Sierra Nevada, Cascade Mountains, and North Coast Ranges up to 1,700 m (about 5,600 ft).

SYMPTOMS AND SIGNS: Resin droplets mark puncture sites on twigs and foliage, which fades to yellow in fall, progressively darkening to dark reddish brown by spring. Meandering larval galleries are found in the phloem that do not originate from a

Damage by Insects

Plate 57. Young ponderosa pines killed by *Cylindrocopturus eatoni*, the Pine Reproduction Weevil. Note the density of competing brush vegetation.

Plate 58. *Cylindrocopturus eatoni* pupa. *(arrow)* Bluestained sapwood.

central gallery or chamber. During winter and spring cream-colored, legless larvae are found in cylindrical pupal chambers about 1 mm (.04 in.) in diameter in the xylem. Round adult emergence holes connect the pupal chambers to the bark surface. The sapwood is discolored by bluestain fungi (pl. 57)(pl. 58).

LIFE CYCLE: Adult weevils emerge from infested trees from May through mid-July. The adults feed on foliage and twigs, leaving punctures that bleed minute drops of resin. Females lay single eggs in the bark of the stem, below the current year's growth. The legless, white larvae feed in the phloem of the stem until they reach maturity in September. They then excavate cylindrical chambers in the xylem, where they remain through winter. Transformation to the pupal stage occurs in early spring followed by adult emergence starting in May (pl. 59)(pl. 60).

Damage by Insects

Plate 59. *Cylindrocopturus eatoni* punctures in ponderosa pine foliage.

Plate 60. *Cylindrocopturus eatoni* pupal cells in dead ponderosa pine stem.

SIGNIFICANCE: This insect is a serious pest in young pine plantations. Trees 45 to 100 cm (1.5 to 3.3 ft) tall are most likely to be attacked and killed. Weevil populations commonly build up over several years, progressively killing trees until the plantation no longer meets density standards. Healthy trees with adequate soil moisture are resistant to attack. Serious losses are seen only where some combination of drought, brush competition, and poor soil renders the trees vulnerable to attack.

SIMILAR PESTS: The Douglas-fir Twig Weevil, *Cylindrocopturus furnissi* has a similar life cycle and habits. It infests and kills small branches of Douglas-fir. Trees killed by pocket gophers or *Armillaria* root disease are similar in appearance to those killed by the Pine Reproduction Weevils; however, trees killed by the Pine Reproduction Weevil have intact root systems, whereas those attacked by pocket gophers do not. Those killed by *Armillaria* root disease have mycelial fans and rhizomorphs associated with the roots.

MANAGEMENT OPTIONS: Good site preparation and control of competing vegetation will usually prevent plantations from becoming susceptible to attack by the Pine Reproduction Weevil. Cutting and burning infested trees before the adult weevils emerge have been done to reduce weevil populations. A hybrid pine, Jeffrey × (Jeffrey × Coulter), is resistant to Pine Rreproduction Weevil and suitable for planting on harsh sites. However, the hybrid is not generally available from forest nurseries.

For More Information
Eaton, C. B. 1942. Biology of the weevil *Cylindrocopturus eatoni* Buchanan, injurious to ponderosa and Jeffrey pine reproduction. *Journal of Economic Entomology* 35:20–25.

Smith, R. H. 1960. Resistance of pines to the Pine Reproduction Weevil, *Cylindrocopturus eatoni* Buchanan. *Journal of Economic Entomology* 53:1044–48.

Stevens, R. E. 1971. *Pine Reproduction Weevil.* Forest Pest Leaflet 15. Washington, D.C.: USDA Forest Service.

DOUGLAS-FIR TWIG WEEVIL   *Cylindrocopturus furnissi*
Pl. 61

HOST: Douglas-fir.

DISTRIBUTION: North Coast Ranges and possibly northern Sierra Nevada.

SYMPTOMS AND SIGNS: Infested branches and terminal shoots die, fading from yellow to reddish brown. Legless larvae are present in galleries under the bark during fall and winter months. L-shaped pupal chambers in dead branches are associated with a bluestain fungus after the insects have emerged. The main stems of lightly infested young trees may present reddish-brown necrotic patches; the stems of heavily infested trees may present swollen areas (pl. 61).

LIFE CYCLE: Adult weevils are predominantly brown, and about 2.5 mm (.1 in.) long. They emerge from infested shoots in June and July. They have light and dark gray spots on the dorsum and light gray spots on the ventral side. Weevils feed on the inner bark of small branches, leaving small, irregular pits in the bark surface. During August and September females deposit oval, white eggs individually in pits excavated in stems and small branches. Most eggs are deposited on four-year-old internodes and hatch in six to 10 days. Larvae excavate feeding galleries in the phloem of the shoots, usually girdling them in the process. When larvae reach maturity in spring they are about 3 mm (.12 in.) long, white with a tan head capsule, legless, and exhibit the typical curved shape of weevil larvae. Mature larvae excavate cylindrical pupal chambers in the sapwood of small twigs or in both the bark and the sapwood

Damage by Insects

Plate 61. Douglas-fir twigs killed by *Cylindrocopturus furnissi*, the Douglas-fir Twig Weevil.

of larger shoots. Transformation of pupae to adults requires about three weeks, but the new adults remain in the pupal chambers for another month before emerging. Under favorable conditions adults may survive the winter and resume laying eggs the next summer. Larvae of all sizes can overwinter.

SIGNIFICANCE: The Douglas-fir Twig Weevil kills lateral and terminal shoots on young Douglas-fir trees. Growth may be retarded and when terminal shoots are killed stem defects may result. On occasion, small trees are killed. Damage is greatest during drought periods and on dry sites.

SIMILAR PESTS: Twig beetles, *Pityophthorus* spp., also kill small branches of Douglas-fir, but the L-shaped pupal chambers and associated bluestain are a distinctive characteristic of Douglas-fir Twig Weevil damage.

MANAGEMENT OPTIONS: Vigorously growing trees resist attack or tolerate infestation without serious damage. Proper site preparation and control of competing vegetation will greatly reduce susceptibility to this insect. Trees 4.5 to 6 m (15 to 20 ft) tall are no longer susceptible to appreciable damage by this weevil.

For More Information

Furniss, R. L. 1942. Biology of *Cylindrocopturus furnissi* Buchanan on Douglas-fir. *Journal of Economic Entomology* 35:853–59.

Oregon Department of Forestry. 2000. Douglas-fir Twig Weevil *(Cylindrocopturus furnissi)*. Forest Health Note. Salem, OR: Oregon

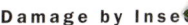

Department of Forestry. http://www.odf.state.or.us/fa/FH/fhn/twigwevil.pdf, (accessed August 2002).

LODGEPOLE TERMINAL WEEVIL — *Pissodes terminalis*
Pls. 62, 63

HOST: Lodgepole pine.

DISTRIBUTION: Sierra Nevada and Cascade Mountains in California; rare or absent on the coast.

SYMPTOMS AND SIGNS: The terminal shoot of an infested lodgepole pine is killed, with full-length needles turning yellow at first and later reddish brown. The interior of the shoot is mined and filled with loose, granular, brown frass (pl. 62)(pl. 63).

LIFE CYCLE: Female weevils excavate oviposition pits in elongating terminal shoots of lodgepole pine in June and early July. One to three eggs are deposited in each pit. Young larvae feed initially in the cambial tissue near the egg pits, making a spiral mine. Older larvae tunnel into the center of the shoot, consuming the pith and loosely filling the gallery with granular, brown frass. The larvae are yellowish brown and legless, with a brown head capsule. They mature in late summer and transform to the pupal stage within the mined shoots. Transformation to the adult stage occurs in fall, although some insects overwinter as mature larvae or pupae. Adults may leave the shoots to overwinter in the duff or remain in the dead shoots.

Plate 62. Lodgepole pine terminal shoot killed by *Pissodes terminalis*, the Lodgepole Pine Terminal Weevil.

Plate 63. *Pissodes terminalis* larvae in a lodgepole pine terminal shoot.

Damage by Insects

SIGNIFICANCE: This insect selectively infests the terminal shoots of lodgepole pines. The infested terminal shoots are killed and the trees replace them with lateral shoots. Lodgepole pines with multiple tops, commonly found in the Sierra Nevada, are largely the result of repeated infestation by this weevil.

SIMILAR PESTS: The White Pine Weevil, *Pissodes strobi*, infests the terminal shoots of Sitka spruce, usually killing the top two internodes. The Western Pine Shoot Borer, *Eucosma sonomana*, feeds within living terminal shoots of lodgepole pine, usually without killing them. The larval mine of the Western Pine Shoot Borer is tightly packed with pitch-soaked frass rather than loose granular frass.

MANAGEMENT OPTIONS: Treatments to reduce damage by this insect are not usually attempted.

For More Information

Cameron, E. A., and R. W. Stark. 1989. Variations in the life cycle of the Lodgepole Terminal Weevil, *Pissodes terminalis* Hopping (Coleoptera: Curculionidae), in California. *The Canadian Entomologist* 121:793–801.

Stark, R. W., and D. L. Wood. 1964. The biology of *Pissodes terminalis* Hopping (Coleoptera: Curculionidae) in California. *The Canadian Entomologist* 96:1208–18.

SEQUOIA PITCH MOTH *Synanthedon sequoiae*
Pls. 64–67

HOSTS: Douglas-fir and most pine species.

DISTRIBUTION: Throughout California.

SYMPTOMS AND SIGNS: Masses of resin are found on the trunk and branches of trees. The pitch masses are concentrated at branch nodes and junctions of limbs and bole. Brown to reddish brown frass pellets are incorporated in the pitch masses. Empty pupal skins protruding from the pitch masses are a good diagnostic character for pitch moth identification (pl. 64)(pl. 65).

LIFE CYCLE: Adults appear in mid-summer. They are black and yellow with clear wings, closely resembling a small wasp. Females deposit eggs singly in bark crevices and in wounds on tree trunks and branches. Larvae are yellowish white with a brown head capsule. They bore into the bark or enter an existing wound and begin feeding on the phloem. Each larva remains in one location,

Damage by Insects

Plate 64. Resin masses at feeding sites of *Synanthedon sequoiae*, the Sequoia Pitch Moth, on a ponderosa pine trunk.

Plate 65. Pupal skin of *Synanthedon sequoiae* protruding from a pitch mass after adult emergence.

excavating an irregular cavity under the bark. A mass of pitch up to 7.5 cm (3 in.) across accumulates on the bark surface, concealing the larva and the entrance to its feeding cavity. Larvae reach maturity after two years and pupate in a chamber within the pitch mass. The pupa protrudes from the pitch mass, allowing the adult to emerge without becoming trapped in the sticky pitch (pl. 66).

Plate 66. *Synanthedon sequoiae* larva in ponderosa pine branch.

TO STEMS AND LARGER BRANCHES

Damage by Insects

Plate 67. Broken treetop and dead branches resulting from repeated *Synanthedon sequoiae* infestation.

SIGNIFICANCE: Continued feeding by pitch moth larvae creates a persistent wound. Several generations of larvae may continue to feed at the same site, creating a weak point in the affected branch or tree trunk. Accumulations of pitch increase the fire hazard in plantations with heavy infestations. The moths are attracted to wounded trees, and are therefore more abundant in urban forests (pl. 67).

SIMILAR PESTS: The Douglas-fir Pitch Moth, *S. novaroensis*, is very similar in habits and appearance; however, the moths are black and orange rather than black and yellow. The Douglas-fir Pitch Moth feeds on Douglas-fir, spruce, ponderosa pine, and lodgepole pine. The pitch tubes of bark beetles, especially those of the Red Turpentine Beetle, *Dendroctonus valens*, resemble the pitch masses of the Sequoia Pitch Moth. Also, stem cankers caused by the pitch canker fungus, *Fusarium circinatum,* can be confused with pitch masses produced by the Sequoia Pitch Moth. The insect-caused pitch masses can also stream down the stem on sun-exposed bark and thus may be misdiagnosed as a pitch canker stem infection. Removal of a patch of bark under a bark beetle pitch tube will reveal the characteristic pattern of a bark beetle egg gallery rather than the small, irregular feeding cavity of pitch moth larva. If larvae are present, legless bark beetle larvae are readily distinguished from pitch moth larvae, which have thoracic legs and abdominal prolegs. The presence of adult bark beetles and bluestain fungi also clearly distinguishes bark beetle infestations from pitch moth damage.

MANAGEMENT OPTIONS: No chemical treatments are specifically aimed at this insect, which can occur at high density in urban environments. Removing the pitch masses, together with the larvae, from highly valued trees may help. Pruning is not recommended, as the adult insect is attracted to open wounds.

For More Information

Koehler, C. S., G. W. Frankie, W. S. Moore, and V. R. Landweher. 1983. Relationship of infestation by the Sequoia Pitch Moth (Lepidoptera: Sesiidae) to Monterey pine trunk injury. *Environmental Entomology* 12:979–81.

Nowak, D. J., and J. R. McBride. 1992. Differences in Monterey pine pest populations in urban and natural forests. *Forest Ecology and Management* 50:133–44.

Powers, R. F., and W. E. Sundahl. 1973. Sequoia Pitch Moth: A new problem in fuel break construction. *Journal of Forestry* 71:338–39.

CALIFORNIA FIVE-SPINED IPS *Ips paraconfusus*
Pls. 68–73

HOSTS: Bishop pine, Coulter pine, gray pine, Jeffrey pine, knobcone pine, lodgepole pine (rarely), Monterey pine, ponderosa pine, sugar pine, and Torrey pine.

DISTRIBUTION: Common west of the Sierra Nevada and Cascade Mountain crests and in Bishop pine, gray pine, knobcone pine, Monterey pine, and ponderosa pine in the Coast Ranges.

SYMPTOMS AND SIGNS: Young pines, about 2.5 to 25 cm (1 to 10 in.) in diameter, are killed, as are the tops of larger trees. During periods of extreme drought and during outbreaks, trees up to 65 cm (26 in.) in diameter may also be killed. Usually, the first evidence of infestation is the typical color change of conifer foliage: from green to lime-green to yellow and then to reddish brown. Small pitch tubes and reddish brown boring dust in the bark crevices and cobwebs on the main stem are symptomatic of infestation. Infestation of snow- or wind-broken branches and of trees and logging debris is indicated by piles of reddish brown boring dust in the crevices and on the ground beneath the branches or logs. On removal of bark, Y-shaped maternal galleries that parallel the grain, and the presence of adult beetles, give positive evidence of the species causing the infestation. The wing cases (elytra) of these dark brown bark beetles have a scooped-out appearance at the end (posterior declivity), with five spines on each side (pl. 68)(pl. 69)(pl. 70)(pl. 71).

LIFE CYCLE: This species of bark beetle has two to six generations per year, depending on latitude and elevation; three or four generations per year occur at about 1,300 m (4,300 ft) in the central

Damage by Insects

Plate 68. Ponderosa pine saplings killed by *Ips paraconfusus*, the California Five-spined Ips.

Sierra Nevada east of Sacramento. Adults overwinter beneath the bark of infested trees. The California Five-spined Ips is one of the earliest colonizers of broken limbs and tops in the spring. Males tunnel through the outer bark into the phloem. A small chamber is then excavated, during which aggregation pheromones are

Plate 69. Top kill of ponderosa pine caused by *Ips paraconfusus*.

Plate 70. Boring dust marking the entry tunnel of an *Ips paraconfusus* beetle in a pine trunk.

Damage by Insects

Plate 71. Nuptial chamber and egg galleries of *Ips paraconfusus* in phloem.

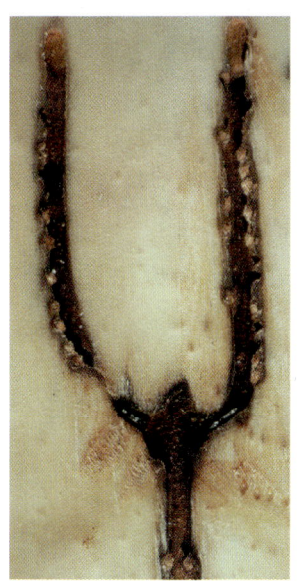

Plate 72. Eggs and egg gallery of *Ips paraconfusus* in phloem.

produced. Each male is joined by up to three females, and each female excavates an egg gallery in the phloem and outer xylem. *Ips* spp. are known as "engraver" beetles because the maternal gallery is excavated only 1 to 3 mm (.04 to .12 in.) into the sapwood. The resultant Y-shaped gallery is not packed with frass. Eggs are laid in niches out in the gallery wall. Larvae excavate small galleries in the phloem that are packed with frass and, at first, are perpendicular to the maternal galleries. At high densities, galleries produced by older larvae turn parallel to the maternal galleries. Each maternal gallery extends 10 to 15 cm (4 to 6 in.). The parental adults emerge after about two weeks and attack new hosts (pl. 72)(pl. 73).

SIGNIFICANCE: Multiple generations per year allow ample opportunity for outbreaks. At outbreak levels, this species can cause considerable tree mortality and, thus, it is an important pest in timber-producing regions. Outbreaks are usually short-lived, not lasting more than one year, although outbreaks have been

Damage by Insects

Plate 73. Pupae and callow adults of *Ips paraconfusus*.

known to last two to three years during periods of extreme drought. Beetle densities can increase in logging debris, leading to the infestation of nearby living trees. Trees with tops killed by *I. paraconfusus* are often subsequently killed by Western Pine Beetle, *D. brevicomis*, Mountain Pine Beetle, *D. ponderosae*, or both. The California Five-spined Ips can kill ornamental pines of very high aesthetic value, especially planted Monterey pines and ponderosa pines in native stands. In urban areas, the cost of removal of killed pines is often a major expense to home owners.

SIMILAR PESTS: Several species of bark beetles and wood-boring beetles infest and kill young trees with a diameter greater than 6 cm (2.5 in.) and the tops of older trees. The distinctive Y-shaped egg galleries of the California Five-spined Ips distinguish it from other bark beetles and wood borers, with the exception of the Pine Engraver, *I. pini*. To distinguish these species, note that California Five-spined Ips adults have five spines on each side of the posterior elytral declivity, whereas Pine Engraver adults have only four.

MANAGEMENT OPTIONS: Cut, lop, and scatter branches from treetops remaining after logging. Exposure to the sun will kill broods beneath the bark. Chip or burn logging debris, or pile it beneath clear plastic exposed to the sun. Elevated temperatures will kill broods beneath the bark. Insecticides can be applied to the bark of highly valued trees to prevent infestation.

For More Information

McNee, W. R., D. L. Wood, and A. J. Storer. 2000. Pre-emergence feeding in bark beetles (Coleoptera: Scolytidae). *Environmental Entomology* 29:495–501.

McPheron, L. J., S. J. Seybold, A. J. Storer, D. L. Wood, T. Ohtsuka, and I. Kubo. 1997. Effects of enantiomeric blend of verbenone on response of *Ips paraconfusus* to naturally produced aggregation pheromone in the laboratory. *Journal of Chemical Ecology* 23: 2825–39.

Schultz, D. E., and W. D. Bedard. 1987. *California Five-spined Ips*. Forest Insect and Disease Leaflet 102. Washington, D.C.: USDA Forest Service.

PINE ENGRAVER *Ips pini*
Pl. 74

HOSTS: Jeffrey pine, limber pine, lodgepole pine, and ponderosa pine.

DISTRIBUTION: Common in the southern Cascade Mountains, northern Sierra Nevada, high-elevation forests of the Sierra Nevada, and mountain ranges of southern California.

SYMPTOMS AND SIGNS: Young pines, about 2.5 to 25 cm (1 to 10 in.) in diameter, are killed. During periods of extreme drought and during outbreaks, trees up to 65 cm (26 in.) in diameter are also killed; in addition, the tops of large trees are killed. Usually, the first evidence of infestation is the typical color change of conifer foliage: from green to lime-green to yellow to reddish brown. Small pitch tubes and reddish brown boring dust in bark crevices and on cobwebs clinging to the main stem are symptomatic of infestation. Infestation of snow- or wind-broken branches and trees and logging debris is indicated by piles of reddish brown boring dust in crevices and on the ground beneath the branches or logs. On removal of bark, the gallery pattern typically consists of four maternal galleries—there may be as many as seven—radiating from one nuptial chamber. Larvae are creamy white, legless, with brown heads. Pine Engraver adults are distinctive in appearance: young (or callow) adults are light yellow whereas mature adults are brown, and the wing cases (elytra) have a scooped out appearance at the end (posterior declivity), with four spines on each side (pl. 74).

LIFE CYCLE: The Pine Engraver produces one or two generations per year, depending on latitude and elevation. Adults usually overwinter beneath the bark of infested trees or in the duff layer. Males emerge in April or May and excavate tunnels through the outer bark and into the phloem of host trees. A small nuptial chamber is

Damage by Insects

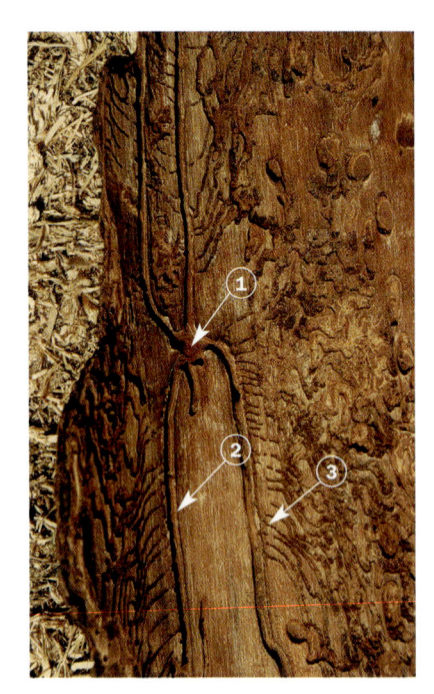

Plate 74. Gallery pattern of *Ips pini*, the Pine Engraver: (1) Nuptial chamber, (2) egg gallery, and (3) larval galleries.

then excavated, during which aggregation pheromones are released. Each male is joined by four to eight females; after mating, each female excavates an egg gallery about 10 to 15 cm (4 to 6 in.) long and etching the sapwood not more than 1 to 3 mm (.04 to .12 in.) deep; hence the name "engraver." The nuptial chamber and the egg galleries are not packed with frass, because the adults push frass out the entrance tunnel. Larvae develop in frass-packed galleries that are oriented perpendicular to the egg galleries. At high densities galleries produced by older larvae turn parallel to the maternal galleries. Parental adults emerge after about three to four weeks and attack new hosts.

SIGNIFICANCE: At outbreak levels, the Pine Engraver can cause considerable tree mortality and, thus, it is an important pest in timber-producing regions. Outbreaks are usually short-lived, not lasting more than one year, although outbreaks have been known to last two to three years during periods of extreme drought.

Population densities of this engraver beetle can increase in logging debris, leading to the infestation of nearby living trees. Trees with tops killed by Pine Engraver are often subsequently killed by Western Pine Beetle, *D. brevicomis,* Mountain Pine Beetle, *D. ponderosae* or Jeffrey Pine Beetle, *D. jeffreyi.* In urban areas, especially among native stands of lodgepole pine, Jeffrey pine, and ponderosa pine, the Pine Engraver can kill ornamental pines of very high aesthetic value. The cost of removal of killed pines is often a major expense to home owners.

SIMILAR PESTS: Young trees with a diameter greater than 6 cm (2.5 in.), and the tops of older trees, are killed by several species of bark and wood-boring beetles. The multiple egg galleries radiating from a central nuptial chamber of the Pine Engraver distinguish it from other bark beetles and wood borers, with the exception of the California Five-spined Ips. To distinguish these species, recall that the adult Pine Engraver have four spines on each side of the posterior declivity, whereas California Five-spined Ips adults have five.

MANAGEMENT OPTIONS: Cut and scatter branches from treetops remaining after logging. Exposure to the sun will kill broods beneath the bark. Chip, lop, or burn logging debris, or pile it beneath clear plastic exposed to the sun. Elevated temperatures will kill broods beneath the bark. Insecticides can be applied to the bark of highly valued trees to prevent infestation.

For More Information

Miller, D. R., K. E. Gibson, K. F. Raffa, S. J. Seybold, S. A. Teale, and D. L. Wood. 1997. Geographic variation in response of Pine Engraver, *Ips pini,* and associated species to the pheromone, lanierone. *Journal of Chemical Ecology* 23:2013–31.

Robertson, I. C. 1998. Flight muscle changes in male Pine Engraver beetles during reproduction: The effects of body size, mating status and breeding failure. *Physiological Entomology* 23:75–80.

Sartwell, C. R., R. F. Schmitz, and W. J. Buckhorn. 1971. *Pine Engraver* Ips pini *in the western states.* Forest Pest Leaflet 122. Washington, D.C.: USDA Forest Service.

MOUNTAIN PINE BEETLE   *Dendroctonus ponderosae*
Pls. 75–77

HOSTS: In California, principally coulter pine, lodgepole pine, ponderosa pine, sugar pine, western white pine, and whitebark pine.

Damage by Insects

DISTRIBUTION: Sierra Nevada, southern Cascade Mountains, Coast Ranges, and southern California mountain ranges, usually above 1,700 m (about 5,600 ft).

SYMPTOMS AND SIGNS: Foliage of infested trees changes from green to yellow to reddish brown to brown. Trees infested early in the season turn yellow to red in the same season, while trees infested later in the season turn yellow to red the following year. Egg galleries excavated by the females follow the grain of the main stem and are 30 to 90 cm (12 to 36 in.) long, with a short bend 2 to 5 cm (.75 to 2 in.) from the entrance tunnel. Pitch tubes and reddish brown boring dust removed by adults are often present in the bark crevices of infested trees. Sapwood is discolored by bluestain fungi carried in by attacking beetles. Mountain Pine Beetle adults are black, cylindrical, and hard shelled with a broadly rounded posterior and vary between 4 and 7.5 mm (0.16 and 0.30 in.) in length (pl. 75)(pl. 76)(pl. 77).

Plate 75. Lodgepole pine killed by *Dendroctonus ponderosae*, the Mountain Pine Beetle.

Plate 76. Western white pine killed by *Dendroctonus ponderosae* after damage by lightning strike. *(arrow)* Scar caused by lightning.

Damage by Insects

Plate 77. Pitch tubes marking entry points of *Dendroctonus ponderosae* on a lodgepole pine trunk.

LIFE CYCLE: Overwintering adults emerge in June from infested trees and attack other trees; this may continue into October. At higher elevations flights peak in June to early August. There is generally only one generation of this bark beetle per year, but at lower latitudes and elevations there may be a second generation and part of a third. Both larvae and adults overwinter under the bark of host trees.

SIGNIFICANCE: Mountain Pine Beetle is the principal insect pest of mature and overmature ponderosa pine, lodgepole pine, sugar pine, and western white pine, although younger trees, from 10 to 12 cm (4 to 5 in.) in diameter, can be killed. Mountain Pine Beetle typically colonizes trees weakened by lightning, fire, wind, and drought. Outbreaks of this bark beetle can kill millions of trees covering many hectares.

SIMILAR PESTS: Other bark beetles infest and kill the same tree species as the Mountain Pine Beetle; they may coexist in their pine hosts. Mountain Pine Beetle can be distinguished from other bark beetles by its long, straight egg gallery. Western Pine Beetle, *D. brevicomis*, makes a sinuous egg gallery and engraver beetles, *Ips* species, excavate multibranched egg galleries radiating from a central nuptial chamber.

TO STEMS AND LARGER BRANCHES

MANAGEMENT OPTIONS: As with other species of bark beetles, removal of trees weakened by drought, competition, dwarf mistletoe, root disease, fire, and wind before infestation by Mountain Pine Beetle will lower the risk of high-level tree mortality. Removing infested trees before emergence of this bark beetle in June may lower the risk of future infestations. Insecticides can be applied to the bark of highly valued trees to prevent infestation.

For More Information

Cerezke, H. F. 1995. Egg gallery, brood production, and adult characteristics of Mountain Pine Beetle, *Dendroctonus ponderosae* Hopkins (Coleoptera: Scolytidae), in three pine hosts. *The Canadian Entomologist* 127:955–65.

Olsen, W. K., J. M. Schmid, and S. A. Mata. 1996. Stand characteristics associated with Mountain Pine Beetle infestations in ponderosa pine. *Forest Science* 42:310–27.

Six, D. L., and T. D. Paine. 1998. Effects of mycangial fungi and host tree species on progeny survival and emergence of *Dendroctonus ponderosae* (Coleoptera: Scolytidae). *Environmental Entomology* 27:1393–1401.

Waters, W. E. 1985. The pine–bark beetle ecosystem: A pest management challenge, in *Integrated Pest Management in Pine–Bark Beetle Ecosystems* (W. E. Waters, R. W. Stark, and D. L. Wood, eds.). New York: John Wiley & Sons. 1–48.

Wilson, I. M., J. H. Borden, R. Gries, and G. Gries. 1996. Green leaf volatiles as antiaggregants for the Mountain Pine Beetle, *Dendroctonus ponderosae* Hopkins (Coleoptera: Scolytidae). *Journal of Chemical Ecology* 22:1861–75.

JEFFREY PINE BEETLE *Dendroctonus jeffreyi*
Pls. 78–80

HOST: Jeffrey pine.

DISTRIBUTION: Sierra Nevada, southern Cascade Mountains, and southern California mountain ranges, usually above 1,700 m (about 5,600 ft).

SYMPTOMS AND SIGNS: Symptoms are similar to those produced by Mountain Pine Beetle, *D. ponderosae*. Foliage of infested trees changes from green to yellow to reddish brown to brown. The foliage of infested trees usually turns yellow to red in the year following infestation. Egg galleries excavated by the females follow the grain of the main stem and are 30 to 90 cm (12 to 36 in.) long, with a short bend 2 to 5 cm (.75 to 2 in.) from the entrance

Damage by Insects

Plate 78. Jeffrey pines killed by *Dendroctonus jeffreyi*, the Jeffrey Pine Beetle, after root damage caused by high water levels.

tunnel. Pitch tubes and reddish brown boring dust removed by adults are often present in the bark crevices and cobwebs attached to the trunks of infested trees. Sapwood is discolored by bluestain fungi carried in by attacking beetles. Jeffrey Pine Beetle adults are black, cylindrical, and hard shelled with a broadly rounded posterior end, and are practically indistinguishable from adult Mountain Pine Beetles (pl. 78)(pl. 79)(pl. 80).

LIFE CYCLE: Overwintering adults emerge in June from infested trees and attack other trees; this may continue into October. At higher elevations flights peak in June to early August. There is one generation per year in the northern part of its range and two

Plate 79. Pitch tubes at entry point of *Dendroctonus jeffreyi*: (arrow) Adult beetle.

TO STEMS AND LARGER BRANCHES

Damage by Insects

Plate 80. *Dendroctonus jeffreyi* egg and larval galleries.

generations per year in the southern part. Both late larvae and adults overwinter under the bark of infested trees.

SIGNIFICANCE: This bark beetle is the principal insect pest of overmature and mature Jeffrey pine. Young trees over 30 cm (12 in.) in diameter are also frequently killed. Like Mountain Pine Beetle, Jeffrey Pine Beetle prefers to colonize trees weakened by lightning, disease, fire, wind, or drought.

SIMILAR PESTS: Engraver beetles, *Ips* species, infest young Jeffrey pines and the tops of older trees. Unlike the egg galleries of Jeffrey Pine Beetle, which are straight, unbranched, and parallel to the wood grain, the egg galleries of engraver beetles have multiple branches radiating from a central nuptial chamber.

MANAGEMENT OPTIONS: As with other species of bark beetles, removal of weakened trees before infestation by this species will lower the risk of high-level tree mortality. Removing infested trees before emergence of the Jeffrey Pine Beetle in June may lower the risk of future infestations. Insecticides can be applied to the bark of highly valued trees to prevent infestation.

For More Information

Paine, T. D., J. G. Millar, C. C. Hanlon, and J. S. Hwang. 1999. Identification of semiochemicals associated with Jeffrey Pine Beetle, *Dendroctonus jeffreyi*. *Journal of Chemical Ecology* 25:433–53.

Six, D. L., T. D. Paine, and J. D. Hare. 1999. Allozyme diversity and gene flow in the bark beetle, *Dendroctonus jeffreyi* (Coleoptera: Scolytidae). *Canadian Journal of Forest Research* 29:315–23.

Smith, R. H. 1971. *Jeffrey Pine Beetle*. Forest Pest Leaflet 11. Washington, D.C.: USDA Forest Service.

Damage by Insects

WESTERN PINE BEETLE — *Dendroctonus brevicomis*
Pls. 81–86

HOSTS: Coulter pine and ponderosa pine.

DISTRIBUTION: Sierra Nevada, southern Cascade Mountains, Coast Ranges, and southern California mountain ranges.

SYMPTOMS AND SIGNS: Pale green to yellow needles that later turn red and reddish brown throughout the canopy are usually the first noticeable evidence of infestation. The foliage of trees killed late in the season will turn yellow to red in the next season. White pitch tubes, 2.5 cm (1 in.) in diameter and few in number, suggest attack, but not necessarily infestation, by this bark beetle. More numerous reddish brown pitch tubes, less than 1.25 cm (.5 in.) in length and diameter, and reddish brown boring dust in the bark crevices and cobwebs attached to the bark, indicate an infestation by Western Pine Beetle. In winter, woodpeckers searching for larvae make small holes in the trunks of trees with green foliage, and the dislodged bark often appears as piles of chips around the root collar. Bluestain fungi carried in by the adults may discolor the sapwood around the egg and larval galleries. Western Pine Beetle adults are brown to black, cylindrical, and hard-shelled, with a broadly rounded posterior end, and vary between 3 and 5 mm (.12 and .20 in.) in length (pl. 81)(pl. 82)(pl. 83).

LIFE CYCLE: Females tunnel through a bark crevice into the phloem, where they excavate winding, criss-crossing egg galleries tightly packed with frass. They deposit eggs in small niches excavated in the phloem around the egg galleries. Larvae hatch from these eggs and feed on the phloem for about a week. This feeding activity results in small, threadlike galleries between the sapwood

Plate 81. Ponderosa pines killed by *Dendroctonus brevicomis*, the Western Pine Beetle.

TO STEMS AND LARGER BRANCHES

Damage by Insects

Plate 82. Female *Dendroctonus brevicomis* on a pitch tube around an entrance tunnel on a ponderosa pine trunk.

Plate 83. Bark removal by woodpeckers feeding on *Dendroctonus brevicomis* larvae.

and phloem. Larvae then turn into the outer bark to complete their development. Adults emerge through the outer bark, creating many small circular holes in the bark surface. There are one to four generations per year, depending on latitude and elevation (pl. 84)(pl. 85).
SIGNIFICANCE: The Western Pine Beetle is the most destructive insect pest of ponderosa pine and Coulter pine in California, infesting

Left: Plate 84. *Dendroctonus brevicomis* egg galleries in the phloem layer under ponderosa pine bark.

Below: Plate 85. Egg galleries of *Dendroctonus brevicomis* on ponderosa pine: Note the discoloration of the sapwood surrounding the galleries, caused by bluestain fungi.

Damage by Insects

Plate 86. Extensive tree mortality caused by *Dendroctonus brevicomis* in a ponderosa pine forest.

trees weakened by smog, annosum root disease, blackstain root disease, fire, drought, and lightning. Dead trees may appear singly or in small groups when Western Pine Beetle attacks at low population levels. At high population levels, groups of 10 to 20 trees are killed (pl. 86).

SIMILAR PESTS: Engraver beetles, *Ips* species, and Mountain Pine Beetle, *D. ponderosae,* also infest and kill ponderosa pine and Coulter pine. Western Pine Beetle is easily distinguished from them by its distinctive winding, criss-crossing egg galleries. The Mountain Pine Beetle makes a straight egg gallery parallel to the wood grain, and engraver beetles make multibranched egg galleries that radiate from a central nuptial chamber.

MANAGEMENT OPTIONS: Stand management practices such as thinning and maintaining a mixture of species help mitigate against the high-level tree mortality associated with large areas of dense, mature pines. Insecticides can be applied to the bark of highly valued trees to prevent infestation.

For More Information

Bertram, S. L., and T. D. Paine. 1994. Influence of aggregation inhibitors (verbenone and ipsdienol) on landing and attack behavior of *Dendroctonus brevicomis* (Coleoptera: Scolytidae). *Journal of Chemical Ecology* 20:1617–29.

———. 1994. Response of *Dendroctonus brevicomis* LeConte (Coleoptera: Scolytidae) to different release rates and ratios of aggregation semiochemicals and the inhibitors verbenone and ipsdienol. *Journal of Chemical Ecology* 20:2931–41.

Dahlsten, D. L., D. L. Rowney, and R. N. Kickert. 1997. Effects of oxidant air pollutants on Western Pine Beetle (Coleoptera: Scolytidae) populations in southern California. *Environmental Pollution* 96:415–23.

Damage by Insects

DeMars, C. J., and B. H. Roettgering. 1982. *Western Pine Beetle.* Forest Insect and Disease Leaflet 1. Washington, D.C.: USDA Forest Service.

Everitt, J. H., J. V. Richerson, J. Karges, M. A. Alaniz, M. R. Davis, and A. Gomez. 1997. Detecting and mapping Western Pine Beetle infestations with airborne videography, global positioning system and geographic information system technologies. *Southwestern Entomologist* 22:293–300.

Hobson, K. R., J. R. Parmeter, Jr., and D. L. Wood. 1994. The role of fungi vectored by *Dendroctonus brevicomis* LeConte (Coleoptera: Scolytidae) in occlusion of ponderosa pine xylem. *The Canadian Entomologist* 126:277–82.

Waters, W. E. 1985. The pine–bark beetle ecosystem: A pest management challenge, in *Integrated Pest Management in Pine–Bark Beetle Ecosystems* (W. E. Waters, R. W. Stark, and D. L. Wood, eds.). New York: John Wiley & Sons. 1–48.

RED TURPENTINE BEETLE *Dendroctonus valens*
Pls. 87–91

HOSTS: Probably all pines in California, but most common in Jeffrey pine, lodgepole pine, Monterey pine, ponderosa pine, and sugar pine.

DISTRIBUTION: Sierra Nevada, southern Cascade Mountains, North Coast Ranges, and southern California mountain ranges. Red Turpentine Beetle is commonly found in urban plantings of Monterey pine and in urbanized native Monterey pine, ponderosa pine, and Jeffrey pine forests.

SYMPTOMS AND SIGNS: Large pitch masses, 2 to 5 cm (.75 to 2 in.) wide, are found on the lower 0 to 3 m (0 to 9 ft) of the trunk and on large, exposed roots. Also, earlier attacks are indicated by the presence of small, whitish gray granules of crystallized resin on the soil at the root collar. Deep holes in the bark of the lower trunk, made by woodpeckers, indicate the presence of Red Turpentine Beetle beneath the bark. Trees with light green to yellow or red foliage will exhibit Red Turpentine Beetle galleries beneath the bark of the lower trunk. Females excavate a vertical gallery about 2 cm (.75 in.) wide and up to 30 cm (12 in.) long in the phloem. Galleries can also occur in larger roots. Eggs are laid in groups of up to 100 along the gallery wall. Larvae feed communally and excavate a large cavity in the phloem. A black-staining fungus carried by the adults discolors the sapwood adjacent to

Damage by Insects

the egg gallery and around the excavated chamber. The reddish brown Red Turpentine Beetle adult has a broadly rounded posterior end, averages 8 mm (0.3 in.) in length, and is the largest of all bark beetles in its genus (pl. 87)(pl. 88)(pl. 89)(pl. 90)(pl. 91).

Left: Plate 87. Pitch tubes of *Dendroctonus valens*, the Red Turpentine Beetle, on a Monterey pine trunk.

Below: Plate 88. *Dendroctonus valens* pitch tubes on an exposed Monterey pine root.

Left: Plate 89. Adult *Dendroctonus valens*.

TO STEMS AND LARGER BRANCHES

Damage by Insects

Plate 90. *D. valens* larvae in feeding chamber.

Plate 91. *D. valens* pupa: Note blackstain fungus.

LIFE CYCLE: Red Turpentine Beetle may produce from one generation every two years to three generations every year, depending on elevation and latitude. In the central Sierra Nevada at 1,300 m (4,300 ft) there is usually one generation per year. Adults emerge in spring and immediately attack trees and stumps; they then reemerge and attack other trees and stumps throughout summer and early fall, at which time larvae and young adults usually begin overwintering beneath the bark. However, in warm, extended fall periods, new brood adults may emerge and infest additional trees and stumps from recently cut trees.

SIGNIFICANCE: Red Turpentine Beetle is attracted to wounded trees and freshly cut stumps, which release certain terpene hydrocarbons attractive to this insect. Trees attacked over the course of several years are weakened but rarely killed by Red Turpentine Beetle. Rather, attacked trees are more frequently killed by Western Pine Beetle, *D. brevicomis* (ponderosa pine), Mountain Pine Beetle, *D. ponderosae* (sugar pine and ponderosa pine), and the California Five-spined Ips, *Ips paraconfusus* (Monterey pine). Stumps are readily infested and may produce populations that attack nearby living trees. Red Turpentine Beetle attacks are often associated with pines infected with blackstain root disease and, in the case of Monterey pine, pitch canker.

SIMILAR PESTS: Other bark beetles infest the same tree species and are often present simultaneously; however, attacks by the other insects do not usually extend downward to the stump and root levels. In addition, the larvae of other bark beetle species feed in distinct, individual tunnels rather than in a communal feeding cavity. The reddish brown adult Red Turpentine Beetle is much larger than other bark beetles, which are dark brown or black. Red Turpentine Beetles predispose pines to infestation by other tree-killing bark beetles.

MANAGEMENT OPTIONS: Avoid wounding trees. A pesticide may be applied to the bark on the lower trunk of highly valued trees to prevent further attacks by Red Turpentine Beetle and lower the risk of tree mortality caused by *Dendroctonus* spp. and *Ips* spp. Covering the lower trunk with window screen is an effective alternative to pesticides.

For More Information

Hobson, K. R., D. L. Wood, L. G. Cool, P. R. White, T. Ohtsuka, I. Kubo, and E. Zavarin. 1993. Chiral specificity in responses by the bark beetle *Dendroctonus valens* to host kairomones. *Journal of Chemical Ecology* 19:1837–46.

Smith, R. H. 1971. *Red Turpentine Beetle*. Forest Pest Leaflet 55. Washington, D.C.: USDA Forest Service.

Svihra, P. 1995. Prevention of Red Turpentine Beetle attack by sevimol and Dragnet. *Journal of Arboriculture* 21:221–24.

DOUGLAS-FIR BEETLE *Dendroctonus pseudotsugae*
Pls. 92, 93

HOST: Douglas-fir.

DISTRIBUTION: Throughout the range of Douglas-fir in the central to northern Sierra Nevada, southern Cascade Mountains, and Coast Ranges.

SYMPTOMS AND SIGNS: The foliage of trees killed by this bark beetle turns from yellow to red to reddish brown. The foliage of trees killed later in the season will turn yellow to red in the next season. Resin streaming from the upper trunk, and reddish brown boring dust in the bark crevices and clinging to cobwebs, indicate attack by the Douglas-fir Beetle. By removing pieces of bark, egg and larval galleries typical of this species may be seen: a straight, vertical egg gallery in the phloem, with larval galleries perpendicular to, and alternating in groups on each side of, the maternal gallery. The galleries are typically filled with frass. Adults are cylindrical, hard shelled, with a broadly rounded posterior end, and are 4.4 to 7.0 mm (0.16 to 0.26 in.) in length. They are dark brown to black with reddish elytra (pl. 92)(pl. 93).

Damage by Insects

Plate 92. Douglas-fir tree killed by *Dendroctonus pseudotsugae*, the Douglas-fir Beetle.

Plate 93. *Dendroctonus pseudotsugae* larvae in larval galleries extending outward from the egg gallery.

LIFE CYCLE: There is one generation per year. Both adults and larvae overwinter: overwintering adults emerge and attack new host trees from April to early June, whereas adults developing from overwintering larvae attack trees in late June and July. Females tunnel through the bark of susceptible trees and release a pheromone that attracts other adults. Twenty to 100 eggs are laid in groups in grooves cut in the phloem on alternating sides of the egg gallery wall. The larval galleries fan out from these egg grooves.

SIGNIFICANCE: Douglas-fir Beetle often infests trees stressed by drought or injured by fire, logging, or wind. Overmature trees weakened by other insect attacks or succumbing to root disease are another favored target. Healthy trees are normally resistant; they are threatened only in the face of high populations of Douglas-fir Beetle. Such increases occur after a prolonged drought or a significant fire or windstorm, or among trees infected with blackstain root disease or *Armillaria* root disease.

SIMILAR PESTS: The Douglas-fir Engraver, *Scolytus unispinosus*, and the Flatheaded Fir Borer, *Melanophila drummondi*, also kill Douglas-fir trees, especially those injured by drought or disease.

Both insects are often present in the tops and limbs of trees killed by the Douglas-fir Beetle. To distinguish these species, the egg galleries must be examined: the Douglas-fir Engraver makes a short, vertical egg gallery with a well-defined nuptial chamber near the center. The larvae of Flatheaded Fir Borers feed in individual mines not associated with an egg gallery. The larvae of the Douglas-fir Beetle feed in galleries that fan out from a long, straight, vertical egg gallery.

MANAGEMENT OPTIONS: Prompt salvage of injured, weakened, or blown-down trees lessens the probability of a population buildup that might threaten neighboring healthy trees. Beetle repellents, or antiattractants, can be deployed on standing and downed trees to prevent colonization by the Douglas-fir Beetle. Insecticides can be applied to the bark of highly valued trees to prevent infestation.

For More Information

Furniss, M. M., and P. W. Orr. 1978. *Douglas-fir Beetle.* Forest Insect and Disease Leaflet 5. Washington, D.C.: USDA Forest Service.

Ross, D. W., and G. E. Daterman. 1995. Efficacy of an antiaggregation pheromone for reducing Douglas-fir Beetle, *Dendroctonus pseudotsugae* Hopkins (Coleoptera: Scolytidae), infestation in high risk stands. *The Canadian Entomologist* 127:805–11.

———. 1997. Using pheromone-baited traps to control the amount and distribution of tree mortality during outbreaks of the Douglas-fir Beetle. *Forest Science* 43:65–70.

———. 1998. Pheromone-baited traps for *Dendroctonus pseudotsugae* (Coleoptera: Scolytidae): Influence of selected release rates and trap designs. *Journal of Economic Entomology* 91: 500–6.

Ross, D. W., and H. Solheim. 1997. Pathogenicity to Douglas-fir of *Ophiostoma pseudotsugae* and *Leptographium abietinum,* fungi associated with the Douglas-fir Beetle. *Canadian Journal of Forest Research* 27:39–43.

DOUGLAS-FIR ENGRAVER *Scolytus unispinosus*

HOSTS: Douglas-fir, and occasionally cedar *(Thuja)* species and fir *(Abies)* species.

DISTRIBUTION: Throughout the range of Douglas-fir in the Sierra Nevada, southern Cascade Mountains, North Coast Ranges, and Central Coast Ranges.

SYMPTOMS AND SIGNS: On living trees, entrance tunnels are identified by resin streaming or small pitch tubes. In logging debris, and in branches and tops of trees killed by the Douglas-fir Beetle, *Dendroctonus pseudotsugae,* reddish brown boring dust marks the entrance tunnel. Young saplings or pole-sized trees infested by the Douglas-fir Engraver exhibit yellow to red foliage. Adults are dark brown to black, cylindrical, and have a sawed-off appearance resulting in a sharply angled-in posterior such that the dorsum is longer than the venter. They average less than 3 mm (0.11 in.) in length.

LIFE CYCLE: Females initiate vertical egg galleries from a centrally located nuptial chamber that is excavated into the outer sapwood. Larvae tunnel horizontally from egg niches cut in the walls of the maternal gallery. There are two generations per year in California, and larvae overwinter.

SIGNIFICANCE: *S. unispinosus* populations build up in slash and during dry periods kill young Douglas-firs in logging areas. Also, young Douglas-firs with blackstain root disease and *Armillaria* root disease are infested by this bark beetle. This species also breeds in the tops and branches of trees killed by the Douglas-fir Beetle.

SIMILAR PESTS: Galleries of this species are often intermixed with those of the Hemlock Engraver, *S. tsugae,* and the Douglas-fir Pole Beetle, *Pseudohylesinus nebulosus,* a secondary bark beetle, and all three are similar in appearance. However, Douglas-fir Engraver has a well-defined nuptial chamber that etches the sapwood.

MANAGEMENT OPTIONS: Timely disposal of logging debris will prevent the buildup of large populations and avoid infestation of adjacent, young Douglas-firs. Insecticides are generally not used to prevent attacks on logging debris and young trees under forested conditions. In urban settings, this option may be considered.

For More Information

McMullen, L. H., and M. D. Atkins. 1962. The life history and habits of *Scolytus unispinosus* LeConte (Coleoptera: Scolytidae) in interior British Columbia. *The Canadian Entomologist* 94:17–25.

FIR ENGRAVER *Scolytus ventralis*
Pls. 94–96

HOSTS: Grand fir, red fir, and white fir.

DISTRIBUTION: Sierra Nevada, Coast Ranges, southern Cascade Mountains, and southern California mountain ranges.

Damage by Insects

Plate 94. White fir killed by *Scolytus ventralis*, the Fir Engraver.

SYMPTOMS AND SIGNS: Reddish brown boring dust appears in bark crevices and on cobwebs, usually indicating a tree is infested. Extensive pitch streaming from entrance holes, and sunken patches of bark, indicate unsuccessful attacks. The Fir Engraver, unlike other bark beetles, excavates horizontal egg galleries, that is, across the grain. Wood surrounding the galleries is soon discolored by the pathogenic fungus, *Trichosporium symbioticum*, introduced by this insect. Attacks often kill branches and the tops of trees. A large number of attacks on the main stem results in death of the entire tree. Usually the foliage turns yellow to red in the year following infestation. In old-growth trees, many new leaders may be produced over the years, giving the crowns the appearance of candelabras. Adult Fir Engravers are black, cylindrical, and have a sawed-off appearance resulting in a sharply angled-in posterior such that the dorsum is longer than the venter. They average about 4.0 mm (0.16 in.) in length (pl. 94)(pl. 95).

Plate 95. Extensive white fir mortality caused by *Scolytus ventralis*.

TO STEMS AND LARGER BRANCHES

Damage by Insects

LIFE CYCLE: Adults tunnel into bark crevices on the main stem of trees usually greater than 10 cm (4 in.) in diameter. If the initial attack is successful, other adults are attracted to the tree. Egg galleries are excavated horizontally, across the grain, in the phloem and sapwood. The sapwood is etched 1 to 2 mm (.04 to .08 in.) by the tunneling female. Two galleries, each 5 to 15 cm (2 to 6 in.) long, are often excavated on each side of the entrance tunnel. Larvae hatch from eggs laid in niches in the gallery wall. Their galleries are oriented vertically on the stem and are 13 to 15 cm (5 to 6 in.) long. Larvae pupate in cells in the phloem and new adults emerge through the bark from these cells. Emergence occurs from June to September, but the greatest numbers emerge in July and August. There are one to two generations per year, depending on latitude and elevation; most often there is one generation per year, with the larvae overwintering in the outer phloem. The biology of the Fir Engraver is unique among tree-killing bark beetles. This species often infests isolated sections of the trunk, where broods fail to complete development and emerge without killing the tree. The patch of cambium destroyed by the feeding larvae is overgrown and remains on the trunk as a brown pocket of resin (or pitch)-impregnated wood. Wetwood often develops in the sapwood beneath the pitch pockets. Pitch pockets, together with tree ring counts, can be used to date Fir Engraver outbreaks (pl. 96).

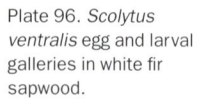

Plate 96. *Scolytus ventralis* egg and larval galleries in white fir sapwood.

SIGNIFICANCE: Trees weakened by root disease, especially annosus root disease, dwarf mistletoe, defoliation by the Douglas-fir Tussock Moth and spruce budworms, high stand density, and drought are infested by the Fir Engraver. Extensive mortality in northern California was related to the extreme drought of 1987–1992. Trees killed during this drought are especially notable in the Lake Tahoe Basin.

SIMILAR PEST: The Silver Fir Beetle, *Pseudohylesinus sericeus,* also infests true firs and excavates a transverse egg gallery similar to that of the Fir Engraver; however, the egg galleries of the Fir Engraver score the sapwood deeply, whereas those of the Silver Fir Beetle are more superficial.

MANAGEMENT OPTIONS: Lower stand density will promote rapid growth of the residual stand. In root-diseased areas, stands can be converted to species not susceptible to annosum root disease. In urban areas, infested trees should be felled and broods killed by removing and destroying the bark. High-risk trees can be removed before infestation by this destructive bark beetle.

For More Information

Ferrell, G. T. 1986. *Fir Engraver*. Forest Insect and Disease Leaflet 13. Washington, D.C.: USDA Forest Service.

Ferrell, G. T., W. J. Otrosina, and C. J. Demars. 1994. Predicting susceptibility of white fir during a drought-associated outbreak of the Fir Engraver, *Scolytus ventralis,* in California. *Canadian Journal of Forest Research* 24:302–5.

Macias-Samano, J. E., J. H. Borden, R. Gries, H. D. Pierce, G. Gries, and G. G. S. King. 1998. Primary attraction of the Fir Engraver, *Scolytus ventralis. Journal of Chemical Ecology* 24:1049–75.

CEDAR BARK BEETLES *Phloeosinus* spp.
Pl. 97

HOSTS: Coast redwood, giant sequoia, incense-cedar, Monterey cypress, Port Orford-cedar, and western juniper.

DISTRIBUTION: Throughout California, wherever suitable hosts are found.

SYMPTOMS AND SIGNS: Reddish brown boring dust is found in the bark crevices of recently killed trees, broken branches, and logging debris. Trees with yellow or red foliage have egg galleries and new broods in the phloem when the bark is removed. Adults tunneling in twigs cause the tips to break and hang from the

Damage by Insects

Plate 97. Egg and larval galleries of cedar bark beetles (*Phloeosinus* sp.) etching the sapwood of a Monterey cypress.

undamaged portion of the twig. Adults are reddish brown to black, cylindrical with a broadly rounded posterior end, and range from 2 to 4 mm (0.08 to 0.16 in.) in length.

LIFE CYCLE: Typically, adults excavate one egg gallery that may be oriented across or parallel to the grain of the wood, depending on species. The egg galleries and egg niches are often deeply etched into the sapwood. There are one to three generations per year (pl. 97).

SIGNIFICANCE: Cedar bark beetles infest twigs, branches, and the main stem of trees weakened by drought, disease, fire, and logging, and they readily colonize broken branches. During the severe drought in the late 1980s and early 1990s these bark beetles were found infesting incense-cedars throughout the Sierra Nevada. Through their tip-feeding habits they may cause extensive flagging on ornamental cedars and redwoods. Some species may transmit the fungus *Seiridium cardinale,* causal agent of cypress canker.

Damage by Insects

SIMILAR PESTS: Bark beetles other than *Phloeosinus* spp. are rarely found colonizing the hosts listed above.

MANAGEMENT OPTIONS: Preventing attacks by these bark beetles with insecticides is attempted only rarely and is confined to urban environments where extensive twig flagging occurs.

For More Information   See general references.

TWIG BEETLES *Pityophthorus* **spp.**
Pls. 98–100

HOSTS: Many coniferous hosts including Douglas-fir, hemlock, pines, spruces, and true firs.

DISTRIBUTION: Throughout California, wherever suitable hosts occur.

SYMPTOMS AND SIGNS: Twig beetles typically colonize shade-weakened branches in the lower canopy but may also colonize branch tips in the upper canopy. These insects can usually be found in lower branches with yellow and red foliage, especially during autumn. They also colonize cones, seedlings, saplings, fallen trees, and uninfested areas on the main stem of trees killed by other bark beetles. Adults are reddish brown to black, cylindrical, and range from 1.5 to 3.0 mm (0.06 to 0.12 in.) in length (pl. 98).

Plate 98. Fading ponderosa pine tips in the upper canopy, caused by twig beetles (*Pityophthorus* spp.).

TO STEMS AND LARGER BRANCHES

Damage by Insects

Right: Plate 99. Adult *Pityophthorus* spp. on the foliage of a Monterey pine.

Below: Plate 100. *Pityophthorus* spp. galleries etched into the sapwood of a 25-mm (1-in.) diameter Monterey pine branch.

LIFE CYCLE: Females excavate several egg galleries in the phloem and outer xylem. These galleries radiate from a nuptial chamber usually initiated by males. These distinctive galleries are found on larger branches, more than 2.5 cm (1 in.) in diameter. On small branches the galleries are found in the pith and surrounding sapwood. One to many generations are produced each year, depending on species, elevation, and latitude (pl. 99)(pl. 100).

SIGNIFICANCE: Twig beetles are generally of little economic significance. However, *P. orarius* kills healthy cone-bearing shoots in Douglas-fir seed orchards, thus reducing a valuable cone crop. Recently, many species have been implicated as vectors of the pitch canker pathogen, *Fusarium circinatum*, in California, where they colonize infected shoots and cones of Monterey pine.

SIMILAR PESTS: Other bark beetle genera are found in similar habitats and have similar appearances and gallery patterns. These include *Myeloborus* spp., *Pityogenes* spp., and *Pityokteines* spp. Species in these genera are rarely pests.

MANAGEMENT OPTIONS: Insecticides are available to reduce losses of cone crops to *P. orarius*.

For More Information

Hoover, K., D. L. Wood, A. J. Storer, J. W. Fox, and W. E. Bros. 1996. Transmission of the pitch canker fungus, *Fusarium subglutinans* f. sp.

pini to Monterey pine, *Pinus radiata,* by cone- and twig-infesting beetles. *The Canadian Entomologist* 128:981–94.

Rappaport, N. G., and D. L. Wood. 1994. *Pityophthorus orarius* Bright (Coleoptera: Scolytidae) in a northern California Douglas-fir seed orchard: Effect of clone, tree vigor, and cone crop on rate of attack. *The Canadian Entomologist* 126:1111–18.

Storer, A. J., T. R. Gordon, D. L. Wood, and P. Bonello. 1997. Current and future impacts of pitch canker disease of pines. *Journal of Forestry* 10 (12):21–26.

Storer, A. J., D. L. Wood, and T. R. Gordon. 1999. Modification of co-evolved insect-plant interactions by an exotic plant pathogen. *Ecological Entomology* 24:238–43.

FLATHEADED WOOD BORERS or METALLIC WOOD-BORING BEETLES

Buprestidae

Pls. 101, 102

HOSTS: All California trees.

DISTRIBUTION: Throughout the forested regions of California.

SYMPTOMS AND SIGNS: Winding, flattened, oval tunnels tightly packed with frass, often arranged in a concentric ring pattern, are found in the phloem and sapwood of infested trees. The white or yellowish larvae are elongate, legless, clearly segmented, with a hardened plate on both the top and bottom of the first segment behind the head. They are called flatheaded wood borers because of their flattened body with the first three segments behind the head two or more times the width of the rest of the body. Flat, oval exit holes in trees or lumber indicate infestation. The adults, called metallic wood-boring beetles, are elongate with short, serrated antennae, and many species have black, bronze, or green metallic colors (pl. 101).

Plate 101. Larva and feeding gallery of the California Flatheaded Wood Borer (family Buprestidae): *(arrow)* Note the wide, flattened thoracic segments.

Damage by Insects

LIFE CYCLE: Females deposit eggs in flattened masses in bark crevices. Newly hatched larvae tunnel through the bark and feed in the phloem, excavating a winding, gradually widening gallery. After a variable time, depending on the species, the larvae turn inward and feed in the sapwood. Larvae reach maturity in fall and may spend winter in the larval or pupal stage. The larvae of some species tunnel into the bark to construct a pupal chamber at maturity. Larvae may survive for many years in dry wood before finally reaching the adult stage. Adults emerge in spring and summer and feed on foliage and twigs before laying eggs. Some species visit flowers and feed on pollen, while others (*Melanophila* spp.) are attracted to smoke from forest fires.

SIGNIFICANCE: Most species of metallic wood-boring beetles infest dead and dying trees. Trees killed by fire, wind, root disease, or bark beetles all provide favorable host material. Metallic wood-boring beetles will also infest cut logs before they are milled into lumber. The holes made by these species reduce the value of the lumber. Larvae can survive in milled lumber, later emerging through finished walls in buildings. Lumber showing flatheaded wood borer tunnels is not acceptable in some export markets. Two species, the California Flatheaded Borer, *Melanophila californica,* and the Flatheaded Fir Borer, *M. drummondi*, infest and kill trees weakened by drought, root disease, or dwarf mistletoe, and *M. drummondi* has been reported to kill apparently healthy trees (pl. 102).

SIMILAR PESTS: Roundheaded wood borers (the larvae of longhorned beetles in the family Cerambycidae) and horntail larvae (larvae of wood wasps in the family Siricidae) tunnel in wood, making galleries similar to those of flatheaded wood borers. The tunnels of roundheaded wood borers are loosely packed with coarser woody material than those of flatheaded wood borers. The tunnels of horntail larvae are tightly packed with fine-grained material but are circular rather than oval in cross section. The wide, flattened segments behind the head of the flatheaded wood borer clearly distinguish it from roundheaded wood borers and horntail larvae.

MANAGEMENT OPTIONS: The use of risk-rating systems to identify and harvest trees likely to be infested with flatheaded wood borers will reduce losses. Prompt salvage of dead and dying trees will reduce damage and reduction in value of lumber. Kiln drying will kill larvae in infested lumber. Highly valued trees weakened by drought or other agents may be treated with insecticides to

Damage by Insects

Plate 102. Weakened ponderosa pines killed by the California Flatheaded Wood Borer, *Melanophila californica*.

prevent attack by flatheaded wood borers and bark beetles. Such treatment will likely prevent infestation by flatheaded wood borers and by roundheaded borers, and wood wasps.

For More Information
Barr, W. F., and E. G. Linsley. 1947. Distributional and biological notes on the species of the subgenus *Melanophila* occurring in western North America (Coleoptera: Buprestidae). *Pan-Pacific Entomologist* 23:162–66.

Lyon, R. L. 1970. *California Flatheaded Borer*. Forest Pest Leaflet 24. Washington, D.C.: USDA Forest Service.

ROUNDHEADED WOOD BORERS or LONGHORNED BEETLES   Cerambycidae
Pls. 103–105

HOSTS: All California trees.

DISTRIBUTION: Throughout the forested regions of California.

SYMPTOMS AND SIGNS: Winding galleries packed with fine to coarse frass, wood chips, or both are found in the phloem and xylem of trees; tunnels in the sapwood have an oval cross section. Larvae (roundheaded wood borers) are elongate, almost legless,

Damage by Insects

Plate 103. Feeding galleries of round-headed wood borers (longhorned beetle larvae; family Cerambycidae) in white fir sapwood. *(arrow)* Smaller galleries to the right are those of the Fir Engraver, *Scolytus vertalis*.

clearly segmented, cylindrical, and white or yellowish in color. The larvae have prominent, strong jaws and a hardened plate on top of the first segment behind the head. In most species the first three segments behind the head are slightly wider than the rest of

Plate 104. Roundheaded wood borer (longhorned beetle larva; family Cerambycidae) and feeding gallery in Douglas-fir sapwood.

Plate 105. Longhorned beetle adult and emergence holes in a white fir trunk.

the body. The adults (longhorned beetles) are elongate beetles with prominent, long antennae. The antennae are at least half the length of the body and for some species (e.g., *Acanthocinus*) the antennae are longer than the body (pl. 103)(pl. 104)(pl. 105).

LIFE CYCLE: Adult females deposit eggs in bark crevices or in slits cut in the bark of weakened or recently killed trees. When the eggs hatch the larvae bore through the bark into the phloem. The larvae feed in the phloem, extending winding mines that become wider as the larvae grow. After several weeks to several months, depending on the species, the larvae turn inward to feed in the sapwood. Most species have a life cycle of one or two years, with some of the largest species requiring three to five years (e.g., *Prionus* spp.). Each mature larva transforms to the pupal stage in a chamber at the end of its feeding tunnel. Adults chew an exit tunnel to emerge from the wood. Adults feed on foliage or twigs and some species visit flowers.

SIGNIFICANCE: Trees weakened by drought or root disease may be killed by attacks of aggressive species such as the Roundheaded Fir Borer, *Tetropium abietis*. The larval tunnels may extend 5 to 7.6 cm (2 to 3 in.) into the wood of the host tree, thus degrading the value of the logs or wood products made from infested trees. Larvae will also infest cut logs before they are milled into lumber, especially if the logs are left in the forest. Lumber with roundheaded wood borer tunnels is unacceptable in some export markets. The larvae of some species survive and continue their development in lumber cut from infested logs. When these insects reach maturity the adults emerge through finished walls and floors in new buildings. The Newhouse Borer, *Arhopalus productus,* frequently found in fire-killed Douglas-fir, and the Firtree Borer, *Semanotus litigiosus,* frequently found in true firs, can cause damage to new buildings. Roundheaded wood borers that are feeding in trees killed by bark beetles sometimes are so numerous and grow so quickly that they destroy a large part of the bark beetle brood. The adults of some species (*Monochamus* spp.) that feed on pine twigs are vectors of *Bursaphelenchus xylophilus,* the Pinewood Nematode, an important pathogen of pines.

SIMILAR PESTS: Flatheaded wood borers (larvae of metallic wood-boring beetles in the family Buprestidae) and horntail larvae (larvae of wood wasps in the family Siricidae) also bore in wood, making tunnels similar to those of roundheaded borers. The larvae of all three groups may coexist in the same tree, with their galleries intermingled. The first three segments behind the head of flat-headed wood borers are more than twice as wide as the rest of the body, whereas those of roundheaded wood borers are only slightly wider than the rest of the body. Horntails have a terminal spine at the end of their bodies and do not feed in the phloem.

MANAGEMENT OPTIONS: Prompt salvage of dead and dying timber will reduce the damage caused by longhorned beetles by limiting the penetration of larval tunnels into the wood. The larvae do not survive in the underwater portion of logs in a mill pond. Kiln drying of lumber kills any larvae present and avoids the problem of beetles emerging from lumber in new buildings. However, lumber used in California markets is usually air-dried for only a short time before use. Highly valued trees weakened by drought or other damage may be treated with insecticides to prevent attack by longhorned beetles.

For More Information

Hanks, L. M. 1999. Influence of the larval host plant or reproductive strategy of Cerambycid beetles. *Annual Review of Entomology* 44:483–505.

Linsley, E. G. 1958. The role of Cerambycidae in forest, urban and agricultural environments. *Pan-Pacific Entomologist* 34:105–24.

———. 1959. Ecology of the Cerambycidae. *Annual Review of Entomology* 4:99–138.

Wickman, B. E. 1965. *Insect caused deterioration of windthrown timber in northern California, 1963–64.* Research Paper PSW 20. Washington, D.C.: USDA Forest Service.

HORNTAILS or WOOD WASPS Siricidae

Pls. 106, 107

HOSTS: All California conifers.

DISTRIBUTION: Throughout the forested regions of California.

SYMPTOMS AND SIGNS: Larvae are cylindrical, yellowish white, and legless with a small, pointed spine on the end of the abdomen. They excavate tunnels in wood that are circular in cross section and packed with fine boring dust. The larval tunnels are confined to the wood and do not start in the phloem. Emerging adults leave circular emergence holes in the bark. The females of most species are metallic, blue-black, or brown with a prominent ovipositor that extends beyond the end of the abdomen. Males lack the ovipositor and the terminal segments of the abdomen

Damage by Insects

Plate 106. Horntail larva in feeding tunnel: *(arrow)* Note the spine at the end of the body.

Plate 107. Horntail emergence holes in a lodgepole pine trunk.

are colored red, orange, or yellow. Both sexes have a hornlike projection on the last abdominal segment of the adults: hence the name horntail (pl. 106)(pl. 107).

LIFE CYCLE: About a dozen species of wood wasps with similar life cycles and habits are found in California. Adults emerge from dead trees in mid to late summer. Females use their long ovipositor to insert eggs through the bark into the wood of recently dead or dying trees. They are strongly attracted to smoke and are often seen depositing eggs in fire-damaged trees before the fire is out. The larvae feed entirely in the wood, requiring two years or more to reach maturity. Mature larvae turn outward to construct a pupal cell near the surface of the wood. Transformation to the adult form is completed in early summer.

SIGNIFICANCE: Larval tunnels may extend 5 to 8 cm (2 to 3 in.) into the wood, which lowers the value of lumber cut from infested trees. The larvae are able to survive and complete their development in lumber cut from infested logs. When these insects reach maturity they may emerge through the finished walls of new buildings. The combination of an extended life cycle and the habit of boring deep within infested trees makes these insects especially likely to be spread in international log or lumber shipments. *Sirex noctilio,* a European species, in combination with a symbiotic fungus, *Amylostereum areolatum,* causes extensive mortality of young pines in New Zealand and Australia. This type of damage does not now occur in California. However, *S. noctilio* and the fungus could be introduced through shipments of logs from Australasia.

SIMILAR PESTS: Flatheaded borers (Buprestidae) and roundheaded borers (Cerambycidae) also bore into wood, making tunnels similar to those of horntail larvae. The larvae of all three types of insect may be present in the same tree, with their galleries intermingled. Flatheaded and roundheaded borers start feeding in the phloem and leave winding galleries there; in addition, flatheaded and roundheaded borers make oval rather than circular tunnels, and do not have a terminal spine at the end of their bodies. Wood wasp emergence holes are very similar to those of *Monochamus* (longhorned beetle) species.

MANAGEMENT OPTIONS: Prompt salvage of dead and dying trees, especially fire-killed trees, will reduce damage by limiting the penetration of larval tunnels into the wood. Kiln drying of lumber will destroy any larvae present and avoids the problem of adult horntails emerging from wood in new buildings.

For More Information

Cameron, E. A. 1967. Notes on *Sirex juvencus californicus* (Hymenoptera: Siricidae) with a description of the male and a key to the California species of *Sirex*. *The Canadian Entomologist* 99:18–24.

Rawlings, G. B. 1984. Recent observations on the *Sirex noctilio* population in *Pinus radiata* forests in New Zealand. *New Zealand Journal of Forestry* 5:411–21.

BIOTIC DISEASES

Biotic Diseases

ELYTRODERMA DISEASE OF PINES   *Elytroderma*
Pls. 108–111   *deformans*

HOSTS: Coulter pine, Jeffrey pine, knobcone pine, lodgepole pine, pinyon pine and ponderosa pine.

DISTRIBUTION: In most California pine forests above about 1,200 to 1,500 m (4,000 to 5,000 ft). *Elytroderma* disease of pines commonly occurs around lakes or along streams.

SYMPTOMS AND SIGNS: Stunting and dying of one-year-old needles, and branch deformation and brooming are the most obvious symptoms of the disease; new growth may be infected but remains green. One-year-old needles die from the tip to the base, beginning in late spring, and turn reddish brown. Shortly thereafter, elongate, black spore-producing bodies of the fungus appear on the surface of the dead needles. Needles turn straw colored as the fruiting bodies mature. The fungus also grows into twigs and branches. Dead, brown flecks of tissue can usually be seen under the inner bark of branches infected for three or more years. Large brooms, resembling those caused by dwarf mistletoe, indicate very old, systemic infections (pl. 108)(pl. 109)(pl. 110)(pl. 111).

Plate 108. Jeffrey pine needle tip mortality caused by *Elytroderma deformans,* causal agent of *Elytroderma* disease of pines.

Biotic Diseases

Left: Plate 109. Large brooms on Jeffrey pine resulting from infection by *Elytroderma deformans*.

Below: Plate 110. Elongate, black fruiting bodies of *Elytroderma deformans* on dead ponderosa pine needles.

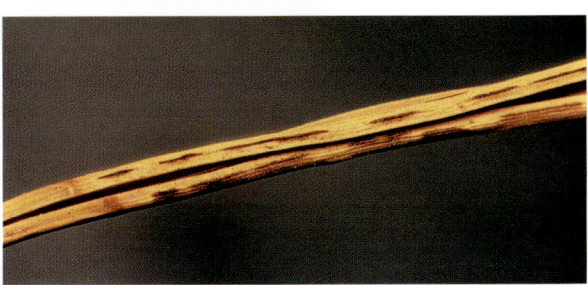

Plate 111. Brown, necrotic flecks in the inner bark of a ponderosa pine twig infected with *Elytroderma deformans*.

NEEDLE DISEASES

Biotic Diseases

LIFE CYCLE: Wind-borne spores released from mature fruiting bodies on needles infect newly developing needles in late spring and early summer when favorable weather permits. The fungus then grows in the needles and into the branch and bud, where new needles are infected the following year. Low levels of infection occur in most years, but when favorable weather conditions coincide with new host growth, epidemics occur.

SIGNIFICANCE: Trees of all sizes and ages are infected with the fungus. Small pines may be severely damaged, particularly when infected in the terminal shoot. Larger, heavily infected trees are weakened and often die from the combined effects of disease, drought, and insect attack.

SIMILAR PESTS: Brooms caused by dwarf mistletoes, *Arceuthobium* spp., and some rust fungi, as well as tip moth infestations, resemble those caused by *Elytroderma* disease of pines.

MANAGEMENT OPTIONS: There is no direct control for this fungus, which develops optimally under cool, humid conditions such as around lakes, streams, and meadows. In highly hazardous sites, where the disease is common and severe, replacement of infected trees with nonhost trees is one way to reduce damage. Pruning large brooms will not effectively prevent spread of the disease.

For More Information

Scharpf, R. F. 1990. Life cycle and epidemiology of *Elytroderma deformans* on pines in California, in *Recent research on foliage diseases* (W. Merrill and M. E. Ostry, eds.). General Technical Report WO-56. Washington, D.C.: USDA Forest Service.

Scharpf, R. F., and R. V. Bega. 1981. *Elytroderma* disease reduces growth and vigor, increases mortality of Jeffrey pines at Lake Tahoe Basin, California. Research Paper PSW-155. Washington, D.C.: USDA Forest Service.

TRUE FIR NEEDLE CAST   *Lirula abietis-concoloris*
Pls. 112–114

HOSTS: Grand fir, noble fir, Pacific silver fir, red fir, and white fir.

DISTRIBUTION: Throughout California, wherever hosts grow. True fir needle cast usually occurs at endemic levels in forests and only occasionally intensifies to high levels of infection when certain weather conditions occur.

SYMPTOMS AND SIGNS: Endemic levels of infection are not easily recognized, as only a few scattered, dead needles appear on

Biotic Diseases

Plate 112. One-year-old needles of white fir killed by *Lirula abietis-concoloris,* causal agent of true fir needle cast.

branches. However, during outbreaks the disease is evident, usually on a single year's complement of needles: the brown, dead needles stand out in contrast to the green, living ones. The dead needles do not drop off readily but remain on branches for a year or more. Elongate, conspicuous, black fruiting bodies of the fungus develop on the lower surface of the dead needles. The fruiting bodies usually appear as a single solid or broken line down the center of the needle (pl. 112)(pl. 113)(pl. 114).

Left: Plate 113. Long, black fruiting bodies of *Lirula abietis-concoloris.*

Below: Plate 114. Open fruiting body of *Lirula abietis-concoloris,* showing the spore-bearing surface.

NEEDLE DISEASES

LIFE CYCLE: Only newly developing needles are infected by this fungus. In most years, just a few needles are infected. However, in some years, when summer rains coincide with the maturation of fruiting bodies, abundant spore dispersal and infection of new foliage occur. Infected needles develop normally and no evidence of the fungus is present until the following year, when needle browning, fruiting body development, and dieback take place. Fruiting bodies and spores are produced once, and only in the year after infection of the newly killed needles.

SIGNIFICANCE: True fir needle cast causes little damage in native forest stands. Endemic levels of infection have no noticeable effect on trees, but during outbreak years small trees and seedlings may suffer some growth reduction. There is no evidence that trees are killed by this disease or are weakened, predisposing them to insect attack. The greatest damage occurs in firs grown as Christmas trees near native fir stands in which the disease occurs. A single outbreak of disease can reduce the quality of a Christmas tree or make it unmerchantable.

SIMILAR PEST: Another fungus, *Virgella robusta,* causes a very similar disease but two rows of fruiting bodies are found on the lower needle surface instead of one. Little is known about its occurrence, distribution, and damage, however.

MANAGEMENT OPTIONS: No attempts have been made to control this disease in native forest stands. In Christmas tree plantations, fungicide sprays have been shown to be effective in preventing infection of young foliage. Locating Christmas tree plantations in areas at least a few hundred yards away from naturally infected stands will probably lessen the chances for disease buildup.

For More Information

McCain, A. H., and R. F. Scharpf. 1987. Control of needle cast of white fir. *California Plant Pathology-Cooperative Extension Service, University of California* 78.

Scharpf, R. F. 1988. Epidemiology of *Lirula abietis-concoloris* on white fir in California. *Plant Disease* 72:855–58.

RED BAND NEEDLE BLIGHT   *Mycosphaerella pini*

Pls. 115–117

HOSTS: Bishop pine, knobcone pine, lodgepole pine, Monterey pine, Monterey × knobcone hybrid pine, ponderosa pine, and western white pine.

Biotic Diseases

DISTRIBUTION: Worldwide in distribution. In California, red band needle blight is limited to native and planted pines growing along the Pacific Coast.

SYMPTOMS AND SIGNS: On needles of all ages, yellow to tan spots at the infection sites are the first symptoms of red band needle blight. These spots enlarge, turn brownish red, and produce around each needle a red band that is diagnostic for the disease. Often the needle base remains green; only the portion near the tip dies. Small, black fruiting bodies eventually develop in the red bands. In extreme cases nearly all the foliage is infected, resulting in trees with only a few live needles (pl. 115)(pl. 116)(pl. 117).

LIFE CYCLE: The fungus has two spore stages: the perfect, or sexual, stage and the imperfect, or asexual, stage. In its perfect stage the fungus is known as *Mycosphaerella pini,* or sometimes *Scirrhia*

Plate 115. Extensive Monterey pine needle death caused by *Mycosphaerella pini,* causal agent of red band needle blight.

Plate 116. *Mycosphaerella pini* infection of Monterey pine.

Plate 117. Fruiting body of *Mycosphaerella pini* on western white pine.

NEEDLE DISEASES

pini; in its imperfect stage the fungus is called *Dothistroma pini.* Spread and buildup of the disease are caused mainly by imperfect stage spores produced in small, black fruiting bodies erupting through the epidermis of red-banded, infected needles. When wet, these fruiting bodies produce abundant spores that are blown or washed onto other needles. The spores germinate and infect only needles—not twigs or branches. This cycle of infection can be repeated more than once as long as moist conditions are present, resulting in severe infection. Dead needles can produce spores for about a year before they fall from the tree.

SIGNIFICANCE: Red band needle blight has caused severe defoliation of some California pines, particularly Monterey pines planted along the northern California coast. Otherwise, the disease is sporadic in occurrence and of little consequence, except perhaps in coastal landscape plantings and Christmas tree plantations.

SIMILAR PESTS: Several other fungi attack and defoliate pines in California. The characteristic red banding on needles distinguishes infections by this fungus from others, however.

MANAGEMENT OPTIONS: No management is usually necessary for red band needle blight. In areas where this disease is the cause of significant damage to ornamental and nursery trees, fungicidal treatments may be used, or clones of resistant pines may be planted.

For More Information

Merrill, W., and M. E. Ostry, eds. 1990. *Recent research on foliage diseases.* General Technical Report WO-56. Washington, D.C.: USDA Forest Service.

Wagener, W. W. 1967. *Red band needle blight of pines: A tentative appraisal for California.* Research Note PSW-153. Washington, D.C.: USDA Forest Service.

DOUGLAS-FIR NEEDLE CAST *Rhabdocline pseudotsugae, R. weirii*
Pls. 118–120

HOSTS: Bigcone Douglas-fir and Douglas-fir.

DISTRIBUTION: Throughout the range of Douglas-fir in California.

SYMPTOMS AND SIGNS: The first symptom, appearing in late fall and early winter, consists of yellowish, 1- to 2-mm (0.04 to 0.08 in.) spots on needles. This is followed by the most characteristic symptoms of Douglas-fir needle cast: reddish brown needle spots and

Biotic Diseases

Plate 118. Douglas-fir foliage showing symptoms of infection by a *Rhabdocline* species, causal agents of Douglas-fir needle cast.

Plate 119. Fruiting bodies of a *Rhabdocline* species.

the elongate, brown, cushion-shaped fruiting bodies that appear in spring on either side of the midrib on the underside of infected needles. Mature fruiting bodies appear as irregular slits in the needle surface, exposing a layer of orange to orange-brown spores. An outbreak of the disease can result in the loss of one or more years' complement of needles (pl. 118) (pl. 119)(pl. 120).

LIFE CYCLE: Douglas-fir needle cast is caused by two species and at least three subspecies of *Rhabdocline*. Because these fungi differ only slightly in their biology they are treated here as one organism. In spring, new needles are infected with windborne spores discharged from mature fruiting bodies

Plate 120. Douglas-fir defoliated by a *Rhabdocline* species.

NEEDLE DISEASES

on previously infected needles. Cool, humid conditions are necessary for infection; wind and rain are also favorable. Douglas-fir needle cast develops in infected needles for one to two years, depending on the fungal species. No obvious stunting or growth abnormalities occur in the infected, developing needles. In spring, fruiting bodies and spores are produced. By the end of summer nearly all the diseased needles are dead and have been cast.

SIGNIFICANCE: This disease occurs at endemic levels in Douglas-fir forests in the West and only occasionally builds up to damaging proportions. In native stands, attack by these *Rhabdocline* fungi usually causes loss of only a single year's needles and little damage to mature, healthy trees. Seedlings and small trees, however, may be more seriously damaged. Successive years of infection can cause growth loss and mortality. The most serious damage from this disease occurs in Christmas tree plantations: loss of even a portion of the foliage on a Christmas tree may make it unmerchantable.

SIMILAR PESTS: Swiss needle cast, caused by *Phaeocryptopus gaeumannii,* causes symptoms that closely resemble the symptoms caused by *Rhabdocline* species. However, the small, black fruiting bodies of the Swiss needle cast fungus are distinctly different from the reddish brown, elongate fruiting bodies of *Rhabdocline* species.

MANAGEMENT OPTIONS: In Christmas tree plantations, early detection and application of fungicides will minimize losses. Management is not necessary in forests.

For More Information

Castagner, G. A., R. S. Byther, and K. L. Riley. 1990. Maturation of apothecia and control of *Rhabdocline* needlecast on Douglas-fir in western Washington, in *Recent research on foliage diseases* (W. Merrill and M. E. Ostry, eds.). General Technical Report WO-56. Washington, D.C.: USDA Forest Service.

BOTRYOSPHAERIA CANKER *Botryosphaeria dothidea*
Pls. 121–123

HOSTS: Coniferous hosts: giant sequoia and incense-cedar. *B. dothidea* actually infects a wide variety of hosts and has received more attention for its infection of fruit trees, particularly apple.

DISTRIBUTION: Common on giant sequoia and, occasionally, incense-cedar when these species are planted outside their native range, particularly in warm, low-elevation regions. *Botryosphaeria* canker

Biotic Diseases

Plate 121. Giant sequoia twigs killed by *Botryosphaeria dothidea*, causal agent of *Botryosphaeria* canker.

has not been found on giant sequoia or incense-cedar in natural forest stands.

SYMPTOMS AND SIGNS: The initial symptom is scattered twig and branch dieback. Recently killed branches are called flags, that is, they retain their needles and appear reddish brown; branches dead for longer periods turn gray-brown and eventually shed their dead needles. Points of infection are typically associated with wounding of some sort, such as recent pruning wounds, cracks, and overall stress. Infected branches develop cankers, which are sunken, necrotic areas that often exude amber-colored resin. With time the fungus can build up and spread throughout much of the crown (pl. 121)(pl. 122)(pl. 123).

LIFE CYCLE: Little is known about the life cycle of this fungus on giant sequoia or incense-cedar. Hot, dry growing conditions are required for development of the disease.

SIGNIFICANCE: This disease is of no significance in natural forest stands. When giant sequoia in particular is planted in hot areas outside its native range, severe damage often occurs. Ornamental plantings and Christmas tree plantations are especially vulnerable to damage.

SIMILAR PESTS: No known diseases produce symptoms similar to those of *Botryosphaeria* canker.

Biotic Diseases

Right: Plate 122. Giant sequoia, branch mortality caused by *Botryosphaeria dothidea*.

Below: Plate 123. Point of infection by *Botryosphaeria dothidea:* (arrow) Note the amber resin drops.

MANAGEMENT OPTIONS: No chemical treatment has been shown to reduce damage from this disease. Giant sequoias that have been heavily irrigated during the hot, dry season tend to have less *Botryosphaeria* canker. Therefore, relieving moisture stress through irrigation may help reduce damage caused by this disease.

For More Information
See general references.

CYTOSPORA CANKER OF TRUE FIRS — *Cytospora abietis*

Pls. 124, 125

HOSTS: Rarely Douglas-fir, true firs.

DISTRIBUTION: Throughout the natural range of true firs in California. *Cytospora* canker has been reported infrequently on Douglas-fir in the extreme northern part of the state.

SYMPTOMS AND SIGNS: Brick-red flagging caused by cankers girdling and killing branches is the most striking symptom of the disease in spring and summer. Later in the year the foliage on these branches dries and turns brown to tan. Other, less easily observed symptoms and signs of infection are the cankers themselves, which are sunken, dead patches of bark tissue; resin exudation at the canker site; small, pimple-like fruiting bodies embedded in the dead bark; and orange, threadlike spore masses exuded from the fruiting bodies when wet (pl. 124)(pl. 125).

LIFE CYCLE: *C. abietis* is a weak parasite that usually attacks only trees that have been weakened by other agents including insects, fire, other diseases, drought, and human activities. One primary agent is dwarf mistletoe (*Arceuthobium* spp.), which predisposes firs to infection by *Cytospora* canker. In some stands of fir nearly a fourth of all branches infected with dwarf mistletoe are also infected with *C. abietis*. The fungus spreads by means of spores, which splash onto surrounding trees and branches when it rains. Small wounds and openings in the bark, such as those made by insects and dwarf mistletoe, are favored sites of infection. The fungus grows within the inner bark and eventually girdles and kills the branch.

Plate 124. Red fir branch mortality caused by *Cytospora abietis*, causal agent of *Cytospora* canker of true firs.

Biotic Diseases

Plate 125. Threadlike spore masses exuding from *Cytospora abietis* fruiting bodies on red fir.

SIGNIFICANCE: *Cytospora* canker is one of the more damaging diseases of true firs in California. Trees of all sizes and ages are affected. Damage is especially heavy in stands also infected with dwarf mistletoe. Trees on poor sites or those on the eastern slopes of the Sierra Nevada suffer severe damage, especially during periods of drought or following insect damage. Crown loss, top dieback, and mortality often result from heavy infection.

SIMILAR PESTS: Branch tips exhibiting yellow to red foliage caused by *Cytospora* canker are distinctive and not readily confused with other diseases of firs.

MANAGEMENT OPTIONS: Reducing levels of dwarf mistletoe in fir stands will in turn reduce the incidence of *Cytospora* canker. Also, any methods that reduce tree stress from drought or other predisposing factors will make firs less susceptible to infection and damage. Avoid planting firs on drought-prone or high-stress sites. Proper spacing and nutrition, and adequate soil moisture, will help lessen the incidence of disease in Christmas tree plantations.

For More Information

Scharpf, R. F. 1969. *Cytospora abietis* associated with dwarf mistletoe on the true firs in California. *Phytopathology* 59:1657–58.

Scharpf, R. F., and H. H. Bynum. 1975. Cytospora *canker of true firs.* Forest Pest Leaflet 146. Washington, D.C.; USDA Forest Service.

PITCH CANKER   *Fusarium circinatum*

Pls. 126–130

HOSTS: At least 12 species of native and introduced pines. Of the native pines, bishop pine and Monterey pine have sustained the most damage. Coulter pine, gray pine, Knobcone pine, ponderosa pine, shore pine, and Torrey pine have also been found with pitch canker, mostly in planted stands. Douglas-fir is the only known nonpine host.

DISTRIBUTION: California coast, from Mendocino County to San Diego County. Until recently, pitch canker was known only as a disease of pines in the southeastern United States. It was first found in California in 1986 on planted Monterey pines near Santa Cruz. The three native Monterey pine forests of California are now heavily infested with this pathogen. These forests are on the Monterey Peninsula, near Point Año Nuevo in San Mateo County, and in Cambria, in San Luis Obispo County. The present distribution of pitch canker ranges from infected bishop pines along the coast in Mendocino County to diseased pines, especially those grown as Christmas trees, in Los Angeles County and San Diego County. The greatest concentration of disease is on planted Monterey pines in the central coastal counties and in the three native forests.

SYMPTOMS AND SIGNS: The most conspicuous symptom of the disease is needle yellowing and wilting, and then red discoloration followed by branch dieback. Abundant resin accumulation usually accompanies dieback. Resinous cankers and resin-soaked, yellowish wood are also noticeable symptoms of the disease. Large branch or bole cankers often produce copious amounts of resin that flows down the trunk or coats lower branches and foliage. The various symptoms of pitch canker can appear at any time of year. The tops of trees and entire trees exhibiting advanced symptoms are often killed by engraver beetles, *Ips* spp., and Red Turpentine Beetle, *Dendroctonus valens*. Positive identification of the disease must be confirmed in the laboratory (pl. 126)(pl. 127)(pl. 128)(pl. 129)(pl. 130).

LIFE CYCLE: *F. circinatum* spores can be wind-borne, but several species of cone, twig, and engraver beetles that infest Monterey pine carry propagules of the fungus and are likely responsible for most transmission of the disease to branch tips, cone whorls, trunk, and branches. Seeds from infected cones are often contaminated with the fungus. The fungus can be internal, on the surface of the seed, or

Biotic Diseases

Plate 126. Monterey pine shoot infected with *Fusarium circinatum*, the causal agent of pitch canker: (arrow) Note discoloration at the infection site.

Plate 127. Monterey pine branch mortality caused by *F. circinatum*: (arrow) Note pitch exudation from the infection site.

Plate 128. Honey-colored, resin-soaked sapwood in an *F. circinatum*-infected Monterey pine branch with the outer bark removed.

both. Isolated, young plantations of pines are believed to have become infected from contaminated seedlings. The climatic conditions that regulate spread and infection are not well understood.

SIGNIFICANCE: It is not known how pitch canker was introduced and became established in California. Many native and exotic pine species and Douglas-fir are susceptible to infection and damage by pitch canker, and therefore this disease should be considered a serious threat to California's forests, even though it is now limited in distribution to the coastal counties. Spread by

Biotic Diseases

Plate 129. Multiple infections by *F. circinatum* on Monterey pine.

Plate 130. Heavy resin flow from a Monterey pine stem canker caused by *F. circinatum*.

insects, seed contamination, infected seedlings, and infected woody material indicates a high probability of eventual introduction and establishment in other areas.

SIMILAR PESTS: Some symptoms caused by a few other insects and diseases may be confused with the yellow to red branch tips caused by pitch canker. The resinous canker between the green and discolored foliage of a pitch canker-infected branch helps distinguish this pathogen from other agents. For example, western gall rust, caused by *Peridermium harknessii*, tip moths, and *Diplodia* blight, caused by *Sphaeropsis sapinea*, on branches, and Sequoia Pitch Moth, *Synanthedon sequoiae*, resin masses can often superficially resemble pitch canker.

MANAGEMENT OPTIONS: Transport of infected wood or trees into new areas should be avoided because fungus-carrying insects in the wood could spread the pathogen to uninfected trees. The pathogen could also be transported to uninfested areas via green waste and thus become associated with insects in the new area. Many exotic pine species, such as Canary Island pine, Aleppo pine, and Italian stone pine, sustain much less damage than Monterey pine. Resistant Monterey pines have been identified and resistant clones are being propagated.

For More Information

Gordon, T. R., D. Okamoto, A. J. Storer, and D. L. Wood. 1997. Susceptibility of five landscape pines to pitch canker disease, caused by *Fusarium subglutinans* f. sp. *pini*. *HortScience* 33:868–71.

Gordon, T. R., A. J. Storer, and D. L. Wood. 2001. The pitch canker epidemic in California. *Plant Disease* 85:1128–39.

McNee, W. R., D. L. Wood, A. J. Storer, and T. R. Gordon. 2002. Incidence of the pitch canker pathogen and associated insects in intact and chipped Monterey pine branches. *The Canadian Entomologist* 134:47–58.

Storer, A. J., T. R. Gordon, D. L. Wood, and P. Bonello. 1997. Current and future impacts of pitch canker disease of pines. *Journal of Forestry* 10 (12):21–26.

DIPLODIA BLIGHT *Sphaeropsis sapinea* or *Diplodia pinea*
Pls. 131–133

HOSTS: Most pines and occasionally Douglas-fir, especially below 600 m (2,000 ft) and along the California coast.

DISTRIBUTION: All low-elevation and coastal areas of California, wherever suitable hosts are found. *Diplodia* blight is found above 600 m (2,000 ft), where wet weather extends into late spring.

SYMPTOMS AND SIGNS: Shoot dieback, affecting primarily the new shoots, is the principal symptom of this fungal disease. Infected shoots remain green until hot weather dries them; they may be curled or bent. The current year's dead needles are usually stunted, as the shoots are often killed before they are fully grown. The bark and wood of infected shoots are resin soaked,

Plate 131. Ponderosa pine branches infected with *Sphaeropsis sapinea*, causal agent of *Diplodia* blight: (1) Infection of first-year growth; (2) infection of second-year growth.

Biotic Diseases

Plate 132. Multiple infections by *Sphaeropsis sapinea* on ponderosa pine.

Plate 133. Resin soaking of a ponderosa pine branch tip infected with *Sphaeropsis sapinea*.

and when the cankers are cut into the tissues vary in color from amber to almost black. Only a small amount, if any, of the previous year's growth is affected, and nonresinous portions of branches support live green needles. Gray needles on shoots killed in previous years may be present. The amount of shoot dieback varies from tree to tree, and within a tree may be more extensive in some parts of the crown than in others. The asexual fruiting bodies, or pycnidia, of this fungus are black and spherical, less than .5 mm (.02 in.) in diameter, and are embedded within the needles or bark of the infected stem. Usually only the beaks of the pycnidia protrude through the surface of infected tissue and resemble numerous small, black pinpoints (pl. 131)(pl. 132)(pl. 133).

LIFE CYCLE: The fungus can infect new needles and shoots, older woody tissue through wounds, and second-year cones. In spring of the year following infection, pycnidia are produced from which spores are released. Infected cones can be found on trees without other types of infection, and spores from cones may provide a source of inoculum for shoot infections. Wet springs can result in disease buildup, and tree stress due to poor site conditions, age, or other agents may increase susceptibility.

CANKER DISEASES

SIGNIFICANCE: *Diplodia* blight is generally less serious than it appears: it is not systemic or inherently chronic, and the majority of diseased trees recover. Mature and overmature trees are most affected, and outbreaks may not recur the following year if there are no late spring rains. Rarely, entire branches, the treetop, or both may be killed. Severe levels of infection on already stressed trees may result in susceptibility to attack by other pests.

SIMILAR PESTS: This disease may be confused with pitch canker caused by *Fusarium circinatum*. Laboratory isolation of the pitch canker pathogen may be necessary to distinguish between these diseases. However, pitch canker generally kills a larger section of the branch tip, does not stain the wood black, and may infect a tree at any time of year. Pinpoint-sized black spots on infected tissue are useful characteristics to identify *Diplodia* blight.

MANAGEMENT OPTIONS: Fungicidal treatments in spring are generally not practical or necessary. To reduce infection through wounds, avoid pruning during wet spring months. Renewed growth in dry years will often conceal previous damage.

For More Information

Owen, D. R. 1999. Diplodia *blight of pines caused by* Sphaeropsis sapinea (Diplodia pinea). Tree Note 23. Sacramento: California Department of Forestry and Fire Protection.

Peterson, G. W. 1981. Diplodia *blight of pines*. Forest Insect and Disease Leaflet 161. Washington, D.C.: USDA Forest Service.

PHOMOPSIS CANKER OF DOUGLAS-FIR

Phomopsis lokoyae or *Diaporthe lokoyae*

Pl. 134

HOST: Douglas-fir.

DISTRIBUTION: Most often found in the northern coastal counties of California. *Phomopsis* canker of Douglas-fir grows occasionally in natural and planted stands elsewhere along the coast and is found only rarely on Douglas-fir inland.

SYMPTOMS AND SIGNS: Branch and, occasionally, leader dieback is the first symptom of the disease. Sunken, reddish brown cankers result when the fungus infects young twigs and then grows into the adjoining branch. The cankers are usually several times longer than they are wide. Branches and tops that are girdled by the fungus turn reddish brown. Trees up to 8 cm (3.1 in.) in diameter are often killed

Biotic Diseases

Plate 134. Reddish to purple, necrotic canker on Douglas-fir caused by *Phomopsis lokoyae,* causal agent of *Phomopsis* canker of Douglas-fir.

by this fungus. Small, black fruiting bodies embedded in the dead bark can often be observed in spring and summer (pl. 134).

LIFE CYCLE: Spread of the fungus is mainly by asexual spores of the imperfect stage, called *Phomopsis lokoyae*. The spores are produced in small (.5 mm, or 0.02 in.), black fruiting bodies embedded in the surface of the cankered bark. The sexual stage, in which form the fungus is called *Diaporthe lokoyae*, is less well understood. It is known that free water, during the period of spore release, is necessary for infection; and yet, paradoxically, severe outbreaks occur more often in trees stressed by prolonged drought. It is probable that wounds on small twigs or branches caused by insects or other agents are the primary infection sites.

SIGNIFICANCE: Small trees are often badly damaged or killed by *Phomopsis* canker. Top dieback resulting from trunk infection is common. Larger trees exhibit branch dieback and some top kill. Trees that survive often have large, dead trunk cankers that fail to heal and may be invaded by secondary insects or decay fungi. Plantations of Douglas-firs grown for the Christmas tree market can be severely damaged by *Phomopsis* canker.

SIMILAR PESTS: *Dermea* canker, caused by *Dermea pseudotsugae,* produces symptoms on Douglas-fir much like those of *Phomopsis* canker. However, the fruiting bodies of *D. pseudotsugae* are two to three times larger than those of *P. lokoyae* and appear as small, black disks on the surface of the dead tissue. The Douglas-fir Engraver, *Scolytus unispinosus,* is often associated with *Dermea* canker.

MANAGEMENT OPTIONS: Methods that maintain tree vigor and reduce stress will help prevent infection with *P. lokoyae*. No chemical control for this disease is available.

For More Information
Funk, A. 1973. Phomopsis (Diaporthe) *canker of Douglas-fir in British Columbia*. Forest Pest Leaflet 60. Victoria: Forest Insect and Disease Survey, Natural Resources Canada, Canadian Forest Service.

DWARF MISTLETOES *Arceuthobium* spp.
Pls. 135–138

HOSTS: In California, there is at least one species of dwarf mistletoe for every genus in the family Pinaceae. Pines, Douglas-fir, and true firs suffer the greatest damage. Dwarf mistletoes are generally given a common name that reflects their principal host, for example, Jeffrey pine dwarf mistletoe.

DISTRIBUTION: From sea level along the coast to more than 3,300 m (10,800 ft) in the alpine forests of the Sierra Nevada. Several species extend north into Oregon and south into Mexico.

SYMPTOMS AND SIGNS: The symptoms produced by dwarf mistletoe vary, depending on the mistletoe species and the host on which it occurs. Large, conspicuous clumps of twigs and foliage, or witches' brooms, are an easily recognized symptom. Brooms are common on Douglas-fir, hemlock, and some pines but less so on true firs. Branch or stem swelling at the point of infection is another clear symptom of dwarf mistletoe. On true firs, large trunk swellings often develop after the parasitic plant has been embedded in the trunk for many years. Old infections, particularly of pines, often develop into open, resinous cankers. The clearest sign of dwarf mistletoe is the plant itself: small, leafless, yellow-green shoots, 2.5 to 25 cm (1 to 10 in.) long, in tufts or scattered along branches (pl. 135)(pl. 136).

LIFE CYCLE: Dwarf mistletoes are small, leafless plants that bear inconspicuous flowers and fruits containing a single seed. Male and female plants are produced separately. The seeds are forcibly discharged and can travel up to 30 m (about 100 ft) before sticking to and infecting new host trees. Typically, young trees in the

Biotic Diseases

Plate 135. Female *Arceuthobium* spp., a dwarf mistletoe, on a Monterey pine branch: Note swelling of the branch.

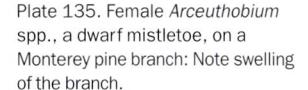

Plate 136. Male *Arceuthobium* spp. plant on Monterey pine.

understory are infected by mistletoe seeds ejected from infected overstory trees. Infection takes place when the radicle of a germinating seed penetrates the thin, green bark of young branch tissue. Within one to three years, at the earliest, young shoots of the mistletoe plant break through the branch surface. Flowering and pollination occur in summer and mature fruits are produced the following year, usually in fall. The shoots are short lived and replaced every few years, whereas the root system grows and expands indefinitely within the living branch (pl. 137)(pl. 138).

SIGNIFICANCE: In California and throughout much of the West, dwarf mistletoes are considered among the most damaging forest disease agents. Mortality, growth reduction, poor wood quality, reduced seed production, and hazardous trees are the consequences of infected forest stands. Infected trees and stands are also predisposed to insect infestation, especially bark beetles and other diseases, particularly during periods of drought or other tree stress.

SIMILAR PESTS: With the exception of a larger, leafy mistletoe that grows on white fir, no other parasitic plants could be confused with dwarf mistletoe. Some other diseases, particularly *Elytroderma*

Biotic Diseases

Plate 137. Female *Arceuthobium* spp. shoot with mature fruits.

Right: Plate 138. Male *Arceuthobium* spp. shoots.

disease of pines, caused by *Elytroderma deformans*, and some rust fungi, cause brooming that could be confused with dwarf mistletoe.

MANAGEMENT OPTIONS: Dwarf mistletoes are obligate parasites. Therefore, the mistletoe dies when the infected part of the host dies or is removed. In the forest various combinations of harvesting of infected trees, thinning, favoring nonsusceptible hosts, and pruning have been used to reduce losses from dwarf mistletoes. Eradication of the pest is not necessary. Keeping the disease at low to moderate levels in the forest is usually enough to prevent unacceptable damage. Highly valued trees, or trees in high-use areas, will require greater levels of management to reduce loss or to minimize hazards.

For More Information

Hawksworth, F. G., and R. F. Scharpf, eds. 1984. *Biology of dwarf mistletoes: Proceedings of the symposium.* General Technical Report RM-111. Fort Collins: Rocky Mountain Forest Range Experiment Station, USDA Forest Service.

Hawksworth, F. G., and D. Wiens. 1996. *Dwarf mistletoes: Biology, pathology, and systematics.* Agriculture Handbook 709. Washington, D.C.: USDA Forest Service.

Scharpf, R. F., and J. R. Parmeter, Jr. 1978. *Proceedings of the symposium on dwarf mistletoe control through forest management.* General Technical Report PSW-31. Washington, D.C.: USDA Forest Service.

Biotic Diseases

JUNIPER MISTLETOES
Pls. 139, 140

Phoradendron juniperinum, P. densum

HOSTS: Cypress and junipers.

DISTRIBUTION: Juniper mistletoe, *Phoradendron juniperinum*, occurs from the eastern slopes of the Cascade Mountains, through the upper elevations of the Sierra Nevada, and into the San Bernardino Mountains of southern California. Dense mistletoe on juniper, *P. densum*, occurs from Baja California north along the eastern slopes of the Sierra Nevada and Cascade Mountains, and along the coast from San Diego to Monterey. Dense mistletoe on cypress, *P. densum*, occurs from the northwest coastal mountains south to Marin County; inland from Butte County north along the western slopes of the Sierra Nevada and Cascade Mountains.

SYMPTOMS AND SIGNS: No brooming or branch deformation is produced by these mistletoes. The clearest sign produced by these mistletoes is the parasitic plant itself: green clusters growing within the crown of the tree, sometimes with such abundance that the crown is partially obscured. The main difference between these two species is that dense mistletoe has conspicuous leaves, whereas the leaves of juniper mistletoe are less than 3 mm (.12 in.) long and very inconspicuous (pl. 139)(pl. 140).

LIFE CYCLE: Seeds are spread primarily by birds, which carry the sticky seeds on their feet and beaks to new branches; seeds that are eaten pass through their digestive tracts unharmed and are deposited on branches with excrement. The sticky seeds adhere to branches, germinate, penetrate thin bark, and develop into new mistletoe plants. Both roots and shoots are produced

Plate 139. *Phoradendron juniperinum,* a juniper mistletoe: Note the inconspicuous leaves, less than 3 mm (.12 in.) long.

MISTLETOES

Biotic Diseases

Plate 140. *Phoradendron densum,* a dense mistletoe, growing on juniper: Note the readily visible leaves.

from the germinating seed, and after several years a large, woody, perennial plant develops. Clumps of shoots and foliage 30 cm (12 in.) or more in diameter are often present in trees.

SIGNIFICANCE: Although both juniper and dense mistletoes often build up to high populations on some trees, the damage caused is usually minimal. Some reduction in growth and vigor probably results from heavy infection, but the parasite seems to have little effect on tree survival or longevity.

SIMILAR PESTS: The only other pests that might be confused with these mistletoes are a few species of rust fungi. Some of these produce brooms that may resemble clusters of mistletoe shoots, but close examination readily distinguishes between mistletoe plants and abnormal host branching caused by rust fungi.

MANAGEMENT OPTIONS: Because these mistletoes cause only limited damage to their hosts, little management has been undertaken. Breaking off mistletoe shoots or pruning infected branches might limit spread and reduce stress somewhat, but these methods should be used only in special circumstances, such as with ornamental trees or in high-use recreational areas.

For More Information

Calvin, C. L., C. A. Wilson, and G. Varughese. 1991. Growth of longitudinal strands of *Phoradendron juniperinum* (Viscaceae) in shoots of *Juniperus occidentalis. Annals of Botany* 67:153–61.

Hawksworth, F. G., and R. F. Scharpf. 1981. Phoradendron *on conifers.* Forest Insect and Disease Leaflet 164. Washington, D.C.: USDA Forest Service.

INCENSE-CEDAR MISTLETOE   *Phoradendron libocedri*
Pl. 141

HOST: Incense-cedar.

DISTRIBUTION: Throughout much of the range of incense-cedar: from Baja California through the mountains of southern California, along the western slopes of the Sierra Nevada, in the southern Cascade Mountains, and into southern Oregon.

SYMPTOMS AND SIGNS: The presence of conspicuous green, pendant, more or less round clumps of leafless shoots is the most obvious sign of incense-cedar mistletoe. Young and small trees are usually free of the parasite, whereas in older trees mistletoe plants are scattered throughout the crown, often at the top. Symptoms include trunk swelling, sometimes associated with the root systems of old, shootless mistletoe plants growing entirely within the trunk. Incense-cedar mistletoe does not stimulate broom formation on infected branches, as some other mistletoes do (pl. 141).

LIFE CYCLE: Incense-cedar mistletoe is a plant that relies entirely on the host for water and nutrients even though the leafless shoots contain chlorophyll and manufacture food for the plant. However, in the absence of shoots, the root system can remain alive in the host for many years. Separate male and female plants are produced. The flowers are very inconspicuous, but in summer the fruits on the female plants are readily visible. Mature fruits contain one seed, are about the size of a pea, and grow in pink clusters. Incense-cedar mistletoe is spread by birds: the seeds stick to their feet or beaks, or are eaten, pass unharmed through the digestive system, and are

Plate 141. Clumps of *Phoradendron libocedri*, incense-cedar mistletoe, in the crown of an incense-cedar tree: (1) and (2) Round clumps of mistletoe plants.

deposited on other branches or trees. Threadlike strands hold seeds tightly on the branch where germination and infection occur. Shoots are initially produced from the seed. Thereafter, new shoots are produced from buds that develop on the enlarging root system. New plants take several years to grow 30 cm (12 in.) or more in diameter.

SIGNIFICANCE: Incense-cedar mistletoe causes no appreciable damage to its host, unless very high populations of the parasite occur in the crown. Slow-growing, old growth trees suffer the most damage, and are weakened substantially by extensive mistletoe infection.

SIMILAR PESTS: The only pest that could be confused with incense-cedar mistletoe is incense-cedar rust, caused by *Gymnosporangium libocedri*. Witches' brooms caused by this rust are usually not pendant and round, like clusters of incense-cedar mistletoe, but more often irregular in shape. The shoots of incense-cedar mistletoe are leafless and therefore the plants closely resemble dwarf mistletoe; however, dwarf mistletoe does not infect incense-cedar.

MANAGEMENT OPTIONS: No management is generally necessary due to the lack of appreciable damage to the host.

For More Information

Hawksworth, F. G., and R. F. Scharpf. 1981. Phoradendron *on conifers*. Forest Insect and Disease Leaflet 164. Washington, D.C.: USDA Forest Service.

Wagener, W. W. 1925. Mistletoe in the lower bole of incense-cedar. *Phytopathology* 15:614–16.

WHITE FIR MISTLETOE *Phoradendron pauciflorum*
Pls. 142, 143

HOST: White fir.

DISTRIBUTION: Sierra Nevada from Placer County south into the mountains of southern California, where it is quite abundant. The colder climate of northern California presumably limits its spread and establishment north of the current range.

SYMPTOMS AND SIGNS: The most obvious sign of white fir mistletoe is the large, ball-shaped clusters of green shoots in the tops of older white firs. Mistletoe plants are less abundant in the lower crown. Dead or forked tops are also associated with older trees that have been infected for many years. Unlike dwarf mistletoe, white fir mistletoe has shoots with conspicuous leaves (pl. 142) (pl. 143).

Biotic Diseases

Above: Plate 142. *Phoradendron pauciflorum* growing from a branch.

Left: Plate 143. *Phoradendron pauciflorum,* white fir mistletoe, at the top of a large white fir.

LIFE CYCLE: The fruit of white fir mistletoe contains a single seed. Seeds are spread almost entirely by birds, who feed on the fruits and carry the sticky seeds on their beaks and feet to other branches and trees. Other seeds pass through the birds' digestive tracts and are deposited on branches at new locations in their feces. About a year or more after germination, the young mistletoe plant produces shoots, flowers, and eventually fruit. Abundant buildup of the parasite in the tops of trees is apparently the result of the habit of many birds to perch on branches in the highest places in trees. Prolonged feeding in these areas intensifies the disease.

SIGNIFICANCE: In general, white fir mistletoe is a problem only in the larger, older firs in a stand. Young or small trees are seldom infected. Growth reduction resulting from intensive infection and, eventually, top dieback are the greatest impacts of this mistletoe. Often insects are responsible for death of the badly weakened tops. Reduction of cone crops and decay entering the trunk from dead tops are also significant losses attributable to white fir mistletoe. Some trees die as a result of extensive mistletoe infection and insect attack, particularly during years of drought.

SIMILAR PESTS: The ragged crown produced by dwarf mistletoe, *Arceuthobium* spp., in older white firs could be confused with the ragged appearance often produced by white fir mistletoe. The occurrence of conspicuous mistletoe plants, alive or dead, usually rules out the presence of dwarf mistletoe. However, in some stands both mistletoes occur in the same trees. Another disease, yellow witches' broom of fir, caused by *Melampsorella caryophyllacearum*, produces large brooms that could resemble clusters of mistletoe. Close examination should reveal the presence or absence of mistletoe shoots and leaves.

MANAGEMENT OPTIONS: Because white fir mistletoe is spread by birds, prevention of the disease in native stands is not practical. Young trees are seldom seriously affected by this mistletoe. Older, heavily infected firs should be harvested before they are killed by insects, drought, or a combination of both. Trees with dead and dying tops will in time undergo extensive trunk decay unless they are harvested. No chemical control is available, and pruning infected branches is not practical.

For More Information

Hawksworth, F. G., and R. F. Scharpf. 1981. Phoradendron *on conifers*. Forest Insect and Disease Leaflet 164. Washington, D.C.: USDA Forest Service.

Wagener, W. W. 1957. The limitation of two leafy mistletoes of the genus *Phoradendron* by low temperatures. *Ecology* 38:142–45.

WHITE PINE BLISTER RUST   *Cronartium ribicola*
Pls. 144–147

HOSTS: In California, sugar pine and western white pine. Bristlecone pine, foxtail pines, limber pine, and whitebark pine are less often infected. All currants and gooseberries (*Ribes* spp.), as alternate hosts, are also susceptible to the white pine blister rust pathogen.

DISTRIBUTION: Widespread and common in California. White pine blister rust was introduced into British Columbia, Canada, in 1910, and has since spread throughout the West. Climatic conditions and certain sites where pine and *Ribes* species grow in close proximity regulate the incidence and intensity of disease.

SYMPTOMS AND SIGNS: In spring, swollen blister-like patches containing masses of orange spores break through the bark of infected pine branches or stems. Earlier symptoms include branch

Biotic Diseases

Plate 144. Spore-producing structure of *Cronartium ribicola*, causal agent of white pine blister rust, on sugar pine.

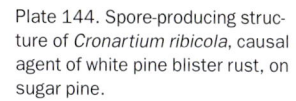

Plate 145. Canker caused by *Cronartium ribicola* on young sugar pine stem.

swelling and a bronze discoloration of the bark of the infected branch, and occasionally a sticky, honey-colored exudate on the swollen branch. Branch killing or flagging occurs after the fungus has completely girdled the branch and killed the cambium. On *Ribes* spp., the alternate host, symptoms include small, yellow pustules or thin, threadlike, orange spore horns on the lower surface of leaves (pl. 144)(pl. 145)(pl. 146)(pl. 147).

LIFE CYCLE: Blister rust has a complicated life cycle involving two hosts and five spore stages, and requiring three to six years to complete. Infection of pines is by wind-borne spores released from infected *Ribes* spp. (sporidial stage), usually growing within 300 m (about 1,000 ft) of the pines. The fungus first infects needles and then grows into the twigs and branches. Two spore stages are produced in the branches: a sexual reproductive stage (pycnial stage) and an asexual, orange "rust" stage (aecial stage) that infects the leaves of *Ribes* spp. The rust spores are windborne, traveling up to 1,300 km (800 miles). Three spore stages develop on *Ribes:* one, repeated over and over again (uredial

Biotic Diseases

Right: Plate 146. Young sugar pine with multiple branch flagging caused by *Cronartium ribicola*.

Below: Plate 147. *Cronartium ribicola* producing spores on the underside of a leaf of an alternate host (*Ribes* sp.).

stage), which intensifies the disease, and the second and third stages produce wind-borne spores that infect the pine hosts (telial and sporidial stages), thus completing the cycle. Climate and weather are major factors that regulate spore production, spread, and infection with white pine blister rust. In general, moist, cool weather in summer and fall favor disease spread.

SIGNIFICANCE: White pine blister rust is the most damaging disease of pines and has cost more to control than any other forest disease in North America. In California alone, millions of trees are damaged or killed each year, and millions of dollars have been spent, probably fruitlessly, on control through eradication of *Ribes* spp. Small trees are particularly vulnerable: only a few branch infections or a single trunk infection can result in tree death. Larger infected trees can also be severely weakened and eventually die or are killed by bark beetles.

SIMILAR PESTS: No similar diseases occur on the white pines in California that could be mistaken for white pine blister rust. A very similar native rust on pinyon pine, *C. occidentale*, also infects *Ribes* spp. as the alternate host. The symptoms of white pine blister rust and pinyon rust on *Ribes* are indistinguishable.

MANAGEMENT OPTIONS: *Ribes* spp. eradication has been discontinued as an effective method of controlling blister rust. Growth of nonhost species on high rust hazard sites is encouraged. Current research and development of rust-resistant pines may lead to a valuable management option in the future.

For More Information

Kinloch, B. B., and D. Dulitz. 1990. *White pine blister rust at Mountain Home Demonstration State Forest: A case study of the epidemic and prospects for genetic control.* Research Paper PSW-204. Washington, D.C.: USDA Forest Service.

Kinloch, B. B., M. Marosy, and M. E. Huddleston, eds. 1992. Sugar pine: Status, roles and values in ecosystems, in *Proceedings of the Symposium by the California Sugar Pine Management Committee, March 30–April 1, 1992.* Publication 3362. Davis: Division of Agriculture and Natural Resources, University of California.

INCENSE-CEDAR RUST *Gymnosporangium libocedri*
Pls. 148–151

HOSTS: Incense-cedar. Incense-cedar rust alternates between incense-cedar and several rosaceous plants, mainly serviceberry (*Amelanchier* spp.). Apple, hawthorn, mountain ash, pear, and quince are also alternate hosts.

DISTRIBUTION: Throughout the range of incense-cedar. It is most common where alternate hosts also occur, often along stream bottoms and in other cool, moist areas.

Biotic Diseases

Plate 148. Brooms caused by *Gymnosporangium libocedri*, causal agent of incense-cedar rust, in the crown of an incense-cedar tree.

SYMPTOMS AND SIGNS: The most conspicuous symptom of incense-cedar rust is the presence of dark green, tight brooms on perennially infected branches. The brooms are often scattered throughout the crown, with the largest and oldest usually in the lower crown. Some grow more than 1 m in diameter. Small, round, reddish cushions that form on the underside of the scalelike needles in spring also indicate infection by incense-cedar rust. Later in the year, when these cushions mature and become wet, they appear as orange, gelatinous masses. On the alternate host, the fungus infects leaves, petioles, and sometimes fruits. Small, cup-shaped fruiting bodies containing orange spores are symptomatic of infection of alternate hosts by this rust (pl. 148)(pl. 149) (pl. 150)(pl. 151).

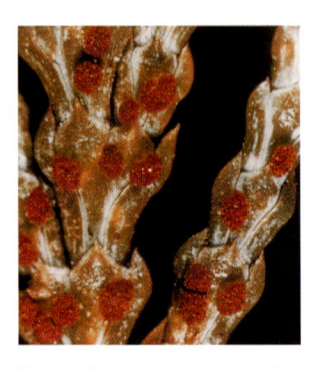

Plate 149. Young, spore-producing structures of *Gymnosporangium libocedri*.

Plate 150. Mature, gelatinous spore stage of *Gymnosporangium libocedri*.

Biotic Diseases

Plate 151. Spore-producing structures of *Gymnosporangium libocedri* on the fruits of western serviceberry.

LIFE CYCLE: Incense-cedar rust, like many rust fungi, requires two unrelated host species to complete its life cycle. The fungus has a stage on incense-cedar that produces spores in spring and summer. These spores infect the young tissues of the alternate host. The fungus grows in the alternate host and eventually produces spores that infect incense-cedar, usually in late summer or fall.

SIGNIFICANCE: Although incense-cedar rust is widespread and fairly common on incense-cedar, the damage is usually minimal. Young infected trees are often deformed and sometimes die, but larger trees are seldom killed, even though some growth reduction and brooming occur. Only when trees are heavily infected or exhibit extensive brooms do they suffer growth reduction and mortality. Abnormal trunk swellings and large knots in lumber caused by large, persistent brooms on branches result in lumber degrade. The rosaceous hosts usually suffer little damage; however, pear orchards in southern Oregon have been defoliated by incense-cedar rust.

SIMILAR PESTS: Incense-cedar mistletoe, *Phoradendron libocedri*, could be confused with incense-cedar rust. The presence of leafless, green shoots growing in more or less round, pendant clusters indicates that the pathogen is incense-cedar mistletoe.

MANAGEMENT OPTIONS: Because of the limited impact on the conifer host, little effort has been made to understand how best to manage incense-cedar rust. Growing incense-cedar near pear orchards or other rosaceous hosts should probably be avoided. Infection and eventual damage can be expected if incense-cedar is planted in or near areas where rust levels are known to be high.

For More Information
See general references.

Biotic Diseases

YELLOW WITCHES' BROOM OF FIR
Pls. 152–154

Melampsorella caryophyllacearum

HOSTS: Grand fir, red fir, subalpine fir, and white fir. Alternate hosts are species in the family Caryophyllaceae, including chickweed *(Stellaria media)* and mouse-ear chickweed *(Cerastium vulgatum)*.

DISTRIBUTION: Throughout most forests in California containing true firs. It is most common in the predominantly true fir stands of the Sierra Nevada and southern Cascade Mountains.

SYMPTOMS AND SIGNS: Yellowish, often large, broomed branches on true firs are the clearest symptom of this disease. The brooms are compact and made up of many short branches bearing short, stubby, yellowish needles. Small tubules are often seen protruding from the surface of these infected needles. Pronounced swelling of the main portion of the broomed branch is also common. Yellowing is most conspicuous in mid-summer and fall. On the alternate host, yellowish spore masses are the most easily recognized sign of the disease. These occur on the leaves, flowers, and stems and are most pronounced in summer, when the host is at the peak of maturation (pl. 152).

LIFE CYCLE: Yellow witches' broom of fir is caused by a rust fungus that requires two unrelated host species to complete its life cycle. The rust on fir causes a perennial infection resulting in large witches' brooms and stem swellings. Spores are produced in summer in structures on infected needles. These spores are wind-borne and infect nearby alternate hosts. A spore stage develops on the surface of the infected alternate host, and the spores are carried by wind to adjacent firs, where infection occurs. Hosts and

Plate 152. *Melampsorella caryophyllacearum,* causal agent of yellow witches' broom of fir, on red fir.

Biotic Diseases

Plate 153. Spore-producing structures (light colored) of *Melampsorella caryophyllacearum* on red fir foliage.

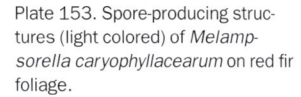

Plate 154. Spore-producing structures of *Melampsorella caryophyllacearum* on the alternate host.

alternate hosts need to be near one another for the rust to complete its life cycle. Outbreaks of the disease are infrequent and apparently regulated by highly specific weather conditions (pl. 153)(pl. 154).

SIGNIFICANCE: True firs of all ages and sizes are susceptible to yellow witches' broom. Broomed branches on young trees often stunt top growth, resulting in deformed and dead trees. Larger trees are seldom killed, but growth reduction, spike tops, trunk swellings, and open wounds leading to stem decay are the usual damage caused by this rust fungus. Large, persistent branches caused by brooms result in large knots and reduced lumber value.

SIMILAR PESTS: With the possible exception of white fir mistletoe, *Phoradendron pauciflorum,* there are no pests in California that could be confused with the disease caused by yellow witches' broom of fir. The yellowish, broomed branches are the distinctive symptom of this disease.

MANAGEMENT OPTIONS: Little has been done to manage yellow witches' broom of fir, for several reasons: the disease usually causes little damage in fir forests, infection occurs only sporadically, and little is known about the epidemiology of the disease. The best way to manage yellow witches' broom of fir at present is to avoid planting or favoring firs in stands where high levels of the rust are present, or where abundant populations of the alternate hosts grow.

For More Information
Peterson, R. S. 1963. *Effects of broom rust on spruce and fir.* Research Paper INT-8. Washington, D.C.: USDA Forest Service.

Ziller, W. G. 1974. *The tree rusts of western Canada.* Canadian Forest Service Publication 1329. Victoria: Department of the Environment, Canadian Forest Service.

WESTERN GALL RUST *Peridermium harknessii*
Pls. 155–157

HOSTS: Nearly all native hard pine species in California; also many exotic hard pines.

DISTRIBUTION: Throughout California, in native forests and on planted trees. Western gall rust is particularly common on lodgepole pine, bishop pine, and shore pine throughout California in their native ranges, and on native and planted Monterey pines wherever they are grown. Exotic Aleppo pine may also be infected with this rust.

SYMPTOMS AND SIGNS: Round to pear-shaped, woody outgrowths, or galls, on branches of trees of all ages, or on the stems of small

Plate 155. *Peridermium harknessii*, causal agent of western gall rust, on lodgepole pine branches.

Biotic Diseases

trees, are the most obvious symptom of infection. These galls grow steadily, sometimes exceeding 30 cm (12 in.) in diameter. Older galls partially circling trunks or large branches sometimes die and become sunken, resinous cankers; those on trunks are called hip cankers. Some brooming or growth of lateral branches also results from infection. In spring and early summer spores are produced. Pustules filled with yellow-orange spore masses erupt from cracks in the galls, which lose their bark, or at the margins of older cankers. Dead branches and branch flagging, often caused by insects invading galls, are common on heavily infected trees (pl. 155)(pl. 156)(pl. 157).

Left: Plate 156. *Peridermium harknessii* infection on the stem of a young ponderosa pine.

Below: Plate 157. Hip canker caused by *Peridermium harknessii* in the trunk of a lodgepole pine.

RUST DISEASES

LIFE CYCLE: Unlike many other rust fungi, the western gall rust fungus does not require an alternate host to complete its life cycle; it is sometimes called pine-to-pine rust for this reason. In spring, spores produced from the galls intensify the disease on the original host, and spread to and infect other pines. Spores may be carried by wind for many miles. Young pines are infected when favorable, usually moist, conditions occur.

SIGNIFICANCE: Western gall rust is a damaging disease in California because it is widespread and common on many pine species. It is the most common native rust found in our pine forests. Young pines, especially those in nurseries and plantations, and pines grown for the Christmas tree market are particularly susceptible to infection and damage. Larger trees are usually severely weakened by numerous galls and killed outright by subsequent bark beetle attack. Larger trees with trunk cankers often break at the canker site, posing a hazard in developed areas.

SIMILAR PESTS: Several other species of rust fungi also infect the hard pines of California. None produce the conspicuous spherical galls of western gall rust, but some produce open cankers on older trunks and stems. On older pines, trunk cankers caused by dwarf mistletoe, *Arceuthobium* spp., could also be confused with those caused by rust fungi.

MANAGEMENT OPTIONS: Management of native pine forests to reduce the impact of western gall rust has been somewhat successful. Favoring nonhost species and controlling stand density may reduce infection and damage in some cases. In highly valued pine stands grown as Christmas trees and in nurseries, managers have had limited success with chemical sprays to prevent infection.

For More Information

Allen, E. A., P. V. Blenis, and Y. Hiratsuka. 1990. Early symptom development in lodgepole pine seedlings infected with *Endocronartium harknessii*. *Canadian Journal of Botany* 68:270–77.

Vogler, D. R., and T. D. Bruns. 1998. Phylogenetic relationships among the pine stem rust fungi (*Cronartium* and *Peridermium* spp.). *Mycologia* 90:244–57.

Vogler, D. R., B. B. Kinloch, F. W. Cobb, and T. L. Popenuck. 1991. Isozyme structure of *Peridermium harknessii* in the western United States. *Canadian Journal of Botany* 69:2434–41.

Biotic Diseases

BROWN STRINGY ROT (INDIAN PAINT FUNGUS)
Echinodontium tinctorium

Pls. 158–160

HOSTS: Common on true firs; less common on hemlock. Douglas-fir and spruce are rarely infected.

DISTRIBUTION: Common in stands of true fir in California, particularly old growth. Indian paint fungus is also found in many forests containing coastal and mountain hemlock.

SYMPTOMS AND SIGNS: The most distinctive symptom produced by advanced Indian paint fungus infection is heartwood decay; the most obvious sign is the presence of conks. Indian paint fungus conks are perennial, dark, hoof-shaped structures up to 30 cm (12 in.) in size. The lower, pore surface consists of irregularly shaped "teeth." The brick-red interior can be powdered and used as a red pigment, hence the name Indian paint fungus. In trees with advanced decay the entire central cylinder of heartwood progresses from yellowish to tan to rust-red in color, and becomes a stringy mass of soft, rotted wood (pl. 158)(pl. 159)(pl. 160).

LIFE CYCLE: Indian paint fungus is spread by airborne spores produced by the fruiting body, or conk. The fungus may infect trees through trunk scars, dead tops, and other wounds. However, it is believed that entry occurs more often through branch stubs on small, suppressed trees. The fungus remains dormant—sometimes for many years—in these stubs until the tree is injured. With air and moisture now available, the fungus becomes active. The fungus causes a white rot of the heartwood. Trees less than about 50 years old suffer little rot because only limited heartwood has developed.

SIGNIFICANCE: Indian paint fungus causes substantial decay in true firs, particularly old growth

Plate 158. Fruiting body, or conk, of *Echinodontium tinctorium,* the Indian paint fungus, on grand fir.

HEART ROTS

Biotic Diseases

Plate 159. Interior of *Echinodontium tinctorium* fruiting body on white fir.

Plate 160. Internal decay of the trunk of a grand fir, caused by *Echinodontium tinctorium*.

firs. In some stands, nearly half the gross volume is decayed. The presence of one or more fruiting bodies on a tree indicates extensive decay. Decay is considered to extend about 5 m (15 ft) below and above any fruiting body. Because fire scars or wounds are not necessary for infection by this fungus, it poses a threat to the management of young growth stands.

SIMILAR PESTS: Although other decay fungi such as *Phellinus pini*, causal agent of white pocket rot, attack some of the same hosts, the brick-red interior of the fruiting body of the Indian paint fungus make misidentification unlikely.

MANAGEMENT OPTIONS: Reducing wounding or injury will help prevent infection and growth of Indian paint fungus. Also, managing stands to increase vigor and prevent suppression will reduce infection and damage caused by the fungus entering branch stubs. Short harvest rotations should also help avoid development of decay in trees.

For More Information

Etheridge, D. H., H. M. Craig, and S. H. Parris. 1972. *Infection of western hemlock by the Indian paint fungus via living branches.* Bimonthly Research Note 28. Ottawa: Department of the Environment, Canada Forest Service.

Kimmey, J. W. 1965. *Rust-red stringy rot.* Forest Pest Leaflet 93. Washington, D.C.: USDA Forest Service.

Wilson, A. D. 1991. Somatic incompatibility in dikaryotic–monokaryotic and dikaryotic pairings of *Echinodontium tinctorium. Canadian Journal of Botany* 12:2716–23.

Biotic Diseases

BROWN CUBICAL ROT (SULFUR FUNGUS) — *Laetiporus sulphureus*
Pls. 161, 162

HOSTS: Douglas-fir, hemlock, pines, spruce, and true firs. Several hardwoods are also hosts of sulfur fungus.

DISTRIBUTION: Common on conifers, particularly true firs, and on hardwoods throughout California.

SYMPTOMS AND SIGNS: The fruiting bodies are the most conspicuous signs of sulfur fungus. They are large, annual, clustered, bracket-type, fleshy structures that, when fresh, are yellow-orange on top and bright sulfur yellow on the lower, pore surface. Old, dried fruiting bodies are usually hard, brittle, and chalky white. Heartwood in advanced decay is reddish brown, cubically cracked and easily crumbled. Often, mycelial felts are present in shrinkage cracks. These white felts may be as much as 6 mm (.24 in.) thick, 7 cm (2.75 in.) wide, and more than 1 m (3.3 ft) long (pl. 161) (pl. 162).

LIFE CYCLE: The life cycle of sulfur fungus is not well known. It is assumed that the fungus enters the heartwood through wounds and fire scars. Sulfur fungus is also known to decay logs, stumps, and dead trees.

SIGNIFICANCE: The sulfur fungus causes substantial loss of timber, particularly among true firs. Because the fungus enters the

Above: Plate 161. Fruiting structure of *Laetiporus sulphureus*, sulfur fungus, showing a dried, chalky margin.

Right: Plate 162. Brown cubical rot caused by *Laetiporus sulphureus*.

HEART ROTS

heartwood mainly through basal wounds and fire scars, timber losses occur mostly in the lower trunk.

SIMILAR PESTS: Several rots of conifers, particularly red-brown butt rot caused by the velvet top fungus, *Phaeolus schweinitzii,* and brown crumbly rot caused by red belt fungus, *Fomitopsis pinicola,* resemble that produced by sulfur fungus; however, the yellow-orange fruiting body of sulfur fungus is distinctive.

MANAGEMENT OPTIONS: Avoiding wounds at the base of trees will reduce the probability of infection by this fungus.

For More Information
Gilbertson, R. L. 1974. *Fungi that decay ponderosa pine.* Tucson: University of Arizona Press.

POCKET DRY ROT
Oligoporus amarus

Pls. 163, 164

HOST: Incense-cedar.

DISTRIBUTION: Throughout the native range of incense-cedar. Pocket dry rot is found more commonly on incense-cedar growing under favorable conditions on the western slopes of the Sierra Nevada and Cascade Mountains rather than in the drier, eastern range of the host.

SYMPTOMS AND SIGNS: The characteristic pockets of rot in the heartwood of incense-cedar are diagnostic of pocket dry rot. The pockets are several times longer than they are wide, and in a cross section of a log often appear to develop in a circular pattern along the annual rings. The wood in the pockets often breaks down into a dry, brown, crumbly residue. The margin between sound wood and the decay pockets remains sharp. Pockets may merge, but the rot is never extensive enough to decay the entire central cylinder of the tree. The presence of fruiting bodies, or conks, on the trunk is certain evidence of extensive rot. Fresh fruiting bodies are soft, annual, bracket-like structures, tan on top and yellow on the lower, spore-producing pore surface. They usually develop in late summer and are present for only a short time; by winter they have either been eaten by insects or have dried and shriveled (pl. 163)(pl. 164).

LIFE CYCLE: Spread and infection are achieved entirely by means of spores produced by the fruiting bodies. Fire scars, wounds, branch stubs, and open knots are all entry points for the fungus. Trees less than 150 years old are relatively free of rot.

Biotic Diseases

Plate 163. *Oligoporus amarus*, causal agent of pocket dry rot, in incense-cedar heartwood.

Plate 164. Fruiting body of *Oligoporus amarus* fungus on the trunk of an incense-cedar.

SIGNIFICANCE: Pocket dry rot is extremely damaging to old growth incense-cedar. It is estimated that about a third of the volume of merchantable timber is lost because of pocket dry rot.

SIMILAR PESTS: There are no other pests of incense-cedar that can be confused with this disease.

MANAGEMENT OPTIONS: The decay caused by pocket dry rot should not be a serious problem in young-growth forests, particularly those less than 100 years old. Prevention of wounds and fire scars should also reduce the incidence and decay caused by this fungus. Trees with pocket dry rot remain structurally sound and constitute no major hazard from breakage in high-use areas.

For More Information

Wagener, W. W., and R. V. Bega. 1958. *Heart rots of incense-cedar*. Forest Pest Leaflet 30. Washington, D.C.: USDA Forest Service.

Biotic Diseases

RED-BROWN BUTT ROT (VELVET TOP FUNGUS)
Phaeolus schweinitzii

Pls. 165, 166

HOSTS: Douglas-fir, incense-cedar, larch, pines, spruce, true firs, and western redcedar; rarely on hemlock.

DISTRIBUTION: Throughout California, wherever hosts grow. Douglas-fir is the host most commonly infected with this decay fungus.

SYMPTOMS AND SIGNS: The presence of this decay fungus is not easily determined in living trees. The leathery fruiting bodies, or conks, when present, are the best means of identifying velvet top fungus and are the best indicators of decay. They may appear from the soil at the base of an infected tree as large, mushroom-shaped bodies, or on the lower trunk of the tree as brackets. The top of the fruiting body is soft and velvety, reddish brown, and encircled by a yellowish margin. Concentric rings are visible on the upper surface. The lower, pore surface varies from dark green to light brown. In fall, the dried fruiting bodies resemble dried cow dung. Only the heartwood is attacked by this fungus. Wood in advanced stages of decay appears reddish brown; it breaks into large cubes and cracks across the grain. Dried, decayed wood can be easily crumbled into fine powder. White fungal mats, or felt, may appear in the cracks. Sometimes infected trees exhibit a pronounced swelling of the lower trunk (pl. 165)(pl. 166).

LIFE CYCLE: Spores produced by the fruiting bodies are wind-borne and infect hosts, usually through basal wounds or fire scars. The fungus is also known to be a root parasite, but there is

Plate 165. Upper surface of *Phaeolus schweinitzii*, the velvet top fungus, fruiting body on Douglas-fir: This conk is growing from the roots of the tree.

Plate 166. Lower, spore-producing surface of *Phaeolus schweinitzii* fruiting body on Douglas-fir.

no evidence of spread from diseased roots to healthy ones as with some of the root disease fungi.

SIGNIFICANCE: Timber losses caused by this decay fungus are important because the valuable wood at the base of trees is often extensively decayed. Also, in campgrounds and other high-use areas, conifers infected with this fungus pose an extreme hazard because they are likely to be uprooted or broken off by wind.

SIMILAR PESTS: In the absence of fruiting bodies the advanced decay caused by velvet top fungus resembles that caused by certain other decay fungi including *Armillaria mellea;* however, infected roots lack the presence of rhizomorphs, and mycelial fans found in *Armillaria* root disease.

MANAGEMENT OPTIONS: Preventing infection by reducing the amount of basal wounding and preventing fire scars is the best method to reduce damage caused by velvet top fungus. In high-use areas, periodic surveys should be conducted to detect and remove infected, high-hazard trees.

For More Information

Partridge, A. D., and D. L. Miller. 1974. *Major wood decays in the inland Northwest*. Natural Resources Series No. 3. Moscow: College of Forestry, Wildlife, and Range Sciences, University of Idaho, and Idaho Research Foundation.

Biotic Diseases

WHITE POCKET ROT or RED RING ROT   *Phellinus pini*
Pls. 167, 168

HOSTS: Douglas-fir, hemlock, pines, spruce, true firs, and rarely incense-cedar.

DISTRIBUTION: Throughout the coniferous forests of California, particularly in northern California. Coastal Douglas-fir is most commonly infected.

SYMPTOMS AND SIGNS: The woody, perennial fruiting bodies are the best indicators of white pocket rot decay. These conks can appear as thin brackets on the tree or as thick, hoof-shaped structures about 5 to 25 cm (2 to 10 in.) wide. The upper fruiting body surface is dark gray to dark brown and usually marked with concentric furrows. The lower pore surface is mazelike and usually rusty brown. Fruiting bodies usually arise from knots or branch stubs. Advanced decay is also a clear symptom of white pocket rot. The decayed, reddish heartwood exhibits small, spindle-shaped pockets containing soft, white fibers. In Douglas-fir and hard pines the decay tends to follow the annual rings and appears in the softer wood formed during the early part of the season's growth (pl. 167)(pl. 168).

Plate 167. Pore surface of a fruiting body of *Phellinus pini*, causal agent of white pocket rot.

Biotic Diseases

Plate 168. Damage caused by *Phellinus pini* in Douglas-fir wood.

LIFE CYCLE: The life cycle of white pocket rot is not well understood. Presumably spores of the fungus infect branch stubs and grow into the heartwood, where the gradual process of decay occurs. Fire scars or wounds are not considered important points of entry for the fungus. Because of the mode of infection, young trees have been found with decay.

SIGNIFICANCE: White pocket rot is one of the most common and serious heart rot problems of California conifers. Old growth trees are often extensively decayed, resulting in the loss or degrade of millions of board feet of lumber, particularly in Douglas-fir. Because young trees can also become infected, white pocket rot may continue to be a serious heart rot problem in future coniferous forests in California.

SIMILAR PESTS: A few other fungi produce pocket rots in coniferous heartwood. The characteristic spindle-shaped pockets containing soft white fibers, and the fruiting bodies, distinguish white pocket rot from all others, particularly in pines and Douglas-fir. The Indian paint fungus, *Echinodontium tinctorium*, has a similar fruiting body, but the interior is brick red.

MANAGEMENT OPTIONS: Little is known about how to manage conifer forests to mitigate the damage caused by white pocket rot. Reducing fire scarring and wounding will probably not have much effect on the incidence and damage caused by this disease. Increasing tree vigor and growth, and shortening the intervals between harvest, will decrease the time the fungus has to cause extensive decay.

For More Information

Blanchette, R. A. 1980. Wood decomposition by *Phellinus (Fomes) pini*: A scanning microscope study. *Canadian Journal of Botany* 58: 1496–1503.

Gilbertson, R. L. 1974. *Fungi that decay ponderosa pine*. Tucson: University of Arizona Press.

ARMILLARIA ROOT DISEASE *Armillaria mellea*
Pls. 169–171

HOSTS: A wide range of hardwood species. However, most conifers in California are also susceptible to *Armillaria* root disease. Reports have concerned primarily Douglas-fir, incense-cedar, and true firs.

DISTRIBUTION: Throughout California. *Armillaria* root disease is especially common in the Coast Ranges, the Central Valley, and the Sierra Nevada up to about 2,000 m (6,500 ft).

SYMPTOMS AND SIGNS: Loss of tree vigor, thinning crown, and eventually death are the above-ground symptoms of *Armillaria* root disease. These symptoms are the result of the progressive death of roots. In advanced stages of the disease, thin, leathery, white to cream-colored mycelial fans, develop between the bark and the sapwood in the lower trunk of the tree or in major roots below the soil. The presence of rhizomorphs is also indicative of *Armillaria* root disease. Rhizomorphs occur in the soil adjacent to roots, on root surfaces, between the bark and wood of infected roots, and within decayed and deteriorating wood (pl. 169). Fruiting bodies, which often

Plate 169. White mycelial fan of *Armillaria mellea*, causal agent of *Armillaria* root disease, on white fir: *(arrow)* Note the white mycelial fan.

Biotic Diseases

Plate 170. Fruiting bodies of *Armillaria mellea* at the root collar.

are used to identify disease-causing fungi, are not usually found with *Armillaria*. When they do occur, usually in late summer and fall, they appear in clusters at the base of infected trees as honey-colored mushrooms 8 cm (3 in.) high with caps 5 to 12 cm (about 2 to 5 in.) across. The underside of the caps bears white to yellowish gills attached to the stalk, and the upper stalk bears a ring of tissue called an annulus. Decayed wood is white, stringy, and often contains rhizomorphs (pl. 170)(pl. 171).

LIFE CYCLE: *A. mellea* is a wood-decaying fungus that lives on roots, stumps, logs, and other woody debris in the forest. The fungus is commonly found growing as epiphytic rhizomorphs

Plate 171. Stringy, white decay caused by *Armillaria mellea* in windthrown white fir.

ROOT DISEASES

on the roots of living trees, particularly hardwoods. On occasion, it attacks the roots of living trees, killing the cambium and eventually decaying the sapwood. Spread of the pathogen is via root-to-root contact, or by rhizomorphs. The maximum extent of rhizomorph growth is approximately 30 to 60 cm (about 12 to 24 in.) from an infected root. Spores produced by the fruiting bodies do not appear to be important in the disease cycle.

SIGNIFICANCE: *Armillaria* root disease is an important disease of conifers in landscape plantings and plantations. It is most often associated with sites that have been converted from hardwoods, particularly oaks. Mortality is primarily centered near infected stumps. In native stands, *Armillaria* root disease in conifers is often found near hardwoods that have died from suppression or other causes (e.g., canker rots), allowing *A. mellea* to increase inoculum on the dying root systems. The fungus then spreads to infect adjacent conifers. Bark beetles may kill infected pines and firs.

SIMILAR PESTS: The above-ground symptoms caused by *Armillaria* root disease are similar to those caused by other root diseases of conifers, but the presence of rhizomorphs, white stringy rot, and mycelial fans distinguishes *Armillaria* root disease from others. Young trees killed by the Pine Reproduction Weevil, *Cylindrocaptorus eatoni* and pocket gophers *Thomomys* spp., are similar in appearance. Those killed by the Pine Reproduction Weevil have intact roots and pupal cells in the wood. The roots of trees killed by pocket gophers are eaten. In addition to *A. mellea*, three other species of *Armillaria* are found in California. Only one of these has been observed to kill conifers, and this species has yet to be given a formal taxonomic name. *A. ostoyae* has not been found in California to date, but is also a very important conifer pathogen.

MANAGEMENT OPTIONS: Recognition of *Armillaria* root disease in a forest is the first step to management. Practices that enhance tree vigor and reduce stress are recommended. Avoid planting on sites with extensive disease in stumps and residual trees, and plant less susceptible species such as ponderosa pine. The most susceptible species are the true firs, Douglas-fir, and western hemlock.

For More Information

Shaw, C. G., III, and G. A. Kile. 1991. Armillaria *root disease*. Agriculture Handbook 691. Washington, D.C.: USDA Forest Service.

ANNOSUM ROOT DISEASE *Heterobasidion annosum*
Pls. 172–175

HOSTS: Nearly all conifers in California. Annosum root disease has also been found on some hardwoods. Two host-specific forms of *H. annosum* are known: a "p-type" that infects pines, hardwoods, incense-cedar, and juniper, and an "s-type" that infects Douglas-fir, giant sequoia, hemlock, spruce, true firs, and western redcedar.

DISTRIBUTION: Most conifer forests of California; especially abundant in true fir forests of the Sierra Nevada and southern Cascade Mountains. Annosum root disease is also common in the pine and mixed-conifer forests of California, particularly where logging has occurred.

SYMPTOMS AND SIGNS: Symptoms and disease development vary depending on the host species attacked. Hemlocks, for example, exhibit no crown symptoms. Infected pines, on the other hand, exhibit slow growth and crown yellowing and thinning. Pine roots are penetrated rapidly, and therefore small pines (seedlings and saplings) are usually killed directly within a few years. With other species, for example, white fir, root death is much slower, and the fungus often grows through the woody root tissue and into the trunk, causing butt rot. In pines and firs, a circular pattern of tree decline and death indicates an annosum root disease infection center, with the most recently diseased trees at the periphery and old, dead trees and often stumps at the center. Perennial fruiting bodies on dead trees and in stumps also indicate annosum root disease. They vary in shape from woody brackets several inches across, usually found inside decaying stumps (especially those of white fir), to small buttons growing between the bark and wood. Root symptoms can also be used for disease diagnosis. In pines the bark separates easily from the wood of recently dead roots, and silvery white flecks of fungus are present at the interface between wood and bark. In firs and other species, a stringy white rot and the presence of fruiting bodies are the most important indications of annosum root disease (pl. 172) (pl. 173)(pl. 174)(pl. 175).

LIFE CYCLE: *H. annosum* exists as both a parasite and a saprophyte. Long-distance spread is by wind-borne spores. Spores produced by the fruiting bodies infect recently cut stumps, trees with basal wounds, and woody material. In pines, the fungus

Biotic Diseases

Plate 172. Young white fir trees killed by *Heterobasidion annosum*, causal agent of annosum root disease: *(arrow)* The cut stump was the source of infection.

Plate 173. *Heterobasidion annosum* fruiting bodies on a Jeffrey pine stump.

grows into a stump and roots as a decay organism until a living root contacts an infected one. The fungus then penetrates and infects the roots of the healthy tree. Infected roots may become resin soaked. In firs and other species, *H. annosum* grows mainly

Above: Plate 174. *Heterobasidion annosum* fruiting body on the root collar of a dead Jeffrey pine seedling.

Left: Plate 175. *Heterobasidion annosum* fruiting body on white fir wood.

ROOT DISEASES

in the woody cylinder of roots and stems and functions primarily as a decay-causing organism.

SIGNIFICANCE: In California, this disease is widespread and damaging to nearly all native conifer species. Because the fungus is known to invade freshly cut stumps, spread and intensification of the disease have increased with timber harvesting and intensive forest management. Once infected, the fungus persists in roots and the soil for many years. Pines suffer the greatest impact as enlarging centers of infection progressively kill trees or render them susceptible to insect attack, especially by bark beetles. Moisture stress, insects, and, in some cases, air pollution contribute to the severity of damage caused by *H. annosum*. Firs and other species are not often killed but are weakened, decayed, and made susceptible to uprooting and attack by insects, especially bark beetles. Infected trees in high-use areas pose a hazard to life and property.

SIMILAR PESTS: Blackstain root disease, caused by *Leptographium wageneri,* and *Armillaria* root disease, caused by *Armillaria mellea,* produce above-ground symptoms similar to those of annosum root disease. Young trees killed by the Pine Reproduction Weevil, *Cylindrocaptorus eatoni,* and gophers, *Thomomys* spp., may resemble those killed by annosum root disease; however, dead roots and silvery white flecks of fungus are present in annosum-killed seedlings and young trees. In the case of damage by gophers, the roots are eaten.

MANAGEMENT OPTIONS: Prevention of infection and spread is the best way to manage annosum root disease of conifers, especially pines. Cut stumps of pine and other hosts can be treated with borax-based registered chemicals. Also, thinning of eastside pines during warm, dry periods in summer inhibits fungal growth and, consequently, tree infection. With conifer species other than pines, prevention of wounds, especially in the lower trunk, is the best way to avoid infection. Mapping of disease centers by global positioning system technology, followed by harvesting, stump treatment or excavation, and planting of resistant trees, are used more often today as the cost of this technology has been reduced.

For More Information

Garbelotto, M., F. W. Cobb, T. D. Bruns, W. J. Otrosina, T. Popenuck, and G. W. Slaughter. 1999. Genetic structure of *Heterobasidion annosum* in white fir mortality centers in California. *Phytopathology* 89:546–54.

Otrosina, W. J., and R. F. Scharpf. 1989. *Research and management of annosus root disease* (Heterobasidion annosum) *in western North America.* General Technical Report 116. Redding: Pacific Southwest Research Station, USDA Forest Service.

Slaughter, G. W., and J. R. Parmeter, Jr. 1995. Enlargement of tree-mortality centers surrounding pine stumps infected by *Heterobasidion annosum* in northeastern California. *Canadian Journal of Forest Research* 25:244–52.

Slaughter, G. W., and D. M. Rizzo. 1999. Past forest management promoted root disease in the Yosemite Valley. *California Agriculture* 53 (3):17–24.

BLACKSTAIN ROOT DISEASE   *Leptographium wageneri*
Pls. 176–179

HOSTS: Douglas-fir (by one host-specific race of *L. wageneri*); Jeffrey pine, ponderosa pine, and a few other pine species (by a second host-specific race of *L. wageneri*); pinyon pine (by a third host-specific race of *L. wageneri*).

DISTRIBUTION: Widespread in California, but a management problem only in certain areas. Some coastal Douglas-fir forests suffer from the disease. Localized pockets of the disease also occur in the pine forests of the Sierra Nevada and southern Cascade Mountains. Several large areas of infection occur on pinyon pines in southern California.

SYMPTOMS AND SIGNS: Loss of tree vigor, crown deterioration, and death are symptoms caused by blackstain root disease. A chocolate-brown to nearly black stain of the sapwood of infected roots is a more distinctive symptom of this disease. The stain usually appears as an area that more or less follows annual rings, in contrast to the dark blue or black, wedge-shaped stain attributed to bluestain fungi. Infected roots are often heavily infiltrated with resin (pl. 176) (pl. 177)(pl. 178)(pl. 179).

Plate 176. *Leptographium wageneri*, causal agent of blackstain root disease, in the lower trunk of a dead ponderosa pine.

Biotic Diseases

Plate 177. *Leptographium wageneri* in the sapwood of Douglas-fir.

LIFE CYCLE: The life cycle of *L. wageneri* is not well understood. The fungus lives within the vascular system of roots and blocks the tree's water supply. The mechanisms of long-distance spread and infection of individual trees are not known. However, certain root-infesting bark beetles and weevils are suspected vectors of this pathogen. Subsequent spread to other trees is by contact of infected roots with healthy ones. Large infection centers develop by spread of the disease from infected to non-infected trees. The fungus remains alive in dead roots for only a few years, and it does not utilize cellulose or decay wood.

SIGNIFICANCE: Although not common in most California forests, severe losses can occur. Heavy growth reduction and mortality

Plate 178. *Leptographium wageneri* in the outer growth rings of a ponderosa pine stump: *(arrow)* Note the black, pitch-soaked wood.

ROOT DISEASES 153

Biotic Diseases

Plate 179. *Leptographium wageneri* in the outer growth rings of young Douglas-fir: *(arrow)* Note that the black-stained areas follow annual rings.

occur in trees in infection centers, and the centers continue to enlarge as long as suitable hosts and favorable conditions are present. This fungus prefers a cool environment, and the stain rarely rises higher than about 2 m (about 7 ft) in pine stems, or 10 m (about 33 ft) in Douglas-fir. Bark beetles kill trees weakened by blackstain root disease. At present, there is no practical method to eliminate the fungus from an infection center.

SIMILAR PESTS: Except for the bluestain fungi, which also stain sapwood, there are no other pests that can be confused with blackstain root disease. The chocolate-brown stain in the sapwood is the distinctive symptom of this root disease. Crown symptoms in diseased trees are similar to those caused by other root diseases.

MANAGEMENT OPTIONS: Planting or favoring nonhost conifer species is the best way to manage this disease. True firs, for example, are immune to blackstain root disease. Maintaining a mixed species stand will also limit the rate of spread between susceptible hosts, and increased spacing between trees will limit root-to-root contact.

For More Information

Diamandis, S., L. Epstein, F. W. Cobb, T. Popenuck, and E. Hecht-Poinar. 1997. Development of *Leptographium wageneri* on root surfaces and other substrata. *European Journal of Forest Pathology* 27:381–90.

Goheen, D. J., and F. W. Cobb. 1980. Infestation of *Ceratocystis wageneri*-infected ponderosa pines by bark beetles (Coleoptera: Scolytidae) in the central Sierra Nevada. *The Canadian Entomologist* 112:725–30.

Harrington, T. C., and F. W. Cobb, Jr. 1988. Leptographium *root disease on conifers*. St. Paul: APS Press.

PORT ORFORD–CEDAR ROOT DISEASE   *Phytophthora lateralis*
Pls. 180, 181
HOST: Port Orford–cedar.
DISTRIBUTION: Most parts of the range of Port Orford–cedar.

Biotic Diseases

Plate 180. Young Port Orford–cedars killed by *Phytophthora lateralis*, causal agent of Port Orford–cedar root disease.

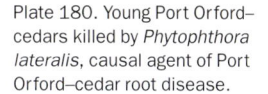

Plate 181. Brown-stained sapwood of Port Orford–cedar infected with *Phytophthora lateralis*.

SYMPTOMS AND SIGNS: Dead and dying trees, characterized by wilting and bronze to brown foliage, are the most conspicuous symptoms of the disease; infected trees are easily windthrown. The disease is caused by a water-borne, fungus-like organism which is spread by water and earth movement. Human activities can greatly enhance spread of Port Orford–cedar root disease. Trees along water courses or along roads are often the first to become diseased. Brown staining of the sapwood in the lower trunk or in major roots is also a common diagnostic symptom (pl. 180) (pl. 181).

LIFE CYCLE: Port Orford–cedar root disease is caused by a nonnative pathogen that is part of a group called water molds. Spores dispersed by water or by the movement of soil infect the fine roots of a host tree. The pathogen then grows through the root system to the root crown, killing the tree. Microscopic fruiting bodies are produced in the dead root tissue. The fruiting bodies produce spores that remain in the tissue, or in the soil, until they are dispersed or contact new host tissue. On contacting live host roots, the spores germinate.

Biotic Diseases

SIGNIFICANCE: Port Orford–cedar root disease is an extremely serious disease of Port Orford–cedar and threatens the management and perhaps survival of the host in many areas. Compounding the seriousness of this disease are cedar bark beetles, *Phoeosinus* spp., which have been observed to increase in windthrown trees and to kill nearby trees.

SIMILAR PESTS: There are no other pests that cause such quick decline and death of native Port Orford–cedar stands. Other water mold pathogens have been shown to damage Port Orford–cedar and other conifers in nurseries and greenhouses.

MANAGEMENT OPTIONS: Preventing the pathogen from spreading and becoming established in new areas is the best control method. Soil or plants should not be moved from areas of infection, and vehicles and equipment used in these areas should be cleaned of all soil before being allowed into new areas. Limiting travel over roads during the wet season will also help prevent spread of the pathogen.

For More Information

Zobel, D. B., L. F. Roth, and G. M. Hawk. 1985. *Ecology, pathology, and management of Port Orford–cedar* (Chamaecyparis lawsoniana). General Technical Report PNW-184. Washington, D.C.: USDA Forest Service.

Hansen, E. M., and P. B. Hamm. 1996. Survival of *Phytophthora lateralis* in infected roots of Port Orford–cedar. *Plant Disease* 80:1075–78.

Hansen, E.M., D.J. Goheen, E.S. Jules, and B. Ullian. 2000. Managing Port Orford–cedar and the introduced pathogen *Phytophthora lateralis*. *Plant Disease* 84:4–14.

ABIOTIC DISEASES

Abiotic Diseases

AIR POLLUTION
Pls. 182, 183

HOSTS: Conifers particularly sensitive to ground-level ozone, the air pollutant most damaging to plants: Jeffrey pine and ponderosa pine. Other conifers showing sensitivity, in order from most to least susceptible: bigcone Douglas-fir, Coulter pine, giant sequoia, incense-cedar, knobcone pine, red fir, sugar pine, and white fir. Monterey pine is also quite sensitive to ozone damage, but some resistant clones have been developed.

DISTRIBUTION: Ozone can move great distances from its point of origin; hence, conifers in the Los Angeles Basin, as well as coniferous forests east and north of the Los Angeles Basin and those in the mountains east of Bakersfield and Fresno, are affected. Potentially damaging levels of air pollutants have been recorded in the greater San Francisco Bay Area and in the Sacramento Valley air basin.

SYMPTOMS AND SIGNS: Chlorotic mottling of needles, shorter needles, and reduced needle retention are the main symptoms of injury. Seriously damaged leaves of deciduous trees in the same area may show distinctive brown edges, called marginal leaf burn. A conifer disease-rating scale ranging from no injury (no chlorotic mottling and at least four years of needle retention) to severe injury (short, yellow, mottled needles and only one or two years of needle retention) has been devised. Additional symptoms of serious damage include loss of apical dominance, that is, suppression of terminal bud growth; reduction in overall shoot growth; increased death of lower and midcrown branches; and diminished cone production (pl. 182) (pl. 183).

ENVIRONMENTAL FACTORS: Damage from air pollution occurs as a result of chronic exposure to the oxidant ozone, the toxic component of photochemical smog. Automobile and industrial emissions are the main causes of smog. Chemical reac-

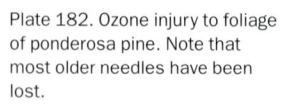

Plate 182. Ozone injury to foliage of ponderosa pine. Note that most older needles have been lost.

Abiotic Diseases

Plate 183. Ponderosa pine showing severe ozone damage. Note reduction in needle length and retention.

tions between emission components, specifically nitrogen dioxide and hydrocarbons, result in the production of ground-level ozone. Conifers are damaged by prolonged exposure to high concentrations of ozone, which usually occurs during the warm, summer months. Several years of exposure to ozone are necessary for symptoms of damage to occur. Continued exposure to ozone often results in tree death. Trees with advanced symptoms are killed by bark beetles.

SIGNIFICANCE: Ozone has seriously weakened conifers in the mountains of southern California. Some damage has also been recorded along the western slopes of the Sierra Nevada. Surviving trees may be more resistant to ozone injury. Current efforts to reduce toxic emissions seem to have reduced the spread of ozone-related damage, but as California's population increases, ozone damage to conifers could become more common.

SIMILAR SYMPTOMS: A number of insects, diseases, and environmental conditions can cause loss of needles and ozone-like mottling, even in areas where ozone is not present. In many of these cases the symptoms diminish over time and the trees appear healthier. With chronic exposure to ozone, symptoms persist, trees continue to decline, and eventually many die, often prematurely from attack by bark beetles.

MANAGEMENT OPTIONS: Government efforts are being made to reduce the levels of toxic air pollutants. These efforts, if successful, will go far in preventing ozone damage to conifers. In the meantime, monitoring of ozone levels, and continuing research into the progression of tree damage and death, plus development of

For More Information

Dahlsten, D. L., D. L. Rowney, and R. N. Kickert. 1997. Effects of oxidant air pollutants on Western Pine Beetle (Coleoptera: Scolytidae) populations in southern California. *Environmental Pollution* 96:415–23.

Miller, P. R. 1992. Mixed conifer forests of the San Bernardino Mountains, California, in *The response of western forests to air pollution* (R. K. Olson, D. Binley, and M. Bohm, eds.). New York: Springer-Verlag. 461–97.

Miller, P. R., G. J. Longbotham, and C. R. Longbotham. 1983. Sensitivity of selected western conifers to ozone. *Plant Disease* 67:1113–15.

DROUGHT

Pls. 184, 185

HOSTS: Douglas-fir, incense-cedar, ponderosa pine, sugar pine, and white fir.

DISTRIBUTION: Throughout the natural and urban forests of California. However, the severity and frequency of drought effects are greatest in the lower elevation forests of the western slopes of the Sierra Nevada, in the dry pine sites of the southern Cascade Mountains, northeastern and southern California, and along the eastern slopes of the Sierra Nevada.

SYMPTOMS AND SIGNS: Conifers respond initially to drought stress by shedding their oldest needles in late summer and fall. These needles turn yellow, then red, and fall from the tree. During periods of more severe moisture stress, needles on the upper branches turn light brown, and this discoloration advances from the tip to the base of each needle, appearing first as a narrow band of yellow, and then as light brown tissue. After exposure to drought for many years, fewer needle whorls are produced, new needles are shorter than average, and branch mortality in both the upper and lower crown becomes evident (pl. 184)(pl. 185).

ENVIRONMENTAL FACTORS: Seedlings and saplings are more susceptible to drought than are larger trees, because of their smaller root systems. Trees growing in thin layers of soil or in dense stands are also more prone to drought effects. Drought reduces tree vigor and thus makes trees more vulnerable to attack by root diseases, other pathogens, and bark beetles.

Abiotic Diseases

Plate 184. Drought damage to young incense-cedars.

Plate 185. Severe drought damage to ponderosa pines in Sierra foothills. Note the surviving gray pines.

SIGNIFICANCE: Trees are weakened and growth is reduced during periods of drought-induced stress. Severe drought occurred in California in the 1920s and early 1930s, early 1960s, 1976–1977, and 1986–1992. All the major 20th century droughts in California resulted in greatly elevated levels of bark beetle-caused tree mortality, especially of ponderosa pine and white fir.

SIMILAR SYMPTOMS: Both severe ozone injury and drought injury cause older needles to die and fall off, leaving "lion's tails"—bare branches with just a tuft of newer needles at the tip. The crowns of trees affected by ozone or drought are shortened because of branch mortality in the lower crown, and overall the crowns have a thinned appearance.

MANAGEMENT OPTIONS: Control of competing vegetation in young plantations will increase the availability of moisture to the trees and alleviate drought stress.

Abiotic Diseases

For More Information
See general references.

FLOODING

HOSTS: All tree species, but among conifers Jeffrey pine and lodgepole pine are especially susceptible.

DISTRIBUTION: Throughout the wildland and urban forests of California.

SYMPTOMS AND SIGNS: Chlorotic foliage, especially in the upper crown and on the branch tips; needle drop; and branch dieback. Flood waters can cause a tree to fall, or strip away enough topsoil that the tree falls later, typically in a windstorm. High water marks may be present on the trunks of trees that have survived past floods. In addition, root crowns may be obscured or overexposed as a result of flooding. The presence of several species of trees and shrubs dying simultaneously in a circumscribed area is also a good indication of flood damage.

ENVIRONMENTAL FACTORS: Flooding depletes the oxygen supply to roots, and therefore this condition is more severe among trees growing in poorly drained, poorly aerated soils, such as those with a high clay content. In addition, construction activity associated with road building and housing development often changes drainage patterns: a new road may function as a dam and cause temporary ponding during heavy rains or spring snow melt. Conifers near lakes and streams, and in lawns that are frequently watered, may be weakened by excess water pooling around their roots.

SIGNIFICANCE: Trees gradually weakened by flooding ultimately die, most often as a result of *Armillaria* root disease, caused by *Armillaria mellea*, and bark beetle infestation.

SIMILAR SYMPTOMS: Chronic ozone exposure and drought also cause the older needles of pines to die and fall off the branch, leaving "lion's tails"—bare branches with just a tuft of newer needles at the tip. The crowns of flooded trees are shortened because of branch mortality in the lower crown, and overall the crowns have a thinned appearance. After the water drains away, the weakened trees will typically succumb to root disease or bark beetle infestation.

For More Information
See general references.

Abiotic Diseases

FROST

Pls. 186–189

HOSTS: Most conifers.

DISTRIBUTION: Throughout California, from time to time.

SYMPTOMS AND SIGNS: Buds, new needles and shoots, older shoots still growing in fall, and terminal leaders are killed. Foliage killed by frost remains on the twigs and turns gray rather than falling off. Typically, the upper crown of small trees, where growth is initiated, is affected first. Sometimes, however, only the lower parts of trees are affected: this occurs among taller trees growing in low-lying areas where cold air accumulates (frost pockets) (pl. 186)(pl. 187)(pl. 188)(pl. 189).

ENVIRONMENTAL FACTORS: Early autumn frosts will kill shoots that are still growing. Late spring frosts will kill buds and newly elongating needles and shoots. Frosts are more common in frost pockets, where cold air accumulates. Native stands are usually well adapted to the local weather patterns, and rarely suffer frost damage. Planted trees in both landscape and forest plantation settings are more likely to have

Plate 186. Frost damage to ponderosa pine.

Plate 187. Frost damage to small Jeffrey pines on the eastern slopes of the Sierra Nevada.

Abiotic Diseases

Above: Plate 188. Jeffrey pine foliage damaged by frost.

Left: Plate 189. Frost damage to young foliage of white fir.

problems. Frost damage to planted trees usually occurs when they start producing new growth in spring, before the end of the cold weather. The underlying cause is often the use of a seed source from a lower elevation or more southerly location.

SIGNIFICANCE: Frost damage may cause complete failures in plantation establishment, requiring repeated site preparation and replanting. In less severe cases, both landscape and plantation trees may suffer deformations and growth reductions.

SIMILAR SYMPTOMS: Frost-killed terminal leaders of pines resemble leaders killed by bark beetles, moths, weevils, drought, or lightning.

MANAGEMENT OPTIONS: Plant tree species adapted to local conditions and utilize local seed sources. In particular, avoid seed sources from more southerly or lower elevation locations.

For More Information
Wagener, W. W. 1960. A comment on cold susceptibility of ponderosa and Jeffrey pines. *Madroño* 15:217–19.

HEAT

HOSTS: Most California conifers. Different species are sensitive to different types of high-temperature extremes.

DISTRIBUTION: Throughout California, wherever susceptible species occur.

SYMPTOMS AND SIGNS: Sudden increases in temperature in May and June can cause tips to droop, turn red-brown, and drop off. Shoots of Douglas-fir and white fir are particularly sensitive to this type of heat damage. Prolonged high temperature at any time, in conjunction with soil moisture deficiency, can cause top dieback in large trees. A generally cool spring followed by a sudden rise in temperature to over 38 degrees Celsius (100 degrees Fahrenheit) causes needle browning in coast redwood, Monterey pine, and juniper.

ENVIRONMENTAL FACTORS: Heat damage occurs with sudden increases in temperature in late spring, prolonged periods of high temperatures, and high temperature extremes following a cool spring.

SIGNIFICANCE: Serious losses occur mainly among trees grown for the Christmas market and other high-value trees.

SIMILAR SYMPTOMS: Bark beetles and Porcupines, *Erethizon dorsatum,* also kill the terminal leaders of pines. Also, tip moths, weevils, and pitch canker caused by *Fusarium circinatum, Diplodia* blight caused by *Sphaeropsis sapinea,* and *Elytroderma* disease of pines caused by *Elytroderma deprmons* kill pine shoots. Sawflies cause fir shoot dieback. Shoots killed by heat have intact but dead foliage; their bark has not been removed and they show no signs of internal insect feeding damage. The simultaneous presence of symptomatic firs, incense-cedar, and pines rules out insects or biotic diseases as the cause.

MANAGEMENT OPTIONS: Planting pines rather than more sensitive Douglas-fir or white fir on hot, dry sites may avoid heat damage.

For More Information
See general references.

HERBICIDES
Pl. 190
HOSTS: All trees.
DISTRIBUTION: All areas in close proximity to herbicide application, or where herbicide drift may occur.
SYMPTOMS AND SIGNS: Symptoms vary according to the type of herbicide used, but may include twisting of needles or shoots, needle necrosis, severe stunting, chlorosis, and yellow-white bleaching.
ENVIRONMENTAL FACTORS: Chemical herbicide applications are usually associated with particular land use practices such as control of brush around dwellings; clearing vegetation from strips of land for fuelbreaks; and along roadside, railroad, and power line right-of-ways. Damage to desirable plants often occurs along the edges of these areas, or as a result of drift of herbicide droplets into

Abiotic Diseases

Plate 190. Damage to incense-cedar from chemical spray along a railroad right-of-way.

neighboring areas. Damage may also occur during weed control operations in plantations (pl. 190).

SIGNIFICANCE: Herbicide damage to trees can be significant, especially in young plantations where vegetation management is taking place. Root-absorbed herbicides usually cause death or lasting damage in the most severe cases. Low doses may cause only defoliation or foliage discoloration.

SIMILAR SYMPTOMS: Other abiotic factors, such as frost, may cause similar damage, but the range of herbicide damage is likely to be restricted to small or discrete areas. The effects of needlecast fungi and tip moth infestation may be confused with herbicide-caused damage. Nearby plants, such as brush species, should also exhibit symptoms, as they were the original target of the herbicide treatment.

MANAGEMENT OPTIONS: Proper care in selecting herbicides, in timing of applications to avoid periods when trees are sensitive, and in avoiding weather conditions that cause spray drift, will reduce herbicide damage.

For More Information

Bentley, J. R., D. A. Blakeman, and S. B. Cooper. 1971. *Recovery of young ponderosa pines damaged by herbicide spraying.* Research Note PSW-252. Washington, D.C.: USDA Forest Service.

SALT
Pls. 191–194

HOSTS: All conifers. Conifers are very susceptible to damage caused by salts used for highway deicing. Associated herbaceous plants and other woody vegetation also show varying degrees of damage.

Abiotic Diseases

DISTRIBUTION: Mainly along the all-weather, mountain roads of California where highway deicing salts are used to melt ice and snow in winter. Damage is also found adjacent to driveways, sidewalks, parking lots, and other areas where salt has been used as a deicing agent. In some coastal areas, salt spray can damage conifers as far as 500 m (1,600 ft) from the beach. Salts from water softeners used to treat waste water can accumulate along septic tank leach lines and cause damage to adjacent trees.

SYMPTOMS AND SIGNS: Progressive, often very uniform yellowing and dieback of conifer foliage facing the road is the most conspicuous symptom of salt spray damage. Dieback begins in winter and is usually most conspicuous in early spring as salts are absorbed and accumulate in the foliage. Nearly all damage occurs within 15 m (50 ft) of roadways that have been treated with salt. Soil salt damage leads to needle tip burn. Other woody vegetation in the immediate area also shows browning and dying of leaf margins. Salt damage can be further diagnosed by foliar analysis to determine whether above-normal levels of sodium and chloride are present in tissues (pl. 191)(pl. 192)(pl. 193)(pl. 194).

Plate 191. Salt damage to Jeffrey pine.

Plate 192. Salt damage symptoms on sugar pine foliage.

Abiotic Diseases

Plate 193. Spiraling pattern of salt damage on roadside white fir.

Plate 194. Salt damage symptoms on manzanita leaf.

ENVIRONMENTAL FACTORS: In general, conifers grow in areas where little salt is present in the environment. As such, they are more sensitive to salt exposure than many other plants. Damage can occur through the uptake of salt solution by roots and also by direct contact of foliage with salt-laden water. Repeated application of salt for deicing, or piling of salt-laden snow around conifers, results in prolonged uptake of salt during the winter and spring months. Direct foliar damage usually results from salt-laden snow being blown onto foliage by snow removal equipment or from piling of salty snow against smaller trees. Excess sodium and chloride ions in the soil are not only toxic to conifer tissues but subject the tree to moisture stress similar to that during exposure to drought.

SIGNIFICANCE: Trees along many highways in California where deicing salts are regularly used have suffered severe damage and high levels of mortality. Highway managers have found that salt application to icy roads is the most economical and efficient method of providing safe, ice-free driving under certain weather conditions. Important decisions need to be made about how best to provide for safe driving without seriously damaging the roadside environment.

SIMILAR SYMPTOMS: A few foliar diseases of conifers, such as *Elytroderma* disease of pines caused by *Elytroderma deprmans,* as well as frost and drought damage, produce symptoms similar to those caused by salt, but the pattern of disease in proximity to roads treated with salt, and excess levels of sodium and chloride in the symptomatic foliage distinguish salt damage from other causal agents. Trees with general chlorosis of the crown due to salt damage are difficult to distinguish from bark beetle-killed trees.

MANAGEMENT OPTIONS: The best way to avoid salt damage to conifers is not to use salt for deicing roadways. Where this is not possible, the prudent use of salt along with an abrasive material such as sand has been found to be an effective, less environmentally damaging method of highway deicing. In some cases it may be necessary to transport salt-laden snow or ice to designated disposal sites in order to avoid damage to highly valued conifers.

For More Information
Scharpf, R. F., and M. D. Srago. 1974. *Conifer damage and death associated with the use of highway deicing salt in the Lake Tahoe Basin of California and Nevada.* Forest Pest Control Technical Report No. 1. San Francisco: USDA Forest Service, California Region.

WINTER BURN and WINTER DRYING
Pl. 195

HOSTS: Douglas-fir, Jeffrey pine, ponderosa pine, and western redcedar.

DISTRIBUTION: Conifers growing at higher elevations and latitudes where freezing temperatures occur for extended periods.

SYMPTOMS AND SIGNS: Winter burn—needles on the sunny side of the tree and above the snow pack turn yellow to reddish brown. After the snow melts, a distinct margin usually exists between the undamaged foliage that was under snow and foliage that was damaged. Winter drying—foliage droops and turns reddish brown in late winter and early spring. Discolored foliage occurs on all sides of the tree (pl. 195).

ENVIRONMENTAL FACTORS: Winter burn occurs when there are sudden temperature changes, such as on sunny days following periods of below-freezing temperatures. This causes irreversible desiccation of the needles. Branches with southern exposure are more likely to be injured under these conditions. Winter drying

Abiotic Diseases

Plate 195. Winter drying of Douglas-fir.

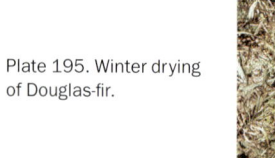

occurs when warm, dry winter winds coincide with frozen soil and reduced water conduction in the tree. As a result, the winds remove water from the needles faster than it can be replaced.

SIGNIFICANCE: Damage is occasionally significant in pine and Douglas-fir plantations.

SIMILAR SYMPTOMS: Gouty Pitch Midge, *Cecidomyia piniinopsis*, causes branch tip mortality in young ponderosa pines, but the presence of pitch-filled cavities along the length of infested shoots distinguishes this damage from winter drying and winter burn. Other tip-infesting insects may cause damage that is superficially similar to winter drying or winter burn. In these cases symptoms will usually be found throughout the crown of the tree, and cutting into symptomatic branches will reveal the feeding damage of the insect. Western gall rust, caused by *Peridermium harknessii*, kills branch tips of pines, but the distinctive galls distinguish this damage from winter burn and winter drying.

MANAGEMENT OPTIONS: Utilizing local seed sources may result in reduced damage from winter burn and winter drying.

For More Information

Grier, C. C. 1988. Foliage loss due to snow, wind and winter drying damage: Its effects on leaf biomass of some western conifer forests. *Canadian Journal of Forest Research* 18:1097–102.

DAMAGE BY LARGER ANIMALS

Damage by Larger Animals

AMERICAN BLACK BEAR *Ursus americanus*
Pls. 196, 197

HOSTS: Conifers damaged most frequently: Coast redwood, Douglas-fir, and Port Orford-cedar. Typical foods include grasses, berries, nuts, tubers, wood fiber, insects, small mammals, eggs, carrion, and garbage.

DISTRIBUTION: In California, throughout the forested areas of the Coast Ranges and the Sierra Nevada.

SYMPTOMS AND SIGNS: Bark is stripped from trees, exposing the sapwood surface. Sometimes the outer bark is more or less intact, but exhibits claw marks. Small seedlings are eaten (pl. 196)(pl. 197).

LIFE CYCLE: American Black Bears breed during the summer months, usually from June to mid-July. Females in good condition will produce two or three cubs early the following year; females typically reproduce every other year. The young are born while the female hibernates, and are cared for by their mother until they are about 1.5 years old and their mother enters estrus once again. American Black Bears become sexually mature at about 3.5 years. Adult females may have a home range of 15 to 50 km^2 (6 to 20 mi^2).

SIGNIFICANCE: American Black Bears can cause extensive damage to young trees by feeding on the inner bark or by clawing off the bark to leave territorial markings.

Above: Plate 196. *Ursus americanus*, American Black Bear, and damaged trees.

Right: Plate 197. *Ursus americanus* feeding damage to young redwood trees.

SIMILAR PESTS: None.

MANAGEMENT OPTIONS: Highly valued properties can be protected by various fencing systems. Sanitation is the best preventive method. Bears can be deterred by various frightening devices. They also may be trapped and removed from the site, or baited and shot where laws permit. State and local officials should be consulted before management activities are undertaken.

For More Information

Giusti, G. A. 1990. Observation of black bear, *Ursus americanus,* feeding damage to Port Orford-cedar, *Chamaecyparis lawsoniana,* in Del Norte County, California. *California Fish and Game* 76:127–28.

Hygnstrom, S. E. 1994. Black bear, in *Prevention and control of wildlife damage* (S. E. Hygnstrom, R. M. Timm, and G. E. Larson, eds.). Lincoln: University of Nebraska Cooperative Extension, University of Nebraska; Washington, D.C.: Animal Damage Control, USDA Animal and Plant Health Inspection Service.

DEER and ELK   *Odocoileus* **spp.,**
Pls. 198–200   *Cervus* **spp.**

HOSTS: A wide variety of vegetation including grasses, forbs, shrubs, conifer seedlings, and twigs and small branches of larger trees.

DISTRIBUTION: Mule Deer, *Odocoileus hemionus*—throughout California; Black-tailed Deer, *O. h. columbianus*—Coast Ranges and northern California; Roosevelt Elk, *Cervus elaphus roosevelti*—northern coastal areas of California. Deer migrate from high elevations in the Sierra Nevada and Coast Ranges to lower elevations in fall and early winter and return to the high country in spring.

SYMPTOMS AND SIGNS: Vegetation is grasped between the lower incisors and the upper palate, which lacks incisors. Twigs and stems are thus ripped or torn, leaving a jagged or fragmented

Plate 198. *Odocoileus hemionus columbianus,* Black-tailed Deer.

Damage by Larger Animals

Plate 199. *Cervus elaphus roosevelti*, Roosevelt Elk.

Plate 200. Deer feeding damage to Douglas-fir shoots. Note that the terminal portion of the shoot and terminal bud have been torn off.

appearance. In late summer, male deer and elk remove the velvet coat from their new antlers by rubbing them against woody vegetation, thus stripping the bark from young trees. Threat displays, in which male deer and elk beat their antlers against vegetation, can demolish young trees (pl. 198)(pl. 199)(pl. 200).

LIFE CYCLE: Deer breed from October to January, depending on latitude. Most females (does) are able to breed by their second fall. The peak of fawn drop is in May and June. Deer are most active during the early morning and evening and they have a home range of about 120 ha (300 acres). The breeding season for elk begins in late summer, when dominant males herd females (cows) into "harems" for breeding. A single calf is born about eight months later. Elk tend to roam over larger areas than deer.

SIGNIFICANCE: Deer and elk can cause significant damage to conifer plantations. Elk strip the bark from young trees, and will eat the bark of larger trees during winter when other food is unavailable. Both deer and elk eat young shoots as well. Seedlings are entirely eaten. Young trees subjected to repeated attacks exhibit greatly retarded growth. Low-elevation plantations in favored deer winter range may be heavily damaged when new growth starts in spring.

Damage by Larger Animals

SIMILAR PESTS: Rabbits and large rodents also cause similar damage to young conifer seedlings and trees. However, rabbits and large rodents leave a clean cut at a sharp, 45 degree angle.

MANAGEMENT OPTIONS: Plantations can be protected with deer fences. Individual seedlings can also be protected with plastic mesh tubes. Chemical repellents can be sprayed on young trees.

For More Information

Craven, S. R., and S. E. Hygnstrom. 1994. Deer, in *Prevention and control of wildlife damage* (S. E. Hygnstrom, R. M. Timm, and G. E. Larson, eds.). Lincoln: University of Nebraska Cooperative Extension, University of Nebraska; Washington, D.C.: Animal Damage Control, USDA Animal and Plant Health Inspection Service.

JACKRABBITS and HARES — *Lepus* spp.
Pls. 201, 202

HOSTS: Conifer seedlings and young trees, forbs, grasses, and low-growing shrubs, especially in replanted, cut-over areas.

DISTRIBUTION: White-tailed Jackrabbit, *Lepus tousendii* and Snowshoe Hare, *Lepus americanus*—northeastern California; Black-tailed Jackrabbit, *Lepus californicus*—throughout most of California.

SYMPTOMS AND SIGNS: Foliage and twigs of seedlings and young trees are eaten. Bark is stripped from seedlings and from the shoots and stems of saplings. Twigs and branches are severed at a 45 degree angle, leaving a clean cut (pl. 201)(pl. 202).

Plate 201. *Lepus californicus,* Black-tailed Jackrabbit.

Plate 202. Top of a ponderosa pine seedling removed by *Lepus californicus:* (arrow) Note the clean, 45 degree cut.

LIFE CYCLE: Females produce up to four litters per year, with two to eight young per litter. Daily movements of 1.6 to 3.2 km (1 to 2 mi) each way are common when food and shelter areas are separated.

SIGNIFICANCE: Rabbits can cause extensive damage to seedlings and saplings in young plantations. Seedlings are entirely consumed. Bark is stripped from the stem of young trees, which subsequently die.

SIMILAR PESTS: Deer and elk can cause similar damage, but leave torn and jagged edges on the vegetation. Porcupines, *Erethizon dorsatum*, also strip bark from trees. Mountain Beaver, *Aplodontia rufa,* and woodrats eat foliage and twigs. Jackrabbits and hares do not climb, and so their feeding damage is confined to the lower 60 cm (2 feet) of the trees.

MANAGEMENT OPTIONS: Jackrabbits and hares can be excluded by wire mesh fences and plastic mesh tubes placed around seedlings. Various chemical repellents are available.

For More Information

Knight, J. E. 1994. Jackrabbits and other hares, in *Prevention and control of wildlife damage* (S. E. Hygnstrom, R. M. Timm, and G. E. Larson, eds.).

Lincoln: University of Nebraska Cooperative Extension, University of Nebraska; Washington, D.C.: Animal Damage Control, USDA Animal and Plant Health Inspection Service.

LIVESTOCK

HOSTS: All conifer seedlings and young trees.

DISTRIBUTION: Wherever livestock are able to invade forested land or are reared on forested land. Extensive areas of forest are used as seasonal livestock pasture. Typically, cattle are pastured at low elevations during the winter months and moved to higher, forested areas in summer. Large flocks of sheep were formerly pastured in the Sierra Nevada forests, but the practice is now confined to limited areas east of the crest of the Sierra Nevada. Goats are not commonly pastured in forested areas of California, but are extremely destructive in areas where they are kept on forested lands. Feral hogs are present in low-elevation coastal areas, where their rooting habits kill young trees.

SYMPTOMS AND SIGNS: Most reported damage involves the trampling of seedlings in plantations younger than 10 years of age. Sheep and goats eat foliage and twigs.

LIFE CYCLE: Most cattle damage occurs during the warmer months when cattle are moved to high-elevation pastures. During winter they are moved to pastures below the forest zone.

SIGNIFICANCE: Livestock can cause significant losses in young plantations of any tree species.

SIMILAR PESTS: Other large mammals may cause trampling, but livestock are indicated on the basis of their presence and their fecal material on the forest floor. Deer and elk eat foliage and shoots but seldom trample young trees.

MANAGEMENT OPTIONS: Fencing of susceptible areas will reduce damage. Also, livestock can be attracted to salt blocks placed in the forest and can then be rounded up and removed if necessary.

For More Information

See recent reports from the Animal Damage Committee to the California Forest Pest Council, published annually by the California Department of Forestry and Fire Protection.

Damage by Larger Animals

MOUNTAIN BEAVER *Aplodontia rufa*
Pls. 203, 204

HOSTS: Douglas-fir, hemlock, red alder, and western redcedar. Bracken fern, forbs, grasses, and sword fern are preferred.

DISTRIBUTION: Throughout northern California in the North Coast Ranges and southern Cascade Mountains, and south through the Sierra Nevada. Populations in the Sierra Nevada are more scattered but can reach high densities in some areas. Mountain Beavers prefer openings in forests or thinned stands with deep soil and abundant understory vegetation.

SYMPTOMS AND SIGNS: Mountain Beavers are primitive rodents, not closely related to true beavers. Adults are about 30 cm (1 ft) long, with blunt snouts and very short tails. Mountain Beavers climb small shrubs and trees as high as 2.5 m (7 ft) or more and cut off branches up to approximately 2 cm (.75 in.) in diameter. Cut stubs on lower tree trunks are indicative of their feeding habits. They also girdle small trees (up to 15 cm [6 in.] in diameter) at the root collar and feed on the bark of the lower stem. Roots of larger trees are also consumed. Small stems are chewed into lengths of about 15 cm (6 in.) and frequently stacked near the entrance to their burrows. On conifer seedlings, angular and rough cuts about .5 to 2 cm (.20 to .79 in.) in diameter appear on the sapwood surface. Stem girdling usually occurs at the base of a tree, and no bark pieces are left on the ground around the tree. The bark is cut smoothly along the edges on the girdled tree trunk (pl. 203)(pl. 204).

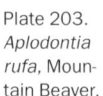

Plate 203. *Aplodontia rufa*, Mountain Beaver.

Plate 204. Terminal shoot and lateral branches of a young Douglas-fir removed by *Aplodontia rufa*: (arrow) Note the angular cuts.

LIFE CYCLE: Mountain Beavers excavate a very large system of underground burrows. Each burrow system has a dome-shaped chamber with a nest, which is excavated about 1 m (3 ft) beneath the ground surface. One or many entrances are present. Each system also has a chamber for food storage and one for fecal deposits. Mating occurs from January to March and birth occurs 30 days later. Densities are limited to about 10 per hectare (four per acre), owing to territorial behavior.

SIGNIFICANCE: Mountain Beavers can be destructive to conifer plantations less than 20 years of age. Small trees are cut at ground level and larger trees are climbed to eat branches. Extensive root girdling kills trees greater than 15 cm (6 in.) in diameter.

SIMILAR PESTS: Stem girdling by American Black Bears, *Ursus americanus*, and porcupines, *Erethizon dorsatum*, may appear similar to Mountain Beavers damage, as bears and porcupines leave bark pieces on the ground around the tree. Deer, jackrabbits, porcupines, and voles also eat foliage, twigs, and bark of seedlings and young trees. Jackrabbits and deer do not climb and therefore their damage is limited to the lower levels of tree.

MANAGEMENT OPTIONS: Small-diameter plastic mesh cages will protect seedlings, and wire mesh cages up to 1 m (3 ft) in diameter are used to protect larger trees.

For More Information

Borrecco, J. E., and R. J. Anderson. 1980. Mountain Beaver problems in the forests of California, Oregon and Washington. *Proceedings of the Vertebrate Pest Conference* 9:135–42.

Campbell, J. E. 1994. Mountain Beavers, in *Prevention and control of wildlife damage* (S. E. Hygnstrom, R. M. Timm, and G. E. Larson, eds.). Lincoln: University of Nebraska Cooperative Extension, University of Nebraska; Washington, D.C.: Animal Damage Control, USDA Animal and Plant Health Inspection Service.

Damage by Larger Animals

POCKET GOPHERS *Thomomys* spp.
Pls. 205–207

HOSTS: Mainly the underground parts of herbs, shrubs and trees, including all California conifers.

DISTRIBUTION: Four species in California with one or more found in all the forested parts of the state.

SYMPTOMS AND SIGNS: The most severe damage occurs during the early phase of forest regeneration as gophers eat the roots and main stems of seedlings, leaving intact dead tops. Fan-shaped mounds of soil mark entrances to tunnel systems of pocket gophers. After snow melt, long winding strips of soil on the surface indicate where pocket gophers had tunnelled through the snow and backfilled the tunnels with soil from underground burrows (pl. 205)(pl. 206).

LIFE CYCLE: Pocket gophers are adapted to burrowing and spend most of their time in extensive underground tunnel systems. Typically only one pocket gopher occupies a burrow system, except during mating and birth of young. Burrow systems consist of a main tunnel with a variable number of lateral tunnels excavated 10 to 45 cm (4 to 18 in.) below the ground surface and may contain up to 180 m (600 ft) of tunnel. Pocket gophers do not hibernate. They become sexually mature in the spring following birth and produce one or two litters with one to 10 young

Above: Plate 205. Pocket gopher, *Thomomys* spp.

Right: Plate 206. Characteristic mound of soil over a *Thomomys* spp. tunnel entrance.

Plate 207. *Thomomys* spp. feeding damage at the base of a young Douglas-fir.

per year. Densities of 42 to 49 per hectare (17 to 20 per acre) are common.

SIGNIFICANCE: Pocket gophers damage trees by eating roots and girdling stems. During the winter they strip twigs and bark from those portions of trees under the snow. They often kill enough trees in the first two or three years after planting that minimal stocking levels are not maintained and replanting is required (pl. 207).

SIMILAR PESTS: Woodrats, *Neotoma* spp., clip off twigs and branches, and debark seedlings and saplings. Trees killed by Pine Reproduction Weevil, *Cylindrocopturus eatoni,* and *Armillaria* root disease caused by *Armillaria mellea* are similar in appearance to those killed by pocket gophers. However, trees killed by the Pine Reproduction Weevil have intact root systems and those killed by *Armillaria* root disease have mycelial fans and rhizomorphs associated with the roots.

MANAGEMENT OPTIONS: Options include trapping and application of toxic grain pellets.

For More Information

Case, R. M. 1994. Pocket gophers, in Prevention and control of wildlife damage (S. E. Hygnstrom, R. M. Timm, and G. E. Larson, eds.). Lincoln: University of Nebraska Cooperative Extension, University of Nebraska; Washington, D.C.: Animal Damage Control, USDA Animal and Plant Health Inspection Service.

Damage by Larger Animals

Engeman, R. M., R. M. Anthony, V. G. Barnes, Jr., H. W. Krupa, and J. Evans. 1998. Double-stocking for overcoming damage to conifer seedlings by pocket gophers. *Crop Protection* 17:687–90.

PORCUPINE *Erethizon dorsatum*
Pls. 208, 209

HOSTS: Herbaceous plants, and inner bark, twigs, and leaves of both young and mature trees, especially Jeffrey pine and ponderosa pine. The bark and inner bark of young trees, especially aspen, cottonwood, ponderosa pine, and willow, are preferred.

DISTRIBUTION: Throughout the coniferous forests of California.

SYMPTOMS AND SIGNS: Tooth marks on the sapwood of young trees indicate feeding by Porcupines. The tops of young ponderosa pines and Jeffrey pines are often killed, especially in northeastern California. During the winter months a Porcupine may live in the same tree for an extended period, progressively eating the bark off several feet of the upper trunk and leaving a pile of bark chips on the ground (pl. 208)(pl. 209).

LIFE CYCLE: Porcupines are principally nocturnal and are active all year. They breed in autumn and give birth to one pup in spring.

Plate 208. Ponderosa pine tops killed by *Erethizon dorsatum*, Porcupine, feeding.

Plate 209. *Erethizon dorsatum* feeding damage to young ponderosa pine.

SIGNIFICANCE: Porcupine damage to young pine plantations can be severe. Top-killed trees produce crooked or multiple stems. Lumber values are reduced in damaged trees.

SIMILAR PESTS: Rabbits and Mountain Beaver, *Aplodontia rufa*, will eat the bark of young trees, but rabbits do not climb and Mountain Beavers do not climb above about 2.5 m (7.5 ft). Western Pine Shoot Borer, *Eucosma sonomana*, and Lodgepole Terminal Weevil, *Pissodes terminalis*, infestations result in forked or crooked stems. Dead stems and branches with tooth marks distinguish Porcupine feeding from Western Pine Shoot Borer and Lodgepole Terminal Weevil damage, where dead shoots retain intact bark.

MANAGEMENT OPTIONS: *E. dorsatum* can be excluded by fencing small areas. Tree trunk guards, poisoning, trapping, and shooting where legal have been effective in reducing damage.

For More Information

Schemnitz, S. D. 1994. Porcupines, in *Prevention and control of wildlife damage* (S. E. Hygnstrom, R. M. Timm, and G. E. Larson, eds.). Lincoln: University of Nebraska Cooperative Extension, University of Nebraska; Washington, D.C.: Animal Damage Control, USDA Animal and Plant Health Inspection Service.

WOODRATS *Neotoma* spp.
Pls. 210, 211

HOSTS: Acorns, cactus, fruits, green vegetation, human food supplies, seeds, shoots, and twigs, depending on the woodrat species.

DISTRIBUTION: Four species occur in California, with one or more living in all the forested regions of the state. Woodrats are more abundant at lower elevations.

SYMPTOMS AND SIGNS: Twigs and branches are clipped off, and seedlings and saplings are debarked. Woodrats are strictly nocturnal and seldom seen. They build houses consisting of large piles of sticks and leaves. The houses often include human-made objects, earning these animals the nickname "pack rat". The houses may be used by successive generations of rats over many years, and reach a height of 1.5 m (4.5 ft)(pl. 210)(pl. 211).

LIFE CYCLE: Nests are constructed from finely shredded plant material and are located inside the house. The house is usually on the ground but may be in trees, or in rocky outcroppings.

Damage by Larger Animals

Right: Plate 210. Redwood stump sprouts killed by woodrat, *Neotoma* spp., feeding.

Below: Plate 211. *Neotoma* spp. feeding damage to Douglas-fir.

Mating occurs in spring, and one to four offspring are produced in one or two litters each year. Woodrats are active throughout the year.

SIGNIFICANCE: Woodrats debark fruit trees and conifer seedlings and saplings, especially redwoods. They are also a nuisance in cabins or vehicles that are infrequently used. Here they shred upholstered furniture and mattresses for use in nests. They transmit human diseases such as plague.

SIMILAR PESTS: Pocket gophers, Mountain Beaver, *Aplodontia rufa,* and Porcupines, *Erethizon dorsatum,* also debark trees and eat twigs and foliage.

MANAGEMENT OPTIONS: Poison baits formulated with anticoagulants and traps are available.

For More Information

Salmon, T. P., and W. P. Gorenzal. 1994. Woodrats, in *Prevention and control of wildlife damage* (S. E. Hygnstrom, R. M. Timm, and G. E. Larson, eds.). Lincoln: University of Nebraska Cooperative Extension, University of Nebraska; Washington, D.C.: Animal Damage Control, USDA Animal and Plant Health Inspection Service.

GUIDE TO DAMAGE BY SYMPTOM LOCATION

On Reproductive Structures

Douglas-fir
Douglas-fir Cone Moth *(Barbara colfaxiana)*
Douglas-fir Seed Chalcid *(Megastigmus spermotrophus)*
Western Conifer Seed Bug *(Leptoglossus occidentalis)*

Pine
Cone beetles *(Conophthorus* spp.*)*
Pine Seedworm *(Cydia miscitata)*
Western Conifer Seed Bug *(Leptoglossus occidentalis)*

On Smaller Branches or Treetops

All conifers
Abiotic diseases
Deer and elk *(Odocoileus* spp., *Cervus* spp.*)*
Jackrabbits and hares *(Lepus* spp.*)*
Livestock
Porcupine *(Erethizon dorsatum)*

Cedar
Botryosphaeria canker *(Botryosphaeria dothidea)*
Incense-cedar mistletoe *(Phoradendron libocedri)*
Incense-cedar rust *(Gymnosporangium libocedri)*

Cypress
Cedar bark beetles *(Phloeosinus* spp.*)*
Juniper mistletoes *(Phoradendron juniperinum, P. densum)*
Silverspotted Tiger Moth *(Lophocampa argentata)*

Douglas-fir
Black Pineleaf Scale *(Nuculaspis californica)*
Cooley Spruce Gall Adelgid *(Adelges cooleyi)*
Douglas-fir needle cast *(Rhabdocline pseudotsugae, R. weirii)*
Douglas-fir Tussock Moth *(Orgyia pseudotsugata)*
Douglas-fir Twig Weevil *(Cylindrocopturus furnissi)*
Dwarf mistletoes *(Arceuthobium* spp.*)*
Phomopsis canker of Douglas-fir *(Phomopsis lokoyae* or *Diaporthe lokoyae)*
Pine Needle Scale *(Chionaspis pinifoliae)*
Pitch canker *(Fusarium circinatum)*
Silverspotted Tiger Moth *(Lophocampa argentata)*
Spruce budworms *(Choristoneura* spp.*)*

Giant sequoia
Botryosphaeria canker *(Botryosphaeria dothidea)*

Hemlock
Silverspotted Tiger Moth *(Lophocampa argentata)*

Juniper
Cedar bark beetles *(Phloeosinus* spp.*)*
Juniper mistletoes *(Phoradendron juniperinum, P. densum)*

Pine
Black Pineleaf Scale *(Nuculaspis californica)*
California Five-spined Ips *(Ips paraconfusus)*
Diplodia blight *(Sphaeropsis sapinea* or *Diplodia pinea)*
Dwarf mistletoes *(Arceuthobium* spp.*)*
Elytroderma disease of pines *(Elytroderma deformans)*
Giant Conifer Aphid *(Cinara ponderosae)*
Gouty Pitch Midge *(Cecidomyia piniinopsis)*
Lodgepole Needleminer *(Coleotechnites milleri)*
Lodgepole Terminal Weevil *(Pissodes terminalis)*
Pandora Moth *(Coloradia pandora)*
Pine Engraver *(Ips pini)*
Pine Needle Scale *(Chionaspis pinifoliae)*
Pine Needle Sheathminer *(Zelleria haimbachi)*
Pitch canker *(Fusarium circinatum)*
Ponderosa Pine Tip Moth *(Rhyacionia zozana)*
Red band needle blight *(Mycosphaerella pini)*

Silverspotted Tiger Moth *(Lophocampa argentata)*
Western gall rust *(Peridermium harknessii)*
Western Pine Beetle *(Dendroctonus brevicomis)*
Western Pine Shoot Borer *(Eucosma sonomana)*
White pine blister rust *(Cronartium ribicola)*

Redwood
Cedar bark beetles *(Phloeosinus* spp.*)*

Spruce
Cooley Spruce Gall Adelgid *(Adelges cooleyi)*

True fir
Balsam Twig Aphid *(Mindarus abietinus)*
Cytospora canker of true firs *(Cytospora abietis)*
Douglas-fir Tussock Moth *(Orgyia pseudotsugata)*
Dwarf mistletoes *(Arceuthobium* spp.*)*
Fir Engraver *(Scolytus ventralis)*
Silverspotted Tiger Moth *(Lophocampa argentata)*
Spruce budworms *(Choristoneura* spp.*)*
True fir needle cast *(Lirula abietis-concoloris)*
White fir mistletoe *(Phoradendron pauciflorum)*
Yellow witches' broom of fir *(Melampsorella caryophyllacearum)*

On Stems, Larger Branches, or Both

All conifers
American Black Bear *(Ursus americanus)*
Mountain Beaver *(Aplodontia rufa)*
Pocket gophers *(Thomomys* spp.*)*
Woodrats *(Neotoma* spp.*)*

Cedar
Incense-cedar mistletoe *(Phoradendron libocedri)*

Cypress
Juniper mistletoes *(Phoradendron juniperinum, P. densum)*

Douglas-fir
Douglas-fir Beetle *(Dendroctonus pseudotsugae)*
Douglas-fir Engraver *(Scolytus unispinosus)*
Dwarf mistletoes *(Arceuthobium* spp.*)*

Phomopsis canker of Douglas-fir *(Phomopsis lokoyae* or *Diaporthe lokoyae)*
Pitch canker *(Fusarium circinatum)*

Juniper
Juniper mistletoes *(Phoradendron juniperinum, P. densum)*

Pine
California Five-spined Ips *(Ips paraconfusus)*
Dwarf mistletoes *(Arceuthobium* spp.*)*
Jeffrey Pine Beetle *(Dendroctonus jeffreyi)*
Mountain Pine Beetle *(Dendroctonus ponderosae)*
Pine Engraver *(Ips pini)*
Pitch canker *(Fusarium circinatum)*
Red Turpentine Beetle *(Dendroctonus valens)*
Sequoia Pitch Moth *(Synanthedon sequoiae)*
Western gall rust *(Peridermium harknessii)*
Western Pine Beetle *(Dendroctonus brevicomis)*
White pine blister rust *(Cronartium ribicola)*

True fir
Dwarf mistletoes *(Arceuthobium* spp.*)*
Fir Engraver *(Scolytus ventralis)*
White fir mistletoe *(Phoradendron pauciflorum)*

Overall Tree Decline

All conifers
Abiotic diseases
Annosum root disease *(Heterobasidion annosum)*
Brown cubical rot or sulfur fungus *(Laetiporus sulphureus)*
Flatheaded wood borers or metallic wood-boring beetles (Buprestidae)
Horntails or wood wasps (Siricidae)
Red-brown butt rot or velvet top fungus *(Phaeolus schweinitzii)*
Roundheaded wood borers or longhorned beetles (Cerambycidae)
White pocket rot or red ring rot *(Phellinus pini)*

Cedar
Cedar bark beetles *(Phloeosinus* spp.*)*
Incense-cedar mistletoe *(Phoradendron libocedri)*

Pocket dry rot of incense-cedar *(Oligoporus amarus)*
Port Orford–cedar root disease *(Phytophthora lateralis)*

Cypress
Juniper mistletoes *(Phoradendron juniperinum, P. densum)*

Douglas-fir
Armillaria root disease *(Armillaria mellea)*
Black Pineleaf Scale *(Nuculaspis californica)*
Blackstain root disease *(Leptographium wageneri)*
Devastating Grasshopper *(Melanoplus devastator)*
Douglas-fir Beetle *(Dendroctonus pseudotsugae)*
Douglas-fir Engraver *(Scolytus unispinosus)*
Douglas-fir needle cast *(Rhabdocline pseudotsugae, R. weirii)*
Douglas-fir Tussock Moth *(Orgyia pseudotsugata)*
Dwarf mistletoes *(Arceuthobium spp.)*
Phomopsis canker of Douglas-fir *(Phomopsis lokoyae* or *Diaporthe lokoyae)*
Pine Needle Scale *(Chionaspis pinifoliae)*
Pitch canker *(Fusarium circinatum)*
Silverspotted Tiger Moth *(Lophocampa argentata)*
Spruce budworms *(Choristoneura spp.)*

Hemlock
Silverspotted Tiger Moth *(Lophocampa argentata)*

Juniper
Juniper mistletoes *(Phoradendron juniperinum, P. densum)*

Pine
Armillaria root disease *(Armillaria mellea)*
Black Pineleaf Scale *(Nuculaspis californica)*
Blackstain root disease *(Leptographium wageneri)*
California Five-spined Ips *(Ips paraconfusus)*
Devastating Grasshopper *(Melanoplus devastator)*
Dwarf mistletoes *(Arceuthobium spp.)*
Elytroderma disease of pines *(Elytroderma deformans)*
Giant Conifer Aphid *(Cinara ponderosae)*
Gouty Pitch Midge *(Cecidomyia piniinopsis)*
Jeffrey Pine Beetle *(Dendroctonus jeffreyi)*
Lodgepole Needleminer *(Coleotechnites milleri)*

Mountain Pine Beetle *(Dendroctonus ponderosae)*
Pandora Moth *(Coloradia pandora)*
Pine Engraver *(Ips pini)*
Pine Needle Scale *(Chionaspis pinifoliae)*
Pine Needle Sheathminer *(Zelleria haimbachi)*
Pine Reproduction Weevil *(Cylindrocopturus eatoni)*
Pitch canker *(Fusarium circinatum)*
Red band needle blight *(Mycosphaerella pini)*
Red Turpentine Beetle *(Dendroctonus valens)*
Silverspotted Tiger Moth *(Lophocampa argentata)*
Western gall rust *(Peridermium harknessii)*
Western Pine Beetle *(Dendroctonus brevicomis)*
White pine blister rust *(Cronartium ribicola)*

True fir
Armillaria root disease *(Armillaria mellea)*
Balsam Twig Aphid *(Mindarus abietinus)*
Black Pineleaf Scale *(Nuculaspis californica)*
Douglas-fir Tussock Moth *(Orgyia pseudotsugata)*
Dwarf mistletoes *(Arceuthobium* spp.*)*
Fir Engraver *(Scolytus ventralis)*
Silverspotted Tiger Moth *(Lophocampa argentata)*
Spruce budworms *(Choristoneura* spp.*)*
True fir needle cast *(Lirula abietis-concoloris)*
White fir mistletoe *(Phoradendron pauciflorum)*

GUIDE TO DAMAGE BY HOST SPECIES

Douglas-fir

Reproductive structures
Douglas-fir Cone Moth *(Barbara colfaxiana)*
Douglas-fir Seed Chalcid *(Megastigmus spermotrophus)*
Western Conifer Seed Bug *(Leptoglossus occidentalis)*

Foliage and shoots
Black Pineleaf Scale *(Nuculaspis californica)*
Cooley Spruce Gall Adelgid *(Adelges cooleyi)*
Devastating Grasshopper *(Melanoplus devastator)*
Douglas-fir needle cast *(Rhabdocline pseudotsugae, R. weirii)*
Douglas-fir Tussock Moth *(Orgyia pseudotsugata)*
Pine Needle Scale *(Chionaspis pinifoliae)*
Silverspotted Tiger Moth *(Lophocampa argentata)*
Spruce budworms *(Choristoneura* spp.*)*

Buds, terminals, and branch tips
Douglas-fir Twig Weevil *(Cylindrocopturus furnissi)*
Dwarf mistletoes *(Arceuthobium* spp.*)*
Phomopsis canker of Douglas-fir *(Phomopsis lokoyae* or *Diaporthe lokoyae)*
Spruce budworms *(Choristoneura* spp.*)*
Twig beetles *(Pityophthorus* spp.*)*

Larger branches and trunk
Douglas-fir Beetle *(Dendroctonus pseudotsugae)*
Douglas-fir Engraver *(Scolytus unispinosus)*
Dwarf mistletoes *(Arceuthobium* spp.*)*
Flatheaded wood borers or metallic wood-boring beetles (Buprestidae)

Horntails or wood wasps (Siricidae)
Phomopsis canker of Douglas-fir *(Phomopsis lokoyae* or *Diaporthe lokoyae)*
Roundheaded wood borers or longhorned beetles (Cerambycidae)
Twig beetles *(Pityophthorus* spp.*)*

Lower trunk and roots
Annosum root disease *(Heterobasidion annosum)*
Armillaria root disease *(Armillaria mellea)*
Blackstain root disease *(Leptographium wageneri)*
Brown cubical rot or sulfur fungus *(Laetiporus sulphureus)*
Red-brown butt rot or velvet top fungus *(Phaeolus schweinitzii)*
White pocket rot or red ring rot *(Phellinus pini)*

Grand Fir

Reproductive structures
Several agents damage cones, but are not included in this guide.

Foliage and shoots
Silverspotted Tiger Moth *(Lophocampa argentata)*
Spruce budworms *(Choristoneura* spp.*)*
True fir needle cast *(Lirula abietis-concoloris)*

Buds, terminals, and branch tips
Cytospora canker of true firs *(Cytospora abietis)*
Dwarf mistletoes *(Arceuthobium* spp.*)*
Spruce budworms *(Choristoneura* spp.*)*
Twig beetles *(Pityophthorus* spp.*)*
Yellow witches' broom of fir *(Melampsorella caryophyllacearum)*

Larger branches and trunk
Dwarf mistletoes *(Arceuthobium* spp.*)*
Fir Engraver *(Scolytus ventralis)*
Horntails or wood wasps (Siricidae)
Roundheaded wood borers or longhorned beetles (Cerambycidae)
Twig beetles *(Pityophthorus* spp.*)*
Yellow witches' broom of fir *(Melampsorella caryophyllacearum)*

Lower trunk and roots
Annosum root disease *(Heterobasidion annosum)*
Armillaria root disease *(Armillaria mellea)*

Brown cubical rot or sulfur fungus *(Laetiporus sulphureus)*
Red-brown butt rot or velvet top fungus *(Phaeolus schweinitzii)*
White pocket rot or red ring rot *(Phellinus pini)*

Red Fir

Reproductive structures
Several agents damage cones, but are not included in this guide.

Foliage and shoots
True fir needle cast *(Lirula abietis-concoloris)*

Buds, terminals, and branch tips
Cytospora canker of true firs *(Cytospora abietis)*
Dwarf mistletoes *(Arceuthobium* spp.*)*
Twig beetles *(Pityophthorus* spp.*)*
Yellow witches' broom of fir *(Melampsorella caryophyllacearum)*

Larger branches and trunk
Dwarf mistletoes *(Arceuthobium* spp.*)*
Fir Engraver *(Scolytus ventralis)*
Horntails or wood wasps (Siricidae)
Roundheaded wood borers or longhorned beetles (Cerambycidae)
Twig beetles *(Pityophthorus* spp.*)*
Yellow witches' broom of fir *(Melampsorella caryophyllacearum)*

Lower trunk and roots
Annosum root disease *(Heterobasidion annosum)*
Armillaria root disease *(Armillaria mellea)*
Brown cubical rot or sulfur fungus *(Laetiporus sulphureus)*
Brown stringy rot or Indian paint fungus *(Echinodontium tinctorium)*
Red-brown butt rot or velvet top fungus *(Phaeolus schweinitzii)*
White pocket rot or red ring rot *(Phellinus pini)*

White Fir

Reproductive structures
Several agents damage cones, but are not included in this guide.

Foliage and shoots
Balsam Twig Aphid *(Mindarus abietinus)*

Douglas-fir Tussock Moth *(Orgyia pseudotsugata)*
Spruce budworms *(Choristoneura* spp.*)*
True fir needle cast *(Lirula abietis-concoloris)*

Buds, terminals, and branch tips
Cytospora canker of true firs *(Cytospora abietis)*
Dwarf mistletoes *(Arceuthobium* spp.*)*
Twig beetles *(Pityophthorus* spp.*)*
White fir mistletoe *(Phoradendron pauciflorum)*
Yellow witches' broom of fir *(Melampsorella caryophyllacearum)*

Larger branches and trunk
Dwarf mistletoes *(Arceuthobium* spp.*)*
Fir Engraver *(Scolytus ventralis)*
Horntails or wood wasps (Siricidae)
Twig beetles *(Pityophthorus* spp.*)*
White fir mistletoe *(Phoradendron pauciflorum)*
Yellow witches' broom of fir *(Melampsorella caryophyllacearum)*

Lower trunk and roots
Annosum root disease *(Heterobasidion annosum)*
Armillaria root disease *(Armillaria mellea)*
Brown stringy rot or Indian paint fungus *(Echinodontium tinctorium)*
Brown cubical rot or sulfur fungus *(Laetiporus sulphureus)*
Red-brown butt rot or velvet top fungus *(Phaeolus schweinitzii)*
White pocket rot or red ring rot *(Phellinus pini)*

Jeffrey Pine

Reproductive structures
Several agents damage cones, but are not included in this guide.

Foliage and shoots
Black Pineleaf Scale *(Nuculaspis californica)*
Diplodia blight *(Sphaeropsis sapinea* or *Diplodia pinea)*
Elytroderma disease of pines *(Elytroderma deformans)*
Pandora Moth *(Coloradia pandora)*
Pine Needle Scale *(Chionaspis pinifoliae)*

Pine Needle Sheathminer *(Zelleria haimbachi)*
Pine Reproduction Weevil *(Cylindrocopturus eatoni)*
Ponderosa Pine Tip Moth *(Rhyacionia zozana)*

Buds, terminals, and branch tips
Diplodia blight *(Sphaeropsis sapinea* or *Diplodia pinea)*
Dwarf mistletoes *(Arceuthobium* spp.*)*
Elytroderma disease of pines *(Elytroderma deformans)*
Ponderosa Pine Tip Moth *(Rhyacionia zozana)*
Twig beetles *(Pityophthorus* spp.*)*
Western gall rust *(Peridermium harknessii)*
Western Pine Shoot Borer *(Eucosma sonomana)*

Larger branches and trunk
California Five-spined Ips *(Ips paraconfusus)*
Dwarf mistletoes *(Arceuthobium* spp.*)*
Flatheaded wood borers or metallic wood-boring beetles (Buprestidae)
Horntails or wood wasps (Siricidae)
Jeffrey Pine Beetle *(Dendroctonus jeffreyi)*
Pine Engraver *(Ips pini)*
Pine Reproduction Weevil *(Cylindrocopturus eatoni)*
Red Turpentine Beetle *(Dendroctonus valens)*
Roundheaded wood borers or longhorned beetles (Cerambycidae)
Sequoia Pitch Moth *(Synanthedon sequoiae)*
Twig beetles *(Pityophthorus* spp.*)*
Western gall rust *(Peridermium harknessii)*

Lower trunk and roots
Annosum root disease *(Heterobasidion annosum)*
Armillaria root disease *(Armillaria mellea)*
Blackstain root disease *(Leptographium wageneri)*
Brown cubical rot or sulfur fungus *(Laetiporus sulphureus)*
Red-brown butt rot or velvet top fungus *(Phaeolus schweinitzii)*
Red Turpentine Beetle *(Dendroctonus valens)*
White pocket rot or red ring rot *(Phellinus pini)*

Lodgepole Pine

Reproductive structures
Cone beetles *(Conophthorus* spp.*)*

Foliage and shoots
Black Pineleaf Scale *(Nuculaspis californica)*
Elytroderma disease of pines *(Elytroderma deformans)*
Lodgepole Needleminer *(Coleotechnites milleri)*
Pine Needle Scale *(Chionaspis pinifoliae)*
Pine Needle Sheathminer *(Zelleria haimbachi)*
Ponderosa Pine Tip Moth *(Rhyacionia zozana)*
Red band needle blight *(Mycosphaerella pini)*
Silverspotted Tiger Moth *(Lophocampa argentata)*

Buds, terminals, and branch tips
Dwarf mistletoes *(Arceuthobium spp.)*
Elytroderma disease of pines *(Elytroderma deformans)*
Lodgepole Pine Terminal Weevil *(Pissodes terminalis)*
Pitch canker *(Fusarium circinatum)*
Ponderosa Pine Tip Moth *(Rhyacionia zozana)*
Twig beetles *(Pityophthorus spp.)*
Western gall rust *(Peridermium harknessii)*
Western Pine Shoot Borer *(Eucosma sonomana)*

Larger branches and trunk
California Five-spined Ips *(Ips paraconfusus)*
Dwarf mistletoes *(Arceuthobium spp.)*
Flatheaded wood borers or metallic wood-boring beetles (Buprestidae)
Horntails or wood wasps (Siricidae)
Mountain Pine Beetle *(Dendroctonus ponderosae)*
Pine Engraver *(Ips pini)*
Pitch canker *(Fusarium circinatum)*
Red Turpentine Beetle *(Dendroctonus valens)*
Roundheaded wood borers or longhorned beetles (Cerambycidae)
Sequoia Pitch Moth *(Synanthedon sequoiae)*
Twig beetles *(Pityophthorus spp.)*
Western gall rust *(Peridermium harknessii)*

Lower trunk and roots
Annosum root disease *(Heterobasidion annosum)*
Armillaria root disease *(Armillaria mellea)*
Brown cubical rot or sulfur fungus *(Laetiporus sulphureus)*
Red-brown butt rot or velvet top fungus *(Phaeolus schweinitzii)*
Red Turpentine Beetle *(Dendroctonus valens)*
White pocket rot or red ring rot *(Phellinus pini)*

Monterey Pine

Reproductive structures
Cone beetles *(Conophthorus* spp.*)*
Western Conifer Seed Bug *(Leptoglossus occidentalis)*

Foliage and shoots
Black Pineleaf Scale *(Nuculaspis californica)*
Diplodia blight *(Sphaeropsis sapinea* or *Diplodia pinea)*
Elytroderma disease of pines *(Elytroderma deformans)*
Pine Needle Scale *(Chionaspis pinifoliae)*
Pine Needle Sheathminer *(Zelleria haimbachi)*
Red band needle blight *(Mycosphaerella pini)*
Silverspotted Tiger Moth *(Lophocampa argentata)*

Buds, terminals, and branch tips
Diplodia blight *(Sphaeropsis sapinea* or *Diplodia pinea)*
Dwarf mistletoes *(Arceuthobium* spp.*)*
Elytroderma disease of pines *(Elytroderma deformans)*
Pitch canker *(Fusarium circinatum)*
Ponderosa Pine Tip Moth *(Rhyacionia zozana)*
Twig beetles *(Pityophthorus* spp.*)*
Western gall rust *(Peridermium harknessii)*

Larger branches and trunk
California Five-spined Ips *(Ips paraconfusus)*
Dwarf mistletoes *(Arceuthobium* spp.*)*
Flatheaded wood borers or metallic wood-boring beetles (Buprestidae)
Horntails or wood wasps (Siricidae)
Pitch canker *(Fusarium circinatum)*
Red Turpentine Beetle *(Dendroctonus valens)*
Roundheaded wood borers or longhorned beetles (Cerambycidae)
Sequoia Pitch Moth *(Synanthedon sequoiae)*
Twig beetles *(Pityophthorus* spp.*)*
Western gall rust *(Peridermium harknessii)*

Lower trunk and roots
Annosum root disease *(Heterobasidion annosum)*
Armillaria root disease *(Armillaria mellea)*
Brown cubical rot or sulfur fungus *(Laetiporus sulphureus)*

Red-brown butt rot or velvet top fungus *(Phaeolus schweinitzii)*
Red Turpentine Beetle *(Dendroctonus valens)*
White pocket rot or red ring rot *(Phellinus pini)*

Ponderosa Pine

Reproductive structures
Cone beetles *(Conophthorus* spp.*)*
Pine Seedworm *(Cydia miscitata)*
Western Conifer Seed Bug *(Leptoglossus occidentalis)*

Foliage and shoots
Black Pineleaf Scale *(Nuculaspis californica)*
Devastating Grasshopper *(Melanoplus devastator)*
Diplodia blight *(Sphaeropsis sapinea* or *Diplodia pinea)*
Elytroderma disease of pines *(Elytroderma deformans)*
Giant Conifer Aphid *(Cinara ponderosae)*
Gouty Pitch Midge *(Cecidomyia piniinopsis)*
Pandora Moth *(Coloradia pandora)*
Pine Needle Scale *(Chionaspis pinifoliae)*
Pine Needle Sheathminer *(Zelleria haimbachi)*
Pine Reproduction Weevil *(Cylindrocopturus eatoni)*
Red band needle blight *(Mycosphaerella pini)*

Buds, terminals, and branch tips
Diplodia blight *(Sphaeropsis sapinea* or *Diplodia pinea)*
Dwarf mistletoes *(Arceuthobium* spp.*)*
Elytroderma disease of pines *(Elytroderma deformans)*
Pitch canker *(Fusarium circinatum)*
Ponderosa Pine Tip Moth *(Rhyacionia zozana)*
Twig beetles *(Pityophthorus* spp.*)*
Western gall rust *(Peridermium harknessii)*
Western Pine Shoot Borer *(Eucosma sonomana)*

Larger branches and trunk
California Five-spined Ips *(Ips paraconfusus)*
Dwarf mistletoes *(Arceuthobium* spp.*)*
Flatheaded wood borers or metallic wood-boring beetles (Buprestidae)
Horntails or wood wasps (Siricidae)
Mountain Pine Beetle *(Dendroctonus ponderosae)*

Pine Engraver *(Ips pini)*
Pine Reproduction Weevil *(Cylindrocopturus eatoni)*
Pitch canker *(Fusarium circinatum)*
Red Turpentine Beetle *(Dendroctonus valens)*
Roundheaded wood borers or longhorned beetles (Cerambycidae)
Sequoia Pitch Moth *(Synanthedon sequoiae)*
Twig beetles *(Pityophthorus spp.)*
Western gall rust *(Peridermium harknessii)*
Western Pine Beetle *(Dendroctonus brevicomis)*

Lower trunk and roots
Annosum root disease *(Heterobasidion annosum)*
Armillaria root disease *(Armillaria mellea)*
Blackstain root disease *(Leptographium wageneri)*
Brown cubical rot or sulfur fungus *(Laetiporus sulphureus)*
Red-brown butt rot or velvet top fungus *(Phaeolus schweinitzii)*
Red Turpentine Beetle *(Dendroctonus valens)*
White pocket rot or red ring rot *(Phellinus pini)*

Sugar Pine

Reproductive structures
Cone beetles *(Conophthorus spp.)*

Foliage and shoots
Black Pineleaf Scale *(Nuculaspis californica)*
Pine Needle Scale *(Chionaspis pinifoliae)*
Pine Reproduction Weevil *(Cylindrocopturus eatoni)*

Buds, terminals, and branch tips
Dwarf mistletoes *(Arceuthobium spp.)*
Ponderosa Pine Tip Moth *(Rhyacionia zozana)*
Twig beetles *(Pityophthorus spp.)*
Western gall rust *(Peridermium harknessii)*
White pine blister rust *(Cronartium ribicola)*

Larger branches and trunk
California Five-spined Ips *(Ips paraconfusus)*
Dwarf mistletoes *(Arceuthobium spp.)*
Flatheaded wood borers or metallic wood-boring beetles (Buprestidae)
Horntails or wood wasps (Siricidae)

Mountain Pine Beetle *(Dendroctonus ponderosae)*
Pine Reproduction Weevil *(Cylindrocopturus eatoni)*
Red Turpentine Beetle *(Dendroctonus valens)*
Roundheaded wood borers or longhorned beetles (Cerambycidae)
Twig beetles *(Pityophthorus* spp.*)*
Western gall rust *(Peridermium harknessii)*
White pine blister rust *(Cronartium ribicola)*

Lower trunk and roots
Annosum root disease *(Heterobasidion annosum)*
Armillaria root disease *(Armillaria mellea)*
Brown cubical rot or sulfur fungus *(Laetiporus sulphureus)*
Red-brown butt rot or velvet top fungus *(Phaeolus schweinitzii)*
Red Turpentine Beetle *(Dendroctonus valens)*
White pocket rot or red ring rot *(Phellinus pini)*

Other Pines

The following affect one or more other pine species in California.

Reproductive structures
Cone beetles *(Conophthorus* spp.*)*
Western Conifer Seed Bug *(Leptoglossus occidentalis)*

Foliage and shoots
Black Pineleaf Scale *(Nuculaspis californica)*
Diplodia blight *(Sphaeropsis sapinea* or *Diplodia pinea)*
Elytroderma disease of pines *(Elytroderma deformans)*
Pine Needle Scale *(Chionaspis pinifoliae)*
Pine Needle Sheathminer *(Zelleria haimbachi)*
Ponderosa Pine Tip Moth *(Rhyacionia zozana)*
Red band needle blight *(Mycosphaerella pini)*
Silverspotted Tiger Moth *(Lophocampa argentata)*

Buds, terminals, and branch tips
Diplodia blight *(Sphaeropsis sapinea)*
Dwarf mistletoes *(Arceuthobium* spp.*)*
Elytroderma disease of pines *(Elytroderma deformans)*
Pitch canker *(Fusarium circinatum)*
Ponderosa Pine Tip Moth *(Rhyacionia zozana)*
Twig beetles *(Pityophthorus* spp.*)*

Western gall rust *(Peridermium harknessii)*
Western Pine Shoot Borer *(Eucosma sonomana)*

Larger branches and trunk
California Five-spined Ips *(Ips paraconfusus)*
Dwarf mistletoes *(Arceuthobium spp.)*
Flatheaded wood borers or metallic wood-boring beetles (Buprestidae)
Horntails or wood wasps (Siricidae)
Mountain Pine Beetle *(Dendroctonus ponderosae)*
Pitch canker *(Fusarium circinatum)*
Red Turpentine Beetle *(Dendroctonus valens)*
Roundheaded wood borers or longhorned beetles (Cerambycidae)
Sequoia Pitch Moth *(Synanthedon sequoiae)*
Twig beetles *(Pityophthorus spp.)*
Western gall rust *(Peridermium harknessii)*
Western Pine Beetle *(Dendroctonus brevicomis)*
White pine blister rust *(Cronartium ribicola)*

Lower trunk and roots
Annosum root disease *(Heterobasidion annosum)*
Armillaria root disease *(Armillaria mellea)*
Blackstain root disease *(Leptographium wageneri)*
Brown cubical rot or sulfur fungus *(Laetiporus sulphureus)*
Red-brown butt rot or velvet top fungus *(Phaeolus schweinitzii)*
Red Turpentine Beetle *(Dendroctonus valens)*
White pocket rot or red ring rot *(Phellinus pini)*

Incense-cedar

Reproductive structures
Several agents damage cones, but are not included in this guide.

Foliage and shoots
Several agents damage foliage and shoots, but are not included in this guide.

Buds, terminals, and branch tips
Botryosphaeria canker *(Botryosphaeria dothidea)*
Cedar bark beetles *(Phloeosinus spp.)*
Incense-cedar mistletoe *(Phoradendron libocedri)*
Incense-cedar rust *(Gymnosporangium libocedri)*

Larger branches and trunk
Cedar bark beetles *(Phloeosinus* spp.*)*
Flatheaded wood borers or metallic wood-boring beetles (Buprestidae)
Horntails or wood wasps (Siricidae)
Incense-cedar mistletoe *(Phoradendron libocedri)*
Roundheaded wood borers or longhorned beetles (Cerambycidae)

Lower trunk and roots
Annosum root disease *(Heterobasidion annosum)*
Pocket dry rot of incense-cedar *(Oligoporus amarus)*
Red-brown butt rot or velvet top fungus *(Phaeolus schweinitzii)*

Port Orford–cedar

Reproductive structures
Several agents damage cones, but are not included in this guide.

Foliage and shoots
Several agents damage foliage and shoots, but are not included in this guide.

Buds, terminals, and branch tips
Cedar bark beetles *(Phloeosinus* spp.*)*

Larger branches and trunk
Cedar bark beetles *(Phloeosinus* spp.*)*
Flatheaded wood borers or metallic wood-boring beetles (Buprestidae)
Horntails or wood wasps (Siricidae)
Roundheaded wood borers or longhorned beetles (Cerambycidae)

Lower trunk and roots
Annosum root disease *(Heterobasidion annosum)*
Port Orford–cedar root disease *(Phytophthora lateralis)*

GLOSSARY

Abdomen The third or posterior region of an insect body. The abdomen usually has 10 segments and bears no true legs, but can have appendages at the apex.

Annulus A ring of fungus tissue encircling the stem of some mushrooms.

Antiattractant A chemical that disrupts the behavioral response of an insect to one or more chemicals.

Asexual Produced vegetatively in plants; without union of sperm and egg in insects.

Bark beetle A beetle in the family Scolytidae that feeds in the cambial region, either in the bark only, or in the bark and xylem, of stems or branches, and spends most of its life cycle there.

Blight Any atmospheric or soil condition, parasite, or insect that kills, withers, or checks the growth of plants.

Bluestain Coloration of wood infected by fungi with blue, brown, or black hyphae; a group of more primitive fungi that cause bluestain.

Board foot A measure of wood volume equal to an unfinished board, $2.54 \times 30.5 \times 30.5$ cm ($1 \times 12 \times 12$ in.).

Bole The trunk or main stem of a tree.

Boring dust Finely divided woody debris produced by insects as they bore or tunnel into the branches and trunks of trees.

Brood All the individuals that are born alive or hatch from an egg mass or group of eggs deposited by one generation of parents and that mature within the same time span.

Broom An abnormal cluster of twigs and branches caused by certain pathogens. Also called Witches' broom.

Brown rot Decay caused by fungi that attack the cellulose in cell walls.

Bud A dormancy structure of shoots.

Butt The base of a tree or the large end of a log.

Butt rot Rot characteristically confined to the lower trunk of a tree.

Callow adult Young adult; usually refers to a new adult bark beetle that is light brown and has not emerged from under the bark.

Cambium The thin layer of cells beneath the bark of woody plants that gives rise to new cells (phloem and xylem) and is therefore responsible for diameter growth.

Canker A localized, often callused, malformation or area of dead or dying tissue on a tree stem or branch.

Canopy The foliar cover in a forest stand.

Caterpillar The larva of a butterfly or moth.

Cellulose A principal chemical component of plant cell walls.

Chlorosis An abnormal yellowing of normally green plant foliage. The adjectival form is *chlorotic*.

Clone A group of genetically identical individuals that have been produced by vegetative means from a single parent.

Cocoon A protective case surrounding an insect pupa.

Cone The seed-bearing structure of conifers, consisting of a central stem, woody or fleshy scales, bracts, and seeds.

Conk The large, often bracket-like fruiting body of wood-destroying fungi.

Cortex Primary plant tissue that separates the vascular system (xylem and phloem) from the epidermis. The adjectival form is *cortical*.

Crawler The first active instar of a scale insect.

Crook An abrupt bend in a tree or log.

Crown The part of the tree that bears live branches and foliage.

Crown closure The point at which crowns of trees begin to overlap in the canopy.

Decay Process or result of degradation of wood by fungi, bacteria, or yeasts.

Decay pocket Pattern of decay characteristic of some fungi, in which wood in a pocket is more extensively degraded than is the surrounding wood.

Defoliation Loss of current or past years' foliage.

Diapause A condition of suspended animation or arrested development during the life cycle of an insect.

Dieback The death of parts of a tree or plant, usually from the top downward and from the outer crown inward.

Dorsum The top or uppermost side of an insect, or other animal. The adjectival form is *dorsal.*

Duff Decaying organic matter on the forest floor.

Earlywood The part of the annual growth ring of wood that is produced early in the season. It is less dense than wood laid down late in the season (latewood).

Egg niche Cavities excavated beneath the bark by female, usually bark beetles or weevils, into which eggs are deposited.

Elytron The leathery front wing, usually of beetles. The plural is *elytra.*

Endemic Term used to describe a pathogen continually infecting a few plants within an area, or a population of potentially damaging organisms that are at low levels.

Epidemic A pathogen periodically infecting a large number of plants in an area and causing significant damage, or a population of organisms that builds up rapidly to highly damaging levels.

Epidermis The outermost layer of cells in plants and animals.

Epiphyte A plant growing on, but not nourished by, another plant.

Fascicle A cluster of leaves, needles, or flowers.

Femur Upper part, or division, of the insect leg.

Flag Dead shoots or branches on live trees, generally with yellow, brown or red needles attached.

Forb A broad-leafed flowering plant as distinguished from the grasses, sedges, etc.

Fork The development of two or more leaders resulting in multiple stems.

Frass Fecal material of an insect. In the case of wood-boring beetles, phloem and xylem fragments occur with the feces.

Fruiting body A fungal organ specialized for producing spores.

Fungus One of a group of organisms considered by some authorities to be lower plants that lack chlorophyll.

Gall A swelling of plant tissue caused by attack of insects or disease-causing organisms.

Gallery Typically, a tunnel or pathway in host tissue in which an insect lives, feeds, mates, deposits eggs, or all four.

Generation One complete cycle of the life history, that is, egg through adult.

Genus An assemblage of species that have a similar character or series of characters. The plural is genera.

Gout An abnormal proliferation of plant tissue.

Gregarious (species) Living in societies or communities, but not social.

Grub The larval form of beetles and wasps.

Hard pine Pines in the hard pine group are in the subgenus *Pinus (Diploxylon)*. Hard pine is a designation used only for convenience, as some hard pine species are almost as soft as soft pines.

Heart rot Decay restricted to the heartwood.

Heartwood The older, nonliving central portion of the tree stem; it is usually darker and more durable than the surrounding sapwood.

Hibernaculum A silken shelter in which some caterpillars hibernate. The plural is hibernacula.

Honeydew   The sugary secretion produced by sap-sucking insects.

Host   An organism that serves as food, shelter, or resting site for an unrelated organism.

Hypha   A microscopic filament of fungal cells. The plural is *hyphae*.

Incipient decay   Early stages of wood decay.

Infection   The process or result of a pathogen invading host tissue.

Inoculum   The spores or tissues of a pathogen that infect a host.

Insecticide   A substance, usually a chemical, that kills insects.

Instar   The period or stage between molts during larval development; the first instar is the stage between the egg and the first molt.

Internode   The space between consecutive portions of the stem from which branches, leaves, or flowers originate (the nodes).

Knot   A section of branch embedded in lumber or other wood product.

Laminate decay   Wood that is decayed more extensively in earlywood than in latewood and tends to separate into sheets or lamina along annual rings.

Larva   An immature form of some insect species that hatches from an egg and differs fundamentally in form from adults.

Lateral shoot   A side shoot from a branch or stem.

Latewood   The part of the annual growth ring of wood that is produced late in the growing season. It is more dense than wood laid down early in the season (earlywood).

Leader   The terminal or topmost shoot.

Lesion   A wound or localized injury, often expressed as broken outer bark tissue.

Life cycle   The series of changes in the life of an organism, including reproduction.

Lignin   A complex polymer that is found between plant cell walls and within plant cell walls, that gives rigidity to the cell.

Looper   A moth larva that is an inchworm or measuring worm.

Metamorphosis The change of body form through which insects pass in developing from egg to adult. In general, there is simple or complete metamorphosis. In the simple form (e.g., grasshoppers) wings develop externally, there is no pupal stage, and immature forms are called nymphs. In the complete form (e.g., beetles) wings develop internally, there is a pupal stage, and immature forms are called larvae.

Midge Adults of a group of small flies in the order Diptera.

Molt The act of shedding the insect exoskeleton.

Mycangial fungi Fungi that are carried in specialized structures found on some insect species, especially beetles.

Mycelial fan A mass of fungal filaments that are arranged in a flat plane, and in a fan shape.

Mycelium The vegetative (nonfruiting) portion of a fungus, composed of hyphae.

Necrosis Death of plant cells, usually resulting in darkening of affected tissue. The adjectival form is necrotic.

Nematode A roundworm with a long, cylindrical, unsegmented body.

Nucleopolyhedrosis virus A particular type of virus that causes disease in insects.

Nuptial chamber Usually refers to the chamber beneath the bark of host trees where mating of bark beetles takes place.

Nymph A young insect that, when it emerges from the egg, resembles the adult but is smaller and has incompletely developed wings and sexual organs.

Overwinter In an insect life cycle, the act of passing the winter period, usually in an inactive state.

Oviposition The act of depositing eggs. Eggs are deposited by means of a specialized organ called an ovipositor.

Oxidants A class of air pollutants that includes ozone and peroxyacetylnitrate (PAN).

Parasite An organism living on and nourished by another living organism.

Parthenogenesis Reproduction without male fertilization.

Pathogen A disease-causing organism.

Pesticide A substance for controlling, preventing, destroying, disabling or repelling a pest.

Petiole The stalk attaching a leaf to the main twig.

Pheromone A substance, secreted to the outside of an insect's body, that serves as a chemical signal between members of the same species. Pheromones are usually airborne compounds and act to bring the sexes together (sex pheromone), to elicit alarm (alarm pheromone) or aggregation (aggregation pheromone), to guide insects to a suitable food source, or to elicit some other response.

Phloem Active, food-conducting tissue of the inner bark of trees and other woody plants, that with age becomes outer bark.

Pitch mass A mass of resin on the surface of a tree.

Pitch pocket A pocket of resin under the bark of a branch or stem.

Pitch tube A mixture of resin, phloem and xylem fragments, and frass on the surface of trees attacked by certain bark beetles.

Pocket rot A rot caused by certain fungi that occurs in distinct, scattered pockets within the heartwood rather than in a distinct column.

Pore The open end of a tube in which spores of certain higher fungi are produced.

Pore surface The surface of a fungal fruiting body from which spores disperse.

Prepupa The insect stage between the end of the larval stage and the beginning of the true pupal stage.

Progeny The offspring or brood from eggs laid by an adult.

Prolegs The fleshy, unjointed legs of caterpillars and some sawfly larvae; false legs.

Propagule Any part of a fungus, such as spores or hyphal fragments, that may result in propagation of the fungus.

Pupa The intermediate stage of insect development between the larval and adult stages.

Pupal chamber A cavity in the wood, phloem, or bark excavated by an insect and in which it pupates.

Pustule A small, sometimes colored, blister-like swelling caused by certain fungi.

Resinosis Reaction of a tree to invasion by pathogens or insects or to abiotic injury, which results in a flow of resin on the outer bark, or in an accumulation of resin within or under the bark.

Rhizomorph A specialized thread or cordlike structure made up of parallel hyphae with a protective covering.

Root collar The location on a plant where the anatomy changes from that of a stem to that of a root, usually at the soil surface.

Root crown Uppermost portion of a root system, where the major roots join together at the base of the stem.

Sapling A young tree that is larger than a seedling but smaller than a pole (the stage proceeding that of a mature tree).

Saprophyte An organism using dead organic material as food.

Sap rot Decay of sapwood.

Sapwood The soft wood just beneath the bark of a tree.

Seedling A young plant recently emerged from a seed.

Semiochemical Any chemical involved in communication between organisms.

Shoot Young aerial outgrowth from a plant body.

Shoot borer An insect that feeds and spends the majority of its life cycle within expanding or mature shoots of host plants.

Sign Evidence of the cause of a symptom. For example, the pest or pathogen itself, or pest- or pathogen-produced materials or structures left behind.

Silviculture The art and science of establishing, growing, and regenerating a forest.

Slash The residue of branches and treetops left on the ground after logging or accumulating as a result of another factor such as storm damage.

Soft pine Any pine in the subgenus *Strobus* (*Haploxylon*).

Species An aggregation of individuals alike in appearance and structure, and that mate and produce fertile offspring.

Spike top The appearance of a tree with a dead terminal or upper portion of the crown.

Spore Reproductive cell (or multiple cells) of a fungus or bacterium that serves the same purpose as the seed of higher plants.

Sporulate To release spores.

Staminate cone Male, pollen-producing, cones.

Stocking Term to describe the number of trees in a given area; stocking density.

Symptom A visual clue indicating changes in the normal growth or appearance, or both, of a host in response to pests, pathogens, environmental factors, or management practices.

Systemic insecticide An insecticide that is carried internally in a plant.

Tarsus The foot. The outermost jointed division of the insect leg.

Terminal growth Growth of the terminal shoot or leader.

Terminal shoot The leader or topmost shoot.

Thin crown A crown that does not have its full complement of needles.

Thorax The second or intermediate region of an insect body; the thorax bears true legs and wings.

Tibia A single-segmented division of an insect leg between the femur and the tarsus.

Top kill Gradual or sudden dieback of the uppermost portions of the crown of a tree.

Top whorl The top circle of branches that originate from one level on the stem.

Trunk The main stem of a tree.

Upper whorl One of the upper circles of branches that each originate from one level on the stem.

Vector An organism, such as an insect, that transmits a pathogen from one host to another.

Venter The underside of an insect or other animal. The adjectival form is *ventral*.

Wetwood A discolored, water-soaked condition of the heartwood of some conifers, presumably caused by bacterial fermentation. It is often associated with a distinctive odor, gas production, and an exudation called slime flux.

White rot Decay caused by fungi that attack both lignin and cellulose in the cell wall.

Windthrow A fallen tree, or part of a tree, as a result of the action of wind. The adjectival form is *windthrown*.

Witches' broom An abnormal cluster of twigs and branches caused by certain pathogens.

Wood borer Usually refers to the larva of a beetle species (especially longhorned beetles and metallic wood-boring beetles) or wood wasp species (family Siricidae) that feeds and spends the majority of its life within the wood of hosts—as opposed to those that feed primarily in the cambial region of the bark.

Xylem The woody tissue of a plant that conducts water and minerals.

GENERAL REFERENCES AND RESOURCES

General References

Dreistadt, S. H., J. K. Clark, and M. L. Flint. 1994. *Pests of landscape trees and shrubs: An integrated pest management guide.* Statewide Integrated Pest Management Project, Publication 3359. Davis: Division of Agriculture and Natural Resources, University of California.

Funk, A. 1981. *Parasitic microfungi of western trees.* BC-X-222. Victoria: Pacific Forest Research Center, Canadian Forest Service.

Furniss, R. L., and V. M. Carolin. 1977. *Western forest insects.* Miscellaneous Publication No. 1339. Washington, D.C.: USDA Forest Service.

Hamm, P. B., S. J. Cambell, and E. M. Hansen, eds. 1990. *Growing healthy seedlings: Identification and management of pests in northwest forest nurseries.* Special Publication 19. Corvallis: Forest Research Laboratory, Oregon State University.

Hedlin, A. F., H. O. Yates, D. Cibrian-Tovar, B. H. Ebel, T. W. Koerber, and E. P. Merkel. 1981. *Cone and seed insects of North American conifers.* Ottawa: Environment Canada, Canadian Forest Service; Washington, D.C.: USDA Forest Service; Chapingo: Universidad Autónoma Chapingo.

Helms, J. A., ed. 1998. *The dictionary of forestry.* Bethesda: Society of American Foresters.

Hygnstrom, S. E., R. M. Timm, and G. E. Larson, eds. 1994. *Prevention and control of wildlife damage.* Lincoln: University of Nebraska Cooperative Extension, University of Nebraska; Washington, D.C.: Animal Damage Control, USDA Animal and Plant Health Inspection Service.

Jameson, E. W., and H. J. Peeters. 1988. *California mammals.* California Natural History Guides 52. Berkeley and Los Angeles: University of California Press.

Johnson, W. T., and H. H. Lyon. 1991. *Insects that feed on trees and shrubs.* 2nd ed. Ithaca: Cornell University Press.

Scharpf, R. F. 1993. *Diseases of Pacific Coast conifers.* Agriculture Handbook 521. Washington, D.C.: USDA Forest Service.

Sinclair, W. A., H. H. Lyon, and W. T. Johnson. 1987. *Diseases of trees and shrubs.* Ithaca: Cornell University Press.

Tainter, F. H., and F. A. Baker. 1996. *Principles of forest pathology.* New York: John Wiley & Sons.

Pest Management Resources

Additional resources may be available in some areas. Telephone numbers for these and the agencies listed below can be found in your telephone directory. In addition, the Internet can be an excellent source of recent information on pests and diseases.

Federal

United States Department of Agriculture (USDA) Forest Service: P.O. Box 96090, Washington, D.C. 20090-6090. Telephone: (202)205-8333. Web site: http://www.fs.fed.us/

State

California Department of Forestry and Fire Protection (CDF): 1416 9th St., P. O. Box 944246, Sacramento, California 94244-2460. Telephone: (916)653-5123. Web site: http://www.fire.ca.gov/ Agriculture and Natural Resources, University of California (ANR): Franklin St., Oakland, California 94607-5200. Web site: http://ucanr.org/

County

University of California Cooperative Extension (UCCE), the outreach arm of the ANR: You'll find local Cooperative Extension offices in each county seat; look under County Government in the telephone directory. Web site: http://ucanr.org/ce.cfm

County Agriculture Commissioner's Offices: Each California county has one. They enforce pesticide use regulation, license pest control operators and advisers, and regulate and inspect shipments and plant products. In a county with intensive agriculture, the commissioner may have several offices and a staff of 100 or more. Web site: http://www.cdpr.ca.gov/docs/counties/caclist.htm

ILLUSTRATION CREDITS

T.W. KOERBER 1–15, 17–26, 29, 30, 32–26, 40, 41, 47, 48, 50–62, 64–67, 69–73, 75, 76, 79, 83, 93, 94, 96, 103–107, 188, 210

USDA FOREST SERVICE 16, 31, 42, 43, 68, 77, 78, 81, 89–92, 98, 101, 102

J.W. DALE 27, 44–46, 49, 97, 212

D.L. DAHLSTEN 28

J.W. FOX 37, 38

A.J. STORER 39, 74, 80, 99, 121–123, 126–130, 135–138

CALIFORNIA DEPARTMENT OF FORESTRY AND FIRE PROTECTION (CDF) 63

CLYDE D. WILLSON 82

R.W. STARK 84

STU WHITNEY 85

DON PERKINS 86, 95

DONALD R. OWEN 87, 131–133

P. SVIHRA 88

P.L. DALLARA 100

R.F. SCHARPF 108–120, 124, 125, 134, 139–179, 184–187, 189, 190–195

J. KLIEJUNAS 180, 181

P.R. MILLER 182, 183

G.A. GIUSTI 196–203, 205, 207, 209, 211

USDA APHIS FILE PHOTO 204

UC COOPERATIVE EXTENSION FILE PHOTO 206, 208

INDEX OF AGENTS

Agents by Common Name

air pollution, 158–160

American Black Bear *(Ursus americanus)*, 172–173

annosum root disease *(Heterobasidion annosum)*, 149–152

Armillaria root disease *(Armillaria mellea)*, 146–148

Balsam Twig Aphid *(Mindarus abietinus)*, 22–24

bark beetles (Scolytidae), 60–81

Black Pineleaf Scale *(Nuculaspis californica)*, 27–28

blackstain root disease *(Leptographium wageneri)*, 152–154

Botryosphaeria canker *(Botryosphaeria dothidea)*, 106–108

brown cubical rot *(Laetiporus sulphureus)*, 139–140

brown stringy rot *(Echinodontium tinctorium)*, 137–138

California Five-spined Ips *(Ips paraconfusus)*, 61–63

cedar bark beetles *(Phloeosinus* spp.), 85–87

cone beetles *(Conophthorus* spp.), 10–13

Cooley Spruce Gall Adelgid *(Adelges cooleyi)*, 24–26

Cytospora canker of true firs *(Cytospora abietis)*, 109–110

deer and elk (*Odocoileus* spp., *Cervus* spp.) 173–175

Devastating Grasshopper *(Melanoplus devastator)*, 44–47

Diplodia blight *(Sphaeropsis sapinea* or *Diplodia pinea)*, 114–116

Douglas-fir Beetle *(Dendroctonus pseudotsugae)*, 79–81

Douglas-fir Cone Moth *(Barbara colfaxiana)*, 7–8

Douglas-fir needle cast *(Rhabdocline pseudotsugae, R. weirii)*, 104–106

Douglas-fir Seed Chalcid *(Megastigmus spermotrophus)*, 13–15

Douglas-fir Tussock Moth *(Orgyia pseudotsugata)*, 38–41

Douglas-fir Twig Weevil *(Cylindrocopturus furnissi)*, 55–56

drought, 160–161

dwarf mistletoes *(Arceuthobium* spp.), 118–120

elk and deer *(Cervus* spp. *Odocoileus* spp.), 173–175

Elytroderma disease of pines *(Elytroderma deformans)*, 98–100

Fir Engraver *(Scolytus ventralis)*, 82–85
flatheaded wood borers (Buprestidae), 89–91
flooding, 162
frost, 163–164

Giant Conifer Aphid *(Cinara ponderosae)*, 21–22
Gouty Pitch Midge *(Cecidomyia pininopsis)*, 30–32

hares and jackrabbits *(Lepus* spp.), 175–177
heat, 164–165
herbicides, 165–166
horntails (Siricidae), 94–96

incense-cedar mistletoe *(Phoradendron libocedri)*, 123–124
incense-cedar rust *(Gymnosporangium libocedri)*, 129–131
Indian paint fungus *(Echinodontium tinctorium)*, 137–138

jackrabbits and hares *(Lepus* spp.), 175–177
Jeffrey Pine Beetle *(Dendroctonus jeffreyi)*, 70–72
juniper mistletoes *(Phoradendron juniperinum, P. densum)*, 121–122

livestock, 177
Lodgepole Needleminer *(Coleotechnites milleri)*, 15–18
Lodgepole Pine Terminal Weevil *(Pissodes terminalis)*, 49
longhorned beetles (Cerambycidae), 91–94

metallic wood-boring beetles (Buprestidae), 89–91
Mountain Beaver *(Aplodontia rufa)*, 178–179
Mountain Pine Beetle *(Dendroctonus ponderosae)*, 67–70

Pandora Moth *(Coloradia pandora)*, 33–35
Phomopsis canker of Douglas-fir *(Phomopsis lokoyae* or *Diaporthe lokoyae)*, 116–118

Pine Engraver *(Ips pini)*, 65–67
Pine Needle Scale *(Chionaspis pinifoliae)*, 29–30
Pine Needle Sheathminer *(Zelleria haimbachi)*, 19–21
Pine Reproduction Weevil *(Cylindrocopturus eatoni)*, 52–55
Pine Seedworm *(Cydia miscitata)*, 9–10
pitch canker *(Fusarium circinatum)*, 111–114
pocket dry rot of incense-cedar *(Oligoporus amarus)*, 140–141
pocket gophers *(Thomomys* spp.), 180–182
Ponderosa Pine Tip Moth *(Rhyacionia zozana)*, 47–50
Porcupine *(Erethizon dorsatum)*, 182–183
Port Orford-cedar root disease *(Phytophthora lateralis)*, 154–156

red band needle blight *(Mycosphaerella pini)*, 102–104
red ring rot *(Phellinus pini)*, 144–146
Red Turpentine Beetle *(Dendroctonus valens)*, 76–79
red-brown butt rot *(Phaeolus schweinitzii)*, 142–143
roundheaded wood borers (Cerambycidae), 91–94

salt, 166–169
Sequoia Pitch Moth *(Synanthedon sequoiae)*, 58–61
Silverspotted Tiger Moth *(Lophocampa argentata)*, 35–37
spruce budworms *(Choristoneura* spp.), 41–44
sulphur fungus *(Laetiporus sulphureus)*, 139–140

true fir needle cast *(Lirula abietisconcoloris)*, 100–102
twig beetles *(Pityophthorus* spp.), 87–89

velvet top fungus *(Phaeolus schweinitzii)*, 142–143

Western Conifer Seed Bug *(Leptoglossus occidentalis)*, 4–6

western gall rust *(Peridermium harknessii)*, 134–136
Western Pine Beetle *(Dendroctonus brevicomis)*, 73–76
Western Pine Shoot Borer *(Eucosma sonomana)*, 50–52
white fir mistletoe *(Phoradendron pauciflorum)*, 124–126
white pine blister rust *(Cronartium ribicola)*, 126–129
white pocket rot *(Phellinus pini)*, 144–146
winter burn and winter drying, 169–170
wood wasps (Siricidae), 94–96
woodrats *(Neotoma* spp.), 183–185

yellow witches' broom of fir *(Melampsorella caryophyllacearum)*, 132–134

Agents by Scientific Name

Adelges cooleyi (Cooley Spruce Gall Adelgid), 24–26
Aplodontia rufa (Mountain Beaver), 178–179
Arceuthobium spp. (dwarf mistletoes), 118–120
Armillaria mellea (*Armillaria* root disease), 146–148

Barbara colfaxiana (Douglas-fir Cone Moth), 7–8
Botryosphaeria dothidea (*Botryosphaeria* canker), 106–108
Buprestidae (flatheaded wood borers or metallic wood-boring beetles), 89–91

Cecidomyia piniinopsis (Gouty Pitch Midge), 30–32
Cerambycidae (longhorned beetles or roundheaded wood borers), 91–94
Cervus spp., *Odocoileus* spp. (elk and deer), 173–175
Chionaspis pinifoliae (Pine Needle Scale), 29–30
Choristoneura spp. (spruce budworms), 41–44
Cinara ponderosae (Giant Conifer Aphid), 21–22
Coleotechnites milleri (Lodgepole Needleminer), 15–18
Coloradia pandora (Pandora Moth), 33–35
Conophthorus spp. (cone beetles), 10–13
Cronartium ribicola (white pine blister rust), 126–129
Cydia miscitata (Pine Seedworm), 9–10
Cylindrocopturus eatoni (Pine Reproduction Weevil), 52–55
Cylindrocopturus furnissi (Douglas-fir Twig Weevil), 55–56
Cytospora abietis (*Cytospora* canker of true firs), 109–110

Dendroctonus brevicomis (Western Pine Beetle), 73–76
Dendroctonus jeffreyi (Jeffrey Pine Beetle), 70–72
Dendroctonus ponderosae (Mountain Pine Beetle), 67–70
Dendroctonus pseudotsugae (Douglas-fir Beetle), 79–81
Dendroctonus valens (Red Turpentine Beetle), 76–79
Diaporthe lokoyae (*Phomopsis* canker of Douglas-fir), 116–118
Diplodia pinea (*Diplodia* blight), 114–116

Echinodontium tinctorium (brown stringy rot), 137–138
Echinodontium tinctorium (Indian paint fungus), 137–138
Elytroderma deformans (*Elytroderma* disease of pines), 98–100
Erethizon dorsatum (Porcupine), 182–183
Eucosma sonomana (Western Pine Shoot Borer), 50–52

Fusarium circinatum (pitch canker), 111–114

Gymnosporangium libocedri (incense-cedar rust), 129–131

Heterobasidion annosum (annosum root disease), 149–152

Ips paraconfusus (California Five-spined Ips), 61–63
Ips pini (Pine Engraver), 65–67

Laetiporus sulphureus (brown cubical rot), 139–140
Laetiporus sulphureus (sulphur fungus), 139–140
Leptoglossus occidentalis (Western Conifer Seed Bug), 4–6
Leptographium wageneri (blackstain root disease), 152–154
Lepus spp. (jackrabbits and hares), 175–177
Lirula abietis-concoloris (true fir needle cast), 100–102
Lophocampa argentata (Silverspotted Tiger Moth), 35–37

Megastigmus spermotrophus (Douglas-fir Seed Chalcid), 13–15
Melampsorella caryophyllacearum (yellow witches' broom of fir), 132–134
Melanoplus devastator (Devastating Grasshopper), 44–47
Mindarus abietinus (Balsam Twig Aphid), 22–24
Mycosphaerella pini (red band needle blight), 102–104

Neotoma spp. (woodrats), 183–185
Nuculaspis californica (Black Pineleaf Scale), 27–28

Odocoileus spp., *Cervus* spp. (deer and elk), 173–175
Oligoporus amarus (pocket dry rot of incense-cedar), 140–141
Orgyia pseudotsugata (Douglas-fir Tussock Moth), 38–41

Peridermium harknessii (western gall rust), 134–136
Phaeolus schweinitzii (red-brown butt rot or velvet top fungus), 142–143
Phellinus pini (red ring rot or white pocket rot), 144–146
Phloeosinus spp. (cedar bark beetles), 85–87
Phomopsis lokoyae (*Phomopsis* canker of Douglas-fir), 116–118
Phoradendron juniperinum, P. densum (juniper mistletoes), 121–122
Phoradendron libocedri (incense-cedar mistletoe), 123–124
Phoradendron pauciflorum (white fir mistletoe), 124–126
Phytophthora lateralis (Port Orford-cedar root disease), 154–156
Pissodes terminalis (Lodgepole Pine Terminal Weevil), 49
Pityophthorus spp. (twig beetles), 87–89

Rhabdocline pseudotsugae, R. weirii (Douglas-fir needle cast), 104–106
Rhyacionia zozana (Ponderosa Pine Tip Moth), 47–50

Scolytidae (bark beetles), 60–81
Scolytus ventralis (Fir Engraver), 82–85
Siricidae (horntails or wood wasps), 94–96
Sphaeropsis sapinea (Diplodia blight), 114–116
Synanthedon sequoiae (Sequoia Pitch Moth), 58–61

Thomomys spp. (pocket gophers), 180–182

Ursus americanus (American Black Bear), 172–173

Zelleria haimbachi (Pine Needle Sheathminer), 19–21

GENERAL INDEX

Page references in **boldface** refer to the main discussion of the agent.

Abies sp., 81–82
Adelges
   cooleyi, **24–26,** pls. 24–26
   piceae, 26
   tsugae, 26
air pollution, **158–160,** pls. 182–183
American Black Bear, **172–173,** pls. 196–197
annosum root disease, 85, **149–152,** pls. 172–175
aphid(s)
   Balsam Twig, **22–24,** pls. 22–23
   Giant Conifer, **21–22,** pl. 21
Aplodontia rufa, **178–179,** pls. 203–204
Arceuthobium spp., **118–120,** pls. 135–138
Arhopalus productus, 93
Armillaria mellea, **146–148,** pls. 169–171
Armillaria root disease, 54, **146–148,** 151, 181, pls. 169–171

Balsam Twig Aphid, **22–24,** pls. 22–23
Balsam Woolly Aphid, 26
Barbara
   colfaxiana, **7–8,** pls. 4–5
   ulteriorana, 8

bark beetles
   Dendroctonus
      brevicomis, **73–76,** pls. 81–86
      jeffreyi, **70–72,** pls. 78–80
      ponderosae, **67–70,** pls. 75–77
      pseudotsugae, **79–81,** pls. 92–93
      valens, **76–79,** pls. 87–91
   Ips
      paraconfusus, **61–65,** pls. 68–73
      pini, 64, **65–67,** pl. 74
   Phloeosinus spp., **85–87,** pl. 97
   Pityophthorus spp., **87–89,** pls. 98–100
   Scolytus
      unispinosus, **81–82**
      ventralis, **82–85,** pls. 94–96
bigcone Douglas-fir, 13–15, 41–44, 104–106, 158
Black Pineleaf Scale, **27–28,** pls. 27–28
blackstain root disease, 151, **152–154,** 190, pls. 176–179
Black-tailed Deer, **173–175,** pl. 198
Black-tailed Jackrabbit, **175–177,** pls. 201–202
Botryosphaeria canker, **106–108,** pls. 121–123
Botryosphaeria dothidea, **106–108,** pls. 121–123

branch tips and terminals: damage.
See also guide to damage by host species
Douglas-fir Twig Weevil, **55–57**, pl. 61
Lodgepole Terminal Weevil, **57–58**, pls. 62–63
Pine Reproduction Weevil, **52–55**, pls. 57–60
Ponderosa Pine Tip Moth, **47–50**, pls. 51–53
Western Pine Shoot Borer, **50–52**, pls. 54–56
brown cubical rot, **139–140**, pls. 161–162
brown stringy rot, **137–138**, pls. 158–160
bud damage
fir, 41–44
ponderosa pine, pl. 50
bug(s), Western Conifer Seed, **4–6**, pls. 1–3
Buprestidae, **89–91**, pls. 101–102

California Five-spined Ips, **61–65**, pls. 68–73
canker diseases
Botryosphaeria canker, **106–108**, pls. 121–123
Cytospora canker of true firs, **109–110**, pls. 124–125
Diplodia blight, **114–116**, pls. 131–133
Phomopsis canker of Douglas-fir, **116–118**, pl. 134
pitch canker, **111–114**, pls. 126–130
cattle, **177**
Cecidomyia piniinopsis, **30–32**, 170, pls. 30–32
cedar bark beetles, **85–87**, 156, pl. 97
cedar(s)
overall tree decline, 190
smaller branch or treetop damage, 186
stem and larger branch damage, 188
Thuja sp., 81–82
western red, 142–143, 149–152, 169–170, 178–179

Cerambycidae, **91–94**, pls. 104–105
Cervus elaphus subsp. *roosevelti*, **173–175**, pl. 199
chalcid(s), Douglas-fir Seed, **13–15**, pls. 11–12
Chionaspis pinifoliae, 28, **29–30**, pl. 29
Choristoneura spp., **41–44**, pls. 45–47
Chrysoperla rufilabris, 24
Cinara
    curvipes, 22
    ponderosae, **21–22**, pl. 21
coast redwood, 85–87, 172–173
Coleotechnites milleri, **15–18**, pls. 13–16
Coloradia pandora, **33–36**, pls. 33–36
cone beetles, **10–13**, pls. 7–10
cone damage. *See also* reproductive structures: damage
    Douglas-fir, 7–8, pls. 4–5
    pines, 10–13, pls. 7–10
Conophthorus radiatae, 12
Conophthorus spp., **10–13**, pls. 7–10
Cooley Spruce Gall Adelgid, **24–26**, pls. 24–26
Cronartium ribicola, **126–129**, pls. 144–147
Cydia
    injectiva, 10
    miscitata, **9–10**, pl. 6
    piperana, 10
Cylindrocopturus
    eatoni, **52–55**, 148, 151, 181, pls. 57–60
    furnissi, 54, **55–57**, pl. 61
cypress(es)
    mistletoe damage, 121–122
    Monterey, 85–87, pl. 97
    overall tree decline, 190
    smaller branch or treetop damage, 186
    stem and larger branch damage, 188
Cytospora abietis, **109–110**, pls. 124–125
Cytospora canker of true firs, **109–110**, pls. 124–125

damage location, determination, 1
deathwatch beetles, 13

224 **GENERAL INDEX**

Dendroctonus
  brevicomis, **73–76**, pls. 81–86
  jeffreyi, **70–72**, pls. 78–80
  ponderosae, **67–70**, pls. 75–77
  pseudotsugae, **79–81**, pls. 92–93
  valens, 60, **76–79**, pls. 87–91
Dermea pseudotsugae, 117–118
Devastating Grasshopper, **44–47**, pls. 48–50
diagnostic steps, 1–2
Diaporthe lokoyae, **116–118**
Dioryctria reniculelloides, 43–44
Dioryctria spp., 8
Diplodia blight, 113, **114–116**, pls. 131–133
Diplodia pinea, **114–116**, pls. 131–133
Douglas-fir. *See also* bigcone Douglas-fir
  annosum root disease, **149–152**, pls. 172–175
  beaver damage, **178–179**, pl. 204
  black bear damage, **172–173**, pls. 196–197
  blackstain root disease, **152–154**, pls. 176–179
  bud damage, **41–44**, pls. 45–46
  cone moth damage, **7–8**, pls. 4–5
  deer feeding damage, pl. 200
  defoliation, **38–41, 44–47**, pl. 40, pl. 45
  Douglas-fir needle cast, **104–106**, pls. 118–120
  drought stress, **160–161**
  dwarf mistletoe damage, **118–120**
  general guide to damage, **192–193**
  overall tree decline, **190**
  Phomopsis canker, **116–118**, pl. 134
  pitch masses, **58–61**, pls. 64–65
  pocket gopher damage, **180–182**, pl. 207
  red-brown butt rot, **142–143**, pls. 165–166
  reproductive structure damage, **186, 192**
  seed chalcid damage, **13–15**, pls. 11–12
  seed orchards, **88**
  Sequoia Pitch Moth, **58–61**, pls. 64–67
  smaller branch or treetop damage, **187**
  stem and larger branch damage, **188–189**
  sulfur fungus, **139–140**, pls. 161–162
  webbing, **35–37**
  Western Conifer Seed Bug damage, **4–6**, pl. 2
  white pocket rot, **144–146**, pls. 167–168
  woodrat feeding damage, **183–185**, pl. 211
Douglas-fir Beetle, **79–81**, pls. 92–93
Douglas-fir Cone Moth, **7–8**, pls. 4–5
Douglas-fir Engraver, 80–81, **81–82**
Douglas-fir needle cast, **104–106**, pls. 118–120
Douglas-fir Pitch Moth, 60
Douglas-fir Seed Chalcid, **13–15**, pls. 11–17
Douglas-fir Tussock Moth, 37, **38–41**, pls. 40–44
Douglas-fir Twig Weevil, 54, **55–57**, pl. 61
drought, **160–161**, pls. 184–185. *See also* moisture stress

Echinodontium tinctorium, **137–138**, pls. 158–160
Elk, Roosevelt, **173–175**, pl. 199
Elytroderma deformans, **98–100**, pls. 108–111
Elytroderma disease of pines, **98–100**, pls. 108–111
Erethizon dorsatum, **182–183**, pls. 208–209
Ernobius spp., 13
Essigella californica, 22
Eucosma sonomana, 49, **50–52**, 58, 183, pls. 54–56

Fir Engraver, **82–85**, pls. 94–96
fir(s)
  Douglas. *See* Douglas-fir
  dwarf mistletoe damage, 118–120

grand, 82–85, 193–194, pl. 158, pl. 160
red, 82–85, 194, pls. 124–125
   annosum root disease (s-type), 149–152, pls. 172, 175
   brown cubical rot, 139–140, pls. 161–162
   Cytospora canker, 109–110, pls. 124–125
   red-brown butt rot, 142–143, pls. 165–166
   smaller branch or treetop damage, 188–189
   stem and larger branch damage, 191
   white pocket rot, 144–146, pls. 167–168
true
   annosum root disease (s-type), 149–152, pls. 172, 175
   brown cubical rot, 139–140, pls. 161–162
   Cytospora canker, 109–110, pls. 124–125
   red-brown butt rot, 142–143, pls. 165–166
   smaller branch or treetop damage, 188–189
   stem and larger branch damage, 191
   white pocket rot, 144–146, pls. 167–168
true fir needle cast, 100–102, pls. 112–114
twig beetle damage, 87–89, pl. 98
webbing, 35–37
white
   annosum root disease (s-type), 149–152, pls. 172, 175
   Armillaria root disease, pls. 169–171
   Balsam Twig Aphid infestation, 22–24, pls. 22–23
   brown cubical rot, 139–140, pls. 161–162
   Cytospora canker, 109–110, pls. 124–125
   drought damage, 160–161, pls. 184–185
   Fir Engraver infestation, 82–85, pls. 94–96
   frost damage, pl. 188
   general guide to damage, 194–195
   Indian paint fungus, pl. 159
   Phoradendron mistletoe damage, **124–126,** pls. 142–143
   red-brown butt rot, 142–143, pls. 165–166
   salt damage, pl. 193
   smaller branch or treetop damage, 188–189
   stem and larger branch damage, 191
   tussock moth infestation, 38–41, pls. 40–41, pl. 44
   white pocket rot, 144–146, pls. 167–168
   white-fir mistletoe, **124–126,** pls. 142–143
   yellow witches' broom infection, **132–134**
Firtree Borer, 93
flatheaded borer, 80–81
Flatheaded Fir Borer, 90
flatheaded wood borers, **89–91,** pls. 101–102
flooding, **162**
foliage and shoots: damage. *See also* guide to damage by host species
   Balsam Twig Aphid, **22–24,** pls. 22–23
   Black Pineleaf Scale, **27–28,** pls. 27–28
   Cooley Spruce Gall Adelgid, **24–26,** pls. 24–26
   Devastating Grasshopper, **44–47,** pls. 48–50
   Douglas-fir Tussock Moth, **38–41,** pls. 40–44
   Giant Conifer Aphid, **21–22,** pl. 21
   Gouty Pitch Midge, **30–32,** pls. 30–32
   Lodgepole Needleminer, **15–18,** pls. 13–16
   Pandora Moth, **33–35,** pls. 33–36
   Pine Needle Scale, **29–30, 192,** pl. 29

Pine Needle Sheathminer, **19–21,** pls. 17–20
Silverspotted Tiger Moth, **35–37,** pls. 37–39
spruce budworms, **41–44,** pls. 45–47
Fomitopsis pinicola, 140
frost damage, **163–164,** pls. 186–189
fungi. *See also* canker diseases; heart rots; needle diseases; root diseases, rust diseases
   Armillaria mellea, **146–148,** pls. 169–171
   Botryosphaeria dothidea, **106–108,** pls. 121–123
   Cronartium ribicola, **126–129,** pls. 144–147
   Cytospora abietis, **109–110,** pls. 124–125
   Diaporthe lokoyae, **116–118,** pl. 134
   Diplodia pinea, **114–116,** pls. 131–133
   Echinodontium tinctorium, **137–138,** pls. 158–160
   Elytroderma deformans, **98–100,** pls. 108–111
   Fusarium circinatum, 12, 60, 88, **111–114,** pls. 126–130
   Gymnosporangium libocedri, **129–131,** pls. 148–151
   Heterobasidion annosum, **149–152,** pls. 172–175
   Indian paint, **137–138,** pls. 158–160
   Laetiporus sulphureus, **139–140,** pls. 161–162
   Leptographium wageneri, **152–154,** pls. 176–179
   Lirula abietis-concoloris, **100–102,** pls. 112–114
   Mycosphaerella pini, **102–104,** pls. 115–117
   Oligoporus amarus, **140–141,** pls. 163–164
   Peridermium harknessii, **134–136,** pls. 155–157
   Phaeolus schweinitzii, **142–143,** pls. 165–166
   Phellinus pini, **144–146,** pls. 167–168
   Phomopsis lokoyae, **116–118,** pl. 134
   Phytophthora lateralis, **154–156,** pls. 180–181
   red belt, 140
   Rhabdocline pseudotsugae, R. weirii, **104–106,** pls. 118–120
   Sphaeropsis sapinea, **114–116,** pls. 131–133
   sulfur, **139–140,** pls. 161–162
   velvet top, **142–143,** pls. 165–166
   Virgella robusta, 102
Fusarium circinatum, 12, 60, 88, **111–114,** pls. 126–130

Giant Conifer Aphid, **21–22,** pl. 21
giant sequoia
   annosum root disease (s-type), 149–152
   Botryosphaeria canker, 106–108, 187, pls. 121–122
   cedar bark beetle infestation, 85–87
goats, **177**
Gouty Pitch Midge, **30–32,** 170, pls. 30–32
green lacewing, 24
guide to damage by host species, 192–203
Gymnosporangium libocedri, 124, **129–131,** pls. 148–151

hares, **175–177,** pls. 201–202
heart rots
   brown stringy rot, **137–138,** pls. 158–160
   brown cubical rot, **139–140,** pls. 161–162
   pocket dry rot of incense cedar, **140–141,** pls. 163–164
   red-brown butt rot, **142–143,** pls. 165–166
   white pocket rot/red ring rot, **144–146,** pls. 167–168
heat damage, **164–165**
hemlock
   annosum root disease (s-type), **149–152,** pls. 172, 175
   beaver damage, **178–179,** pl. 204
   Brown Cubical Rot, **139–140,** pls. 161–162

Silverspotted Tiger Moth damage, **35–37,** pls.38–39
twig beetle infestation, **87–89,** pl. 98
white pocket rot, **144–146,** pls. 167–168
Hemlock Woolly Adelgid, 26
herbicides, **165–166,** pl. 190
Heterobasidion annosum, **149–152,** pls. 172–175
highway deicing salts, **166–169,** pls. 191–194
hogs, feral, **177**
horntail larvae, 90, **94–96,** pls. 106–107

incense-cedar
annosum root disease, **149–152,** pls. 172–175
Armillaria root disease, **146–148,** pls. 169–171
bark beetle infestation, **85–87,** pl. 97
Botryosphaeria canker, **106–108,** pls. 121–123
drought damage, **160–161,** pls. 184–185
general guide to damage, **202–203**
Gymnosporangium libocedri disease, pls. 148–150
herbicide damage, pl. 190
incense-cedar mistletoe, **123–124,** pl. 141
incense-cedar rust, **129–131,** pls. 148–151
pocket dry rot, **140–141,** pls.163–164
red-brown butt rot, **142–143**
Indian paint fungus, **137–138,** pls. 158–160
Ips
paraconfusus, **61–65,** pls. 68–73
pini, 64, **65–67,** pl. 74

jackrabbits, **175–177,** pls. 201–202
Jeffrey Pine Beetle, **70–72,** pls. 78–80
juniper(s), 187
annosum root disease, **149–152,** pls. 172–175
mistletoe damage, **121–122,** pls. 139–140

overall tree decline, 190
smaller branch or treetop damage, 187
stem and larger branch damage, 189
western, 85–87

Laetiporus sulphureus, **139–140,** pls. 161–162
larch, 142–143
Leptoglossus
occidentalis, **4–6,** pls. 1–3
zonatus, 6
Leptographium wageneri, **152–154,** pls. 176–179
Lepus spp., **175–177,** pls. 201–202
Lirula abietis-concoloris, **100–102,** pls. 112–114
livestock, **177,** 186
Lodgepole Needleminer, **15–18,** pls. 13–16
Lodgepole Terminal Weevil, 49, **57–58,** pls. 62–63
longhorned beetles, **91–94,** pls. 103–105
Lophocampa argentata, **35–37,** pls. 37–39

mammals
American Black Bear, **172–173,** pls. 196–197
deer and elk, **173–175,** pls. 198–200
jackrabbits and hares, **175–177,** pls. 201–202
livestock, **177**
Mountain Beaver, **178–179,** pls. 203–204
pocket gophers, 54, 148, **180–182,** pls. 205–207
Porcupine, **182–183,** pls. 208–209
woodrats, **183–185,** pls. 210–211
manzanita leaf, pl. 194
Megastigmus spermotrophus, **13–15,** pls. 11–12
Melampsorella caryophyllacearum, **132–134,** pls. 152–154
Melanophila drummondi, 80–81, 90
Melanoplus devastator, **44–47,** pls. 48–50

metallic wood-boring beetles, **89–91,** pls. 101–102
midge(s), Gouty Pitch, **30–32,** pls. 30–32
Mindarus abietinus, **22–24,** pls. 22–23
mistletoe(s)
 dwarf, 110, **118–120,** pls. 135–138
 incense-cedar, **123–124,** pl. 141
 juniper, **121–122,** pls. 139–140
 white fir, **124–126,** pls. 142–143
moisture stress, 108, 160. *See also* drought
Monterey Pine Cone Beetle, 12
moth(s)
 Douglas-fir Cone, **7–8,** pls. 4–5
 Douglas-fir Tussock, 37, **38–41,** pls. 40–44
 Nantucket Pine Tip, 48–49
 Pandora, **33–35,** pls. 33–36
 Ponderosa Pine Tip, **47–50,** pls. 51–53
 Sequoia Pitch, **58–61,** pls. 64–67
 Silverspotted Tiger, **35–37,** pls. 37–39
Mountain Beaver, **178–179,** pls. 203–204
Mountain Pine Beetle, **67–70,** pls. 75–77
Mule Deer, **173–175**
mycelial fans, pl. 169
Mycosphaerella pini, **102–104,** pls. 115–117

Nantucket Pine Tip Moth, 48–49
needle diseases
 Douglas-fir needle cast, **104–106,** pls. 118–120
 Elytroderma disease of pines, **98–100,** pls. 108–111
 red band needle blight, **102–104,** pls. 115–117
 true fir needle cast, **100–102,** pls. 112–114
Neotoma spp., **183–185,** pls. 210–211
Newhouse Borer, 93
Nuculaspis californica, **27–28,** pls. 27–28

occurrence pattern, 1
Odocoileus hemionus, **173–175**
 subsp. *columbianus,* pl. 198

Oligoporus amarus, **140–141,** pls. 163–164
Orgyia pseudotsugata, 37, **38–41,** pls. 40–44
overall tree decline, 189–191. *See also* tree killing
ozone damage, **158–160,** pls. 182–183

Pandora Moth, **33–35,** pls. 33–36
Peridermium harknessii, **134–136,** pls. 155–157
Phaeocryptopus gaeumannii, 106
Phaeolus schweinitzii, 140, **142–143,** pls. 165–166
Phellinus pini, **144–146,** pls. 167–168
Phloeosinus spp., **85–87,** 156, pl. 97
Phomopsis canker of Douglas-fir, **116–118,** pl. 134
Phomopsis lokoyae, **116–118,** pl. 134
Phoradendron
 densum, **121–122,** pl. 140
 juniperinum, **121–122,** pl. 139
 libocedri, **123–124,** pl. 141
 pauciflorum, **124–126,** pls. 142–143
Phytophthora lateralis, **154–156,** pls. 180–181
Pine Engraver, 64, **65–67,** pl. 74
Pine Needle Scale, 28, **29–30,** pl. 29
Pine Needle Sheathminer, **19–21,** pls. 17–20
Pine Reproduction Weevil, **52–55,** 148, 151, 181, pls. 57–60
pine(s)
 annosum root disease, **149–152,** pls. 173–174
 bark beetle infestations, 61–70, 76–79
 blackstain root disease, **152–154,** pls. 176, 178
 cone beetle damage, 10–13, pls. 7–10
 Coulter, 73–76
 Elytroderma disease of pines, **98–100,** 169, pls. 108–111
 general guide to damage, 201–202
 gray, 52–55
 Jeffrey, 10
  Cydia injectiva, 10
  Dendroctonus jeffreyi damage, **70–72,** pls. 78–80

GENERAL INDEX 229

Elytroderma disease of pines, **98–100,** pls. 108–109
Eucosoma sonomana infestation, **50–52,** pl. 54, pl. 56
flooding damage, 162
frost damage, **163–164,** pl. 187, pl. 189
general guide to damage, 195–196
Pandora Moth damage, **33–35,** pl. 33
Pine Engraver damage, **65–67**
Pine Reproduction Weevil damage, **52–55**
Porcupine damage, **182–183**
salt damage, **166–169,** pl. 191
winter burn and winter drying, **169–170**
limber, 65–67
lodgepole
  Coleotechnites milleri damage, **15–18,** pls. 13–16
  flooding damage, **162**
  general guide to damage, 196–197
  Pine Engraver damage, **65–67**
  Pissodes terminalis infestation, 57–58, pls. 62–63
  western gall rust, 134–136, pl. 155, pl. 157
  wood wasp infestation, **94–96,** pl. 107
Monterey
  D. valens pitch tubes, 76–79, pls. 87–88
  dwarf mistletoe on, 118–120, pls. 135–136
  general guide to damage, 198–199
  pitch canker damage, 111–114, pls. 126–130
  red band needle blight, 102–104, pls. 115–116
  twig beetle damage, 87–89, pls. 99–100
overall tree decline, 190–191
pitch canker disease, **111–114,** pls. 126–130
pitch masses, 58–61
ponderosa
  aphid infestation, **21–22,** pl. 21
  Diplodia blight, 114–118, pls. 131–133
  drought stress, 160–161
  frost damage, 163–164, pl. 186
  general guide to damage, 199–200
  Ips-infestation, 61–64, pls. 68–69
  ozone damage, **158–160,** pls. 182–183
  Pine Reproduction Weevil damage, **52–55,** pls. 57, 59
  Pine Seedworm damage, **9–10,** pl. 6
  Porcupine damage, 182–183, pls. 208–209
  rabbit damage, 175–177, pl. 202
  seed damage, 4, pl. 1
  red band needle blight, **102–104,** pls. 115–117
  red-brown butt rot, **142–143,** pls. 165–166
  reproductive structure damage, 4–6, 9–13, 186, 192
  smaller branch or treetop damage, 187–188
  stem and larger branch damage, 189
  sugar, 52–55, 160–161, 200–201, pls. 144–146, pl. 192
  sulfur fungus, **139–140,** pls. 161–162
  twig damage, 87–89
  webbing, 35–37, pls. 37, 39
  western gall rust disease, 134–136
  western white, 126–129
  white pocket rot, 144–146
Pine Seedworm, **9–10,** pl. 6
Pissodes
  strobi, 58
  terminalis, 49, **57–58,** pls. 62–63
pitch canker, 12, 60, 88, **111–114,** 116, pls. 126–130
pitch masses, 58–61, pl. 64
pitch tubes, 73, pl. 82, pls. 87–88
Pityophthorus spp., 31, 56, **87–89,** pls. 98–100

pocket dry rot, **140–141,** pls. 163–164
pocket gophers, 54, 148, **180–182,** pls. 205–207
Ponderosa Pine Tip Moth, 31, **47–50,** pls. 51–53
Porcupine, **182–183,** pls. 208–209
Port Orford–cedar, 85–87, 172–173, 203
  root disease, **154–156,** pls. 180–181
Pseudohylesinus sericeus, 85

red band needle blight, **102–104,** pls. 115–117
red-brown butt rot, 140, **142–143,** pls. 165–166
red ring rot, **144–146,** pls. 167–168
Red Turpentine Beetle, 60, **76–79,** pls. 87–91
redwood(s), 188, pl. 210. *See also* coast redwood
reproductive structures: damage; *See also* guide to damage by host species
  cone beetles, **10–13,** pl. 7
  Douglas-fir Cone Moth, **7–8,** pls. 4–5
  Douglas-fir Seed Chalcid, **13–15,** pls. 11–12
  Pine Seedworm, **9–10,** pl. 6
  Western Conifer Seed Bug, **4–6,** pls. 1–3
Rhabdocline
  pseudotsugae, **104–106,** pls. 118–120
  weirii, **104–106,** pls. 118–120
Rhyacionia
  frustrana, 48–49
  zozana, 31, **47–50,** pls. 51–53
Rhyacionia spp., 52
Ribes spp., 126–129
Roosevelt Elk, **173–175,** pl. 199
root diseases. *See also* guide to damage by host species
  annosum root disease, 85, **149–152,** pls. 172–175
  Armillaria root disease, 54, **146–148,** pls. 169–171
  blackstain root disease, **152–154,** pls. 176–179
  Port Orford–cedar root disease, **154–156,** pls. 180–181

roundheaded wood borers, 90–91, **91–94,** pls. 103–105
rust diseases
  incense-cedar rust, **129–131,** pls. 148–151
  western gall rust, **134–136,** pls. 155–157
  white pine blister rust, **126–129,** pls. 144–147
  yellow witches' broom of fir, **132–134,** pls. 152–154

salt, **166–169,** pls. 191–194
scale(s), Black Pinelaf, **27–28,** pls. 27–28
Scolytus
  unispinosus, 80–81, **81–82**
  ventralis, **82–85,** pls. 94–96
seed damage. *See also* reproductive structures: damage
  bigcone Douglas-fir, 13–15
  Douglas-fir, 7–8, 13–15
  pines, 10–13
  pines and Douglas-fir, 4–6
  ponderosa pine, 9–10
Semanotus litigiosus, 93
Sequoia Pitch Moth, **58–61,** pls. 64–67
sheep, **177**
Silver Fir Beetle, 85
Silverspotted Tiger Moth, **35–37,** pls. 37–39
Siricidae, **94–96,** pls. 106–107
smog, **158–159**
Snowshoe Hare, **175–177**
Sphaeropsis sapinea, **114–116,** pls. 131–133
spruce budworms, **41–44,** pls. 45–47
Spruce Coneworm, 43–44
spruce(s)
  annosum root disease, **149–152**
  Engleman, pl. 26
  red-brown butt rot, **142–143**
  Sitka, **24–26**
  smaller branch or treetop damage, **188**
  sulfur fungus, **139–140,** pls. 161–162
  twig beetle damage, **87–89**
  white pocket rot, **144–146,** pl. 167

stems and larger branches: damage.
See also guide to damage by host species
- California Five-spined Ips, **61–65,** pls. 68–73
- cedar bark beetles, **85–87,** pl. 97
- Douglas-fir Beetle, **79–81,** pls. 78–80
- Douglas-fir Engraver, **81–82**
- Fir Engraver, **82–85,** pls. 94–96
- flatheaded wood borers, **89–91,** pls. 101–102
- horntails or wood wasps, **94–96,** pls. 106–107
- Jeffrey Pine Beetle, **70–72,** pls. 78–80
- Mountain Pine Beetle, **67–70,** pls. 75–77
- Pine Engraver, **65–67,** pl. 74
- Red Turpentine Beetle, **76–79,** pls. 87–91
- roundheaded wood borers, **91–94,** pls. 103–105
- Sequoia Pitch Moth, **58–61,** pls. 64–67
- twig beetles, **87–89,** pls. 98–100
- Western Pine Beetle, **73–76,** pls. 81–86

sudden oak death
swiss needle cast, 106
symptom location, guide to damage by, 186–191
Synanthedon
  novaroensis, 60
  sequoiae, **58–61,** pls. 64–67

Thomomys spp., **180–182,** pls. 205–207
Thuja sp., 81–82

trampling damage, 177
tree killing. *See also* overall tree decline
true fir needle cast, **100–102,** pls. 112–114
twig beetles, 186
  Pityophthorus spp., 31, 56, **87–89,** pls. 98–100

Ursus americanus, **172–173,** pls. 196–197

Virgella robusta, 102

Western Conifer Seed Bug, **4–6,** pls. 1–3
western gall rust, **134–136,** pls. 155–157
Western Pine Beetle, **73–76,** pls. 81–86
Western Pine Shoot Borer, 49, **50–52,** 58, 183, pls. 54–56
western serviceberry, pl. 151
white fir mistletoe, **124–126,** pls. 142–143
white pine blister rust, **126–129,** pls. 144–147
White Pine Weevil, 58
white pocket rot, **144–146,** pls. 167–168
White-tailed Jackrabbit, **175–177**
winter burn/winter drying, **169–170,** pl. 195
woodrats, **183–185,** pls. 210–211
wood wasps, **94–96,** pls. 106–107

yellow witches' broom of fir, **132–134,** pls. 152–154

Zelleria haimbachi, **19–21,** pls. 17–20

ABOUT THE AUTHORS

David L. Wood is professor of the graduate school and professor emeritus at the University of California, Berkeley. Thomas W. Koerber is retired research entomologist at the Pacific Southwest Forest and Range Experiment Station, USDA Forest Service, and currently consulting entomologist at Entomological Services Co., Berkeley, California. Robert F. Scharpf is retired research plant pathologist at the Pacific Southwest Forest and Range Experiment Station, USDA Forest Service. Andrew J. Storer is assistant professor of forest insect ecology at the School of Forest Resources and Environmental Science at Michigan Technological University.

Series Design:	Barbara Jellow
Design Enhancements:	Beth Hansen
Design Development:	Jane Tenenbaum
Composition:	TechBooks
Text:	9/10.5 Minion
Display:	Franklin Gothic Book and Demi
Printer and binder:	Everbest Printing Company

CALIFORNIA NATURAL HISTORY GUIDES

"It's always good to read a new California Natural History Guide; these little books are small enough to fit into a pocket, inexpensive, and authoritative." —*Sunset*

"A series of excellent pocket books, carefully researched, clearly written, and handsomely illustrated." —*Los Angeles Times*

The California Natural History Guide series is the state's most authoritative resource for helping outdoor enthusiasts and professionals appreciate the wonderful natural resources of their state. If you would like to receive more information about the series or other books on California natural history, please fill in this card and return it to the University of California Press or register online at www.californianaturalhistory.com.

Name _____

Address _____

City/State/Zip _____

Email _____

Which book did this card come from? _____

_____

Where did you buy this book? _____

_____

What is your profession? _____

UNIVERSITY OF CALIFORNIA PRESS
www.ucpress.edu

POST OFFICE WILL NOT DELIVER WITHOUT POSTAGE

Return to:
University of California Press
Attn: Natural History Editor
2120 Berkeley Way
Berkeley, California 94720

CHILD SEXUAL ABUSE:
A DESCRIPTIVE & TREATMENT STUDY

ELIZABETH MONCK
ARNON BENTOVIM
GILLIAN GOODALL
CAROLINE HYDE
REBEKAH LWIN
ELAINE SHARLAND
WITH ANNE ELTON

STUDIES IN CHILD PROTECTION

LONDON: HMSO

© Crown Copyright 1996

Applications for reproduction should be made to HMSO Copyright Unit, St Clements House, 2–16 Colegate, Norwich NR3 1BQ

ISBN 0 11 321792 7

Acknowledgements

We wish to record our thanks to several groups of people who have helped with this study.

We wish to acknowledge first the help we received from the Child Sexual Abuse clinical team, in the Department of Psychological Medicine, Hospital for Sick Children; second, the advice we have received in the analysis from Dr Marjorie Smith; third, the support and advice of the Advisory Group at the Department of Health; fourth, the help of the teachers and social workers who provided information; and finally, the cooperation of the families and children who took part.

We also wish to record our thanks to the project secretaries: Beverley Kahati, who saw us through the first six months, and Carole Barnham, who carried on from that point; both made important contributions to running the work.

The study was supported with a three-year grant to Dr Arnon Bentovim from the Department of Health, and three-month extension to Elizabeth Monck to lead on the analysis and writing of the report.

PARTICIPANTS IN THE STUDY
Research team:
Mrs Elizabeth Monck, BA, Senior Research Officer
Ms Rebekah Lwin, BSc, MPhil, Research Officer
Ms Gillian Goodall, BSc, Research Officer
Ms Elaine Sharland, BA, Research Officer

Clinical associates:
Dr Arnon Bentovim, MB,BS, FRCPsych., Consultant Psychiatrist
Dr Caroline Hyde, BSc, DCH, MB.BS, MRCPsych, Senior Registrar

Contents

Chapter 1	**Introduction**	1
	Contributions to the increased recognition of child sexual abuse	1
	The effects of abuse—	2
	Evidence from retrospective studies	2
	The effects of abuse observed in children at disclosure and subsequently	3
	Some secondary effects of detection procedures	5
	The need for evaluation of services in child sexual abuse	6
Chapter 2	**Aims and hypotheses**	8
Chapter 3	**The Sample**	9
Chapter 4	**Method**	11
	Research interviews at referral	11
	Research interviews after treatment	12
	Random allocation to treatment	12
	Family treatment	12
	Group treatment	13
	Professional network support and treatment	14
Chapter 5	**Measures**	15
	Standardised instruments	16
	Additional research measures	17
	Sexualised behaviour	17
	Information about the abuse	18
	Clinical measures	19
Chapter 6	**The findings in the descriptive study**	23
	The study population	23
	Completed interviews	24
	Personal and family characteristics of the children	25
	Gender	25
	Age at referral to CSA hospital team	26
	Socio-economic status of parents	27

Living arrangements of the abused children at disclosure and at referral to hospital	28
The children's legal status at the time of referral	28
Menarche	29
Abuse characteristics	29
Age of onset	29
Age at cessation of abuse	30
Duration of abuse	30
The number of abusers	31
Relationship of child to the abuser	31
Type of abuse	32
Children's feelings about the abuse	34
The number of incidents	34
The perpetrator's strategies for preventing dislcosure	34
Physical abuse	35
Disclosure characteristics	35
Choice of whom to disclose to	35
Circumstances leading to disclosure	36
Non-abusing parent's or carer's belief that the child was telling the truth about the abuse	39
Caring parent's reactions to disclosure or discovery	40
Carrying the responsibility for the abuse	41
Psychological impact of the abuse	41
The Children's Depression Inventory	41
The Child Self-esteem Inventory	42
Mother's/carer's report of the children's symptoms	46
Sexualised behaviour of the abused child	47
Child's report of own symptoms	48
Teachers' reports of behaviour in school	50
Protective factors	52
Characteristics of the primary carers	52
Mother's marital status	52
Mother's/carers' childhood experiences	53
Mothers'/carers' psychological state	54
Mothers'/carers' self-esteem	55

Chapter 7   **Discussion of findings in the descriptive study**   57

Introduction	57
Gender ratio	58
Age, onset and the duration of abuse	58
The abuse and the abusers	60
Evidence of physical abuse	62

	The children's feelings about the abuse and the circumstances of disclosure	62
	Mothers'/carers' reactions and support at disclosure	63
	The children's symptoms	64
	The mothers' negative outlook	67
	Sexualised behaviour in the abused children	67
	The mothers'/carers	68
Chapter 8	**Conclusions from the descriptive study**	70
Part II	**The Treatment Outcome Study**	
Chapter 9	**Introduction**	71
Chapter 10	**Aims, Hypotheses and Procedures**	73
Chapter 11	**The Treated Sample**	74
Chapter 12	**The Treatment Programmes**	75
	Family network treatment	75
	Family with the addition of group work	75
	Collaboration with community-based professionals	75
Chapter 13	**The findings**	77
	Introductory note	77
	Comparison of the treated and untreated populations	77
	Numbers involved in the following analyses	77
Chapter 14	**Assessment of whole treated population: non-clinical measures**	79
	Children's Depression Inventory	80
	The mothers'/carers' reports of the children's symptoms	81
	Teachers' reports of children's behaviour in school	81
	Children's self-reported symptoms	81
	Children's self-esteem inventory	82
	Relationships between mother/carers and abused children before and after treatment	82
	Mothers' General Health Questionnaire and self-esteem scores	83
	External sources of stress for mothers/carers	84
Chapter 15	**Assessment of the whole treated population: clinical measures**	86
	Clinical expectations of improvement in the child	86
	Family treatment aims	86

Chapter 16	**Assessment of treatment: comparison of the two treatment programmes**	88
	Comparison of the initial status of mothers and children in each of the two treatment programmes	88
Chapter 17	**Assessment of treatment: progress within each treatment group**	89
	Family treatment only	89
	a) Outcomes for the abused children: non-clinical standardised measures	89
	b) Outcomes for the mothers/carers: non-clinical standardised measures	89
	c) Clinical outcomes following family treatment	90
	Family treatment with additional groupwork	90
	a) Outcomes for the abused children: non-clinical standardised measures	90
	b) Outcomes for the mothers/carers: non-clinical, standardised measures	91
	c) Clinical outcomes following the additional groupwork	91
Chapter 18	**Assessment of treatment:comparison of the two treatment programmes**	93
	The non-clinical standardised measures of outcome	93
	a) The abused children	93
	b) The mothers/carers	93
	The clinical measures of outcome	94
	a) The mother-child relationship	94
	b) The Family Treatment Aims	95
	c) The mothers/carers in treatment	96
	d) Clinicians' predicted outcome compared with final clinical ratings of children's improvement	97
	e) The influence of mothers/carers belief about abuse on clinical ratings of improvement	98
Chapter 19	**Children's and mothers'/carers' attitudes to treatment**	99
	Children's attitudes to treatment	99
	Children's views of the helpfulness of treatment	100
	Mothers'/carers' attitudes to treatment	101
	Mothers'/carers' reports of the helpfulness of treatment	103
	The effect of commitment to treatment	104

Chapter 20	**Discussion: treatment outcome study**	105
Chapter 21	**Conclusions**	112
Chapter 22	**Directions for future work**	114
	Adopting agreed definitions	114
	Standardising assessments of sexually abused children	114
	Evaluation of treatment programmes	115
	Record-keeping in child sexual abuse cases	116
	Recording the content of treatment programmes	117
	Uncovering the long-term consequences of childhood sexual abuse	117
Chapter 23	**Clinical implications of the study findings**	119
	Taking account of the treatment context	119
	The effects of treatment for the whole sample	120
	Differences between the two treatment groups	124
	Summary implications	127
Appendix	**The sexual abuse treatment programmes**	**128**
	Introduction	128
	Assessment and aim of treatment	130
	Treatment programmes	136
	I. Offenders: Young People and Adults	136
	II. Treatment processes with sexually abused children and adolescents and their families	146
	1. Family treatment	146
	2. Group treatment for non-abusing parents and victims	155
References		168

Chapter 1

Introduction

In Western societies child sexual abuse has probably always existed. The taboo about children being involved in sexual activities, particularly those also involving adults, has been most apparent in the reluctance to acknowledge what was being experienced by many children. This has led to an apparent determination to ignore the need to plan strategies for detecting and treating the victims and the abusers. Over the last two decades, however, this situation has changed dramatically in many Western countries. Studies of child sexual abuse in the UK have emphasised an apparently persistent increase in the numbers of detected and reported cases. The National Society for the Prevention of Cruelty to Children (NSPCC) have published the figures from the Child Protection Registers which they manage on behalf of some social services departments, covering about 9% of the total child population in England and Wales. The estimated rate of sexual abuse cases per thousand under the age of 17 years was 0.08 in 1983, rising to 0.65 in 1987 (Creighton & Noyes, 1989). Bearing in mind that these figures refer only to cases of child sexual abuse which are registered as 'at risk', and not to all detected cases, this is, nevertheless, a startling 8-fold increase in 4 years. Figures compiled by the Department of Health (Personal Social Services) for 1989 and 1990 indicate that for 14% of cases on the registers in England and Wales sexual abuse was the primary reason for registration, and for another 2% it was a joint concern with physical abuse or neglect (Department of Health, 1990). It should be noted that Creighton (1992) has found a recent levelling-off in reported cases of child sexual abuse.

Contributions to the increased recognition of child sexual abuse

Following the greater acceptance that sexual abuse is experienced by many more than a tiny minority of children, and that effective treatments can be devised, many professionals (eg police/social work investigation teams) are now trained in more sophisticated and sensitive ways of detecting the signs of abuse, and of eliciting disclosures from children and teenagers. In the UK, media attention and the work of Childline has undoubtedly helped some children to understand that the abuse they suffer is wrong, and to disclose. The provision of a specialist service for the diagnosis, assessment or treatment of child sexual abuse may also contribute to a local rise in the numbers of cases coming to light

(Hobbs, 1990; Kitchur & Bell, 1989). Nevertheless, there is a marked absence of reliable national figures, and Markowe (1988) has expressed doubt about whether the existing surveys could be used to estimate (UK) national prevalence or incidence figures satisfactorily. Prevalence rates in the US literature appear to be affected by the choice of samples, with the highest rates arising in urban female adults (Russell, 1984), intermediate rates arising in national studies (Committee on Sexual Offenses, 1986), and the lowest rates arising in college student samples (Peters, Wyatt & Finkelhor, 1986). Kelly and her colleagues, in an exploratory study of the prevalence of sexual abuse in college students, have discussed the effects of definition on prevalence rates. Using nine different definitions these authors found rates varying between 5% (for severe abuse) and 59% (for less severe) among young women, and between 2% and 27% among young men (Kelly, Regan & Burton, 1995)

The effects of abuse

Evidence from retrospective studies

The apparently sharp increase in disclosed and confirmed cases in such a relatively short space of time has led to a widespread interest in establishing the characteristics of the sexually abused child, their families and the perpetrators. Most of the early, pioneering work (largely in the USA and Canada) which sought to describe the abused child and the effects of sexual abuse was based on the retrospective accounts of adult men and women recalling sexual abuse in their childhood (eg, Finkelhor, 1979; Fritz, Stoll & Wagner, 1981; Briere, 1987; Briere & Runtz, 1988). It is now recognised that the use of such sources will almost certainly have introduced biases into what is known about the abuse or about the children's circumstances at the time of abuse. Recall, even of traumatic events, can be very selective (Sudman & Bradburn, 1973) and further bias may arise from the fact that the adults who choose to disclose may themselves be a self-selected sample. Some adults abused in childhood (though how many it is not possible to calculate) may prefer never to admit it.

From retrospective studies it appears that a wide range of sexually abusive experiences may lead to later medical and psychiatric problems (Walker, Katon, Hansom et al, 1992). Affective disorders, including depression, suicidal behaviour, and anxiety and fears have sometimes been reported more frequently in abused than non-abused samples (Mullen, Romans-Clarkson, Walton & Herbison, 1988; Stein, Golding Siegal et al, 1988; Peters & Range, 1996). But the evidence for these particular adult outcomes are not always found; for example, Sedney & Brooks (1984) and Peters (1988) found no link between early abuse and later suicidal ideation.

The evidence that childhood sexual abuse leads to later problems of sexual adjustment is also confused. Meiselman (1978), and Gold (1986) have reported that women who had been abused were more likely to report adult sexual problems, but later work by Fromuth (1986) and Greenwald, Leitenberg, Cado & Tarran (1990) suggest there are no such differences. Eating problems, including anorexia and bulimia nervosa, have also been reported in adult survivors (eg, Hall, Tice, Beresford, Wooley & Hall, 1989; Waller, 1991).

Many retrospective studies have focused on adult women. Collings (1995) studied the medium-term effects of sexual abuse on men, using a sample of university students. Consistent with findings from female samples, young adult men who had experienced contact abuse showed significant levels of poor psychological adjustment compared with students with non-abused and non-contact histories. The effect of contact abuse remained significant after allowing for the adverse effects of dysfunctional parenting.

Evidence from children observed at disclosure

Following the early descriptions based on recall, more recent research has sought to establish the characteristics of the sexually abused child at the time of disclosure or referral to statutory agencies (Conte & Schuerman, 1987; Gomes-Schwarz, Horowitz & Cardarelli, 1990). In the UK the study based on data from NSPCC Child Protection registers (Creighton & Noyes, 1989) has provided more detailed information, but only for a selected population of sexually abused children (those on the child protection registers). It is clear that, although there are nationally agreed guidelines about placing children on the Registers, these decisions are governed largely by local policy and practice; the weight given to particular features of each case appears to vary from authority to authority, leading to considerable variation in the proportion of cases which are registered.

Important though this contemporary information will be, even these data may therefore not provide a representative picture of all cases. It also seems likely that it is more than chance which distinguishes the identified from the unidentified cases, and that the former will prove, for example, to have more psychological or behaviour problems. There is some evidence that detected cases tend to be found more frequently in lower socio-economic classes compared with 'cases' identified by adult recall of childhood experiences (Baker & Duncan, 1985), and that identified cases may occur more frequently in disorganized, socially chaotic families (Meiselman, 1978; Finkelhor, 1979). However, even this observation may be an artifact arising from these families being more frequently in contact with child care agencies and therefore more prone to have further problems detected.

Despite the difficulties of acquiring unbiased data, there is abundant evidence of the harmful effects of sexual abuse on the children and adolescents in known cases. A study of children referred to Great Ormond Street between 1981 and 1984 indicated a range of psychological and behavioural problems (Bentovim & Boston, 1988). In this study, however, no systematic assessment of the children was made at the time they entered treatment, nor at the point of follow-up in 1986. For most of the sample, the evidence of their progress was supplied retrospectively by the child's social worker, whose judgements were necessarily subjective as they had been involved in the care and treatment of the child.

However, many other studies have confirmed a bleak picture. Kolko, Mosner & Weldy (1988) found that 29 sexually abused and hospitalised children showed greater sexualized behaviour, fear, anxiety and sadness than non-abused children. Freidrich, Urquiza & Beilke (1986) used objective psychological tests on 85 sexually abused children, and found one third had significantly raised scores on a scale measuring fearfulness, and one third had raised scores on a scale measuring aggressive behaviour.

More recently, in a study of the incidence of sexual abuse during 1987 in Northern Ireland, information on the children and their backgrounds was acquired at the time of detection of the case (The Research Team, 1990). However, the authors note that the quality of information on psychological sequelae tended to vary by reporter, and conclude that the results from this part of their enquiry must be treated with caution. Using measures constructed for the study, it was observed that over 50% of boys and girls showed some form of emotional disorder; 40% or the boys and a third of the girls showed conduct disorder; school difficulties were reported for about one quarter of each sex; sexualized behaviour was observed in about one third; and relationship difficulties were reported for nearly half the boys and 40% of the girls (The Research Team, 1990). In a study of sexually abused children referred to a child psychiatric clinic, Sauzier, Salt & Calhoun (1990) reported that although, as a group, the children manifested more behaviour problems and more stressful emotional reactions than children in the general population, they did not all show the severity of symptoms seen in a comparison group of children attending the clinic for other psychiatric problems.

Some effects of detection procedures

While recognising that the sexually abused children seen in clinics and child protection agencies may not be fully representative of all cases, it is also important to bear in mind that the range of behaviour and psychiatric symptoms they display is not solely the product of the sexual abuse. When an abused child is first seen by specialized agencies s/he will have been through at least some stages of the disclosure/discovery procedures. Aspects of these procedures, and many of the events surrounding disclosure and discovery, can themselves be 'abusive', in the sense of causing considerable stress. For example, the sudden departure of the abuser, contact with police, the sudden removal of the child to a Children's Home, a foster family, or to hospital, the physical examination, the lengthy probing interviews by strangers—all these may mean that the level of distress, exhibited as disturbed behaviour or feelings, may well be as high as or higher than when the abuse was going on (DeVine, 1980; Berliner & Barbieri, 1984; Weiss & Berg, 1982; Wattam, 1992). This is no reflection on the skilful and sensitive handling of the children and their families in many cases; it may be at least partly inherent in the situation of disclosure[1]. It is interesting to note that most of the routine procedures in relation to newly-detected sexual abuse would be described as severely stressful in the literature on the effects of life events (Brown, 1989). These post-abuse events might be expected (even without the impact of the original abuse) to affect the behaviour and psychological state of the child (the effect of stressful life events in childhood has been reviewed by Goodyer, 1990). Any information obtained about the psychiatric state or behaviour of the child after disclosure will produce a picture of the effects of abuse tempered or inflated by the effects of the events surrounding disclosure. It will not produce a picture of the effects of abuse alone.

An additional source of bias about abused children may arise from the use of informants, like social workers, parents and teachers, who already know about the nature of the abuse, and who may therefore strain to provide the picture of disturbed children they believe the researcher seeks. There is an understandable desire to recognise that sexual abuse is a damaging experience. There are no simple answers to these problems: at present it may be sufficient that they are acknowledged.

Research has exposed some myths. For example, it was widely believed that there was an association between child abuse and paternal unemployment. In a study of cases referred to a hospital clinic in Sheffield, UK, it was shown that neither recent nor long-term

---

[1] See Wattam (1992) Chapter 2 for a review of the factors influencing disclosure.

unemployment were risk factors for child abuse (sexual abuse included) in previously stable families with no other social problems. There had been no recent increase in paternal unemployment in the child abuse cases after a sharp local economic recession led to male unemployment rates tripling (Taitz, King, Nicholson & Kessel, 1987).

The need for evaluation of services in child sexual abuse

Until recently, the issues surrounding identification and the development of services to support the sexually abused child and non-abusive family members have tended to take precedence over evaluative studies of treatment outcome in the UK. Over one-third of treatment facilities recorded in the survey by the Department of Health and the National Children's Homes in 1990 had only been established in the previous 5 years (NCH, 1992), and very few included evaluation, although many recognised the need.

The UK study by Mrazek, Bentovim & Lynch (1983) found evidence that the most usual response to the discovery of sexual abuse was the prosecution of the offenders, while treatment was offered to only 11% of the abused children. Following this, the Child Sexual Abuse (CSA) treatment programme at the Hospital for Sick Children, Great Ormond Street was developed by a multi-disciplinary team of child psychiatrists, psychiatric social workers and psychologists, under the direction of Dr Arnon Bentovim. The approach is founded in part on the principles underlying family therapy, including those parts of family systems theory which explains the behaviour of individuals (abuser, abused child and non-abusing family members) in terms of the maintenance of the family system, even when this is described by others as malfunctioning.

The CSA team has also drawn on the pioneering work of Giaretto who advocates the use of therapeutic groups of children and parents (Giaretto, 1981). The work of the CSA team at Great Ormond Street has been widely disseminated by teaching and in written reports (Bentovim, Elton, Hildebrand, Tranter & Vizard, 1988; Bentovim, 1991). Group treatment has been strongly advocated as useful in addressing such major difficulties of the sexually abused child and adolescent as low self-esteem, impaired ability to trust, role confusion (within the family), wrongly allocated blame for the abuse, and impaired self-control (Kitchur & Bell, 1989). Systematic evaluation of the effectiveness of treatment is still, however, the exception rather than the rule, although since the present study was started many treatment programmes have attempted evaluation (see, Monck & New, 1996).

The present study

The present study arose from this previous work of the Child Sexual Abuse (CSA) team and the desire to use the clinical experience as the basis for an assessment of the efficacy of particular treatment programmes. In order to improve the validity of the conclusions, a separate research team was appointed to conduct the evaluation.

Chapter 2

Aims and Hypotheses

The research project presented in this report was supported by the Department of Health. The research was planned in two parts: first a descriptive study of sexually abused children and adolescents referred to the CSA clinical team in the Department of Psychological Medicine (DPM) at The Hospital for Sick Children, London and of their families; second, a study of the effectiveness of the treatment available in the DPM to the 4–16 year old children and their families.

The first part of the study was planned to provide a full description of cases referred to a specialized treatment and assessment day-clinic facility.

The treatment outcome study was designed to test the hypothesis that the addition of group therapy to family network treatment would lead to an improved outcome for abused children[2] and their families compared with family network treatment alone.

Subsidiary hypotheses concerned the following issues:

- symptoms at referral would relate to the severity and/or the duration of the abuse;
- the self-esteem and self-reported depression of the abused children at referral would be related to the severity/duration of abuse;
- the self-esteem and self-reported depression of the abused children would be related to the quality of their relationship with the non-abusing parent(s), and to the reaction of those parents to the disclosure.

---

[2] Throughout this report the words 'child' and 'children' have been used to describe both children and adolescents between the ages of 4 and 16 years; this is purely in order to avoid repetition of the phrase 'children and adolescents' and should not be read as a lack of recognition that there are, or can be, differences in the needs, perceptions etc, of the two age groups.

Chapter 3

The Sample

All the cases referred to the CSA clinical team in the Department of Psychological Medicine (DPM) were included when the following criteria were met:

- abuse defined as anal, oral or genital penetration, or forced stimulation of the abuser's external genitalia by the child, stimulation of the child's external genitalia by the abuser, or mutual masturbation; or fondling of breasts and other parts of the child's body for the sexual satisfaction of the abuser;
- abuse had taken place on more than one occasion in the previous 24 months;
- abuse had been perpetrated by a close family member or a member of the household;
- disclosure had occurred in the 12 months before referral to the DPM;
- the abused child was aged over 4 years and under 16 years at referral to the DPM.

It is important to state that the reasons varied why agencies chose to refer particular cases to the specialist team. It was not the case that the CSA team exclusively saw cases in which the abused child was grossly symptomatic, or in which the abuse had been particularly severe or prolonged. However, the Hospital for Sick Children acts in this field, as in others, as a tertiary referral centre, and it was assumed that the children might not be typical of those seen elsewhere.

A high proportion of the children referred by external agencies receive treatment from the CSA team, while the other half are referred for expert assessment or diagnosis, or for professional advice on treatment delivered by other agencies.

For this reason the sample populations for the descriptive and treatment studies would not necessarily be identical. Before the case was accepted for assessment or treatment by the CSA clinical team, the community professional support had to be in place; this meant that appropriate social work and probation service support had been assigned and preliminary protective work had already been carried out, often by these external agencies. Diagnostic work had been done either by others or by the CSA team at Great Ormond Street, and case conferences, police investigations and, when necessary, criminal action or care proceedings had been initiated. All the professionals involved with the case outside the hospital were invited to a preliminary meeting before treatment began, to carry out initial assessments for treatment. These professionals were in close touch with the clinical team throughout treatment and their consent was required for the case to be included in the research.

All the families and abused children meeting the above criteria entered the descriptive study, while those accepted for treatment entered the treatment outcome study as well (Treatment Outcome Study, Part II, Chapter 9).

Chapter 4

Method

The four members of the research team[3] worked separately from the clinical team, but in order to ensure the necessary co-ordination, joint meetings were held every week throughout the planning stages of the research. While the research was in progress a member of the research team and the project secretary attended the weekly allocation meeting of the clinical team.

Referrals to the CSA clinical team come from a variety of sources, including local authority Social Services Departments (social workers), General Practitioners, the courts, community and hospital paediatricians, and the probation service. As noted above, only about half the children are referred for treatment; for the other cases the agencies are seeking expert advice on management or taking advantage of the diagnostic and assessment skills in the CSA team.

Research interviews at referral

Research interviews were planned to take place at referral for all cases entering the descriptive study and the treatment outcome study (the initial interview), and at the end of treatment for those in the treatment outcome study (the final interview)[4].

The initial research interviews were planned with the abused children aged 6 years and over, with the non-abusing parents or carers who had been responsible for the child at the time of the disclosure and at referral, and with the perpetrators. Interviews with children aged 6 and 7 years lasted about 40 minutes; those with older children and perpetrators each lasted about one and a quarter hours, and that with the caring parents/carers lasted about one and three quarter hours. The youngest children in the research programme who were aged 4 and 5 years, were not interviewed. (Summaries of the contents of the different initial research interviews are given in the next section on Measures—page 16.)

---

[3] Elizabeth Monck, team co-ordinator; Rebekah Lwin; Elaine Sharland and Gillian Goodall.

[4] The end of treatment was expected to be within 12 months of referral. In the event, some children or families continued in treatment after 12 months: for these cases the final interview was held at 12 months.

The initial interviews with the abused child and family members were the same for the cases entering the descriptive and treatment studies. Only after the clinical assessment of a case (and therefore after the initial research interview), was the decision taken by clinicians and the family social worker about whether it was appropriate for the family to join the treatment programmes.

Research interviews after treatment

The researchers had no contact with the families during their treatment, which varied in length from 9 to 12 months. The end of treatment was determined by the clinicians on the CSA team, who notified the research team when they were planning their own final assessment. However, if treatment had not been completed after 12 months, the design of the research meant that the second and final research interview was nonetheless carried out. (It should be noted that for some children or families, clinicians judged that treatment should not end after only 12 months, and further treatment or community support was continued for most families whose group and family treatment plan in the hospital clinic had ended.)

The final interviews were conducted in the same style as the initial interviews. Whenever possible, individuals were interviewed by a researcher who had not seen them at the start of treatment, in order that relatively uncontaminated ratings could be made. Although the original intention had been that the interviewer would also be blind to the treatment, this proved unrealistic as it usually became apparent at an early stage in this final interview.

Random allocation to treatment

Once it had been agreed that a family would take part in treatment they were randomly allocated to one of two treatment modes: family treatment only or family treatment plus group treatment. This allocation was made without the direct involvement of the researchers or the clinical team, and was achieved by blind choice of marked cards drawn from an envelope.

Family treatment

All the families (or individual abused children without family support) who were in the treatment study were offered family network treatment sessions every 4–6 weeks. However, not all members of the family attended every one of these sessions. Sometimes the clinicians perceived that the presence of particular relatives would not help the abused child.

For example, disbelieving and denying parents from whom the child had been removed would probably not be invited to attend treatment, unless there was a definite plan to return the child to that household. Sometimes not all family members were willing or able to attend all of these meetings; some family members failed to attend any of the meetings. Those children who had been removed from the family home into local authority Children's Homes attended sessions with their caseworker, and fostered children and adolescents attended with their foster-parent(s). When there were plans for a child to return to their own family in the long run, then some sessions would involve the former parents as well. Even when a child attended family network meetings with professional or foster carers the therapeutic work was closely patterned on the family meetings held with intact families. [A detailed description of family network treatment is given in the Appendix.]

Group treatment

In those families which had been allocated to group treatment each family member would attend groups which were specific to their role in the family, and to their sex and age. The types of group held in the DPM are detailed Table 1.

Table 1 **Composition and duration of groups attending group treatment**

Composition	Duration
children—mixed-sex up to 6 years	6-8 weeks
8 years	6-8 weeks
10 years	6-8 weeks
adolescent girls aged 10–13	20 weeks
14–15	20 weeks
16–17	20 weeks
adolescent male perpetrators	20 weeks
couples	20 weeks
parents of teenage perpetrators	20 weeks
caretaker parents	20 weeks
adult perpetrators	20+ weeks

The group treatment courses started twice a year in October and March. Consequently, if a family was allocated to group treatment they might start treatment immediately or have to wait up to 4-5 months before a new set of group sessions began. It follows from this that the group treatment took place at different stages within the family treatment. Although this may have affected the impact of group treatment, it was not possible to re-arrange the treatment programme in the interests of the research. A full description of the treatment in the groups and the family work is given in the Appendix.

Professional network support & treatment

In addition to the treatment by the CSA team, all the families with whom the child lived, and any family group to which they might return, were supported in the community by social workers, and when appropriate, by probation officers. These community-based professional workers were always invited to the hospital for the family treatment sessions to report and plan for future work: families did not attend without them. All families in treatment received this network support. The extent and nature of this external support and treatment was recorded by the clinical teams.

Chapter 5

Measures

A literature search at the start of the research period revealed few prospective studies of abused children referred for assessment or treatment, and no instruments which were widely or routinely used, at least in the UK, to assess the effects of sexual abuse at referral, or to assess the progress of children through treatment[5]. Discussions were therefore held with the CSA clinical team, in groups and individually, to establish which measures were appropriate for the study. Inevitably, because the initial interviews were the same for the individuals in the Descriptive and the Treatment studies, measures were chosen which reflected the clinical aims of the CSA team, and which could be repeated at the end of treatment.

The following issues were considered by members of the CSA clinical team to form the core of their treatment aims:

AIM/ISSUE	INDIVIDUALS TO BE TREATED
Appropriate allocation of blame	Whole family
Appropriate anger in family and caring parent	Family & caring parent(s)
Communication between caring parent and abused child(ren)	Caring parent(s) & abused child(ren)
Appropriate affection from perpetrator to abused child	Perpetrator
Prevention of further abuse	Child and parent(s)
Raising self esteem	Abused child and caring parent

In addition, members of the CSA clinical team expressed their conviction that non-abusing mothers and the abused children were often extremely depressed, that the quality of the marital relationship was poor (when the father had been the perpetrator), and that school work frequently suffered while the abuse was going on (Bentovim et al, 1988). None of these issues had been systematically assessed before the research began, but the literature on the effects of abuse supported these points.

With this guidance the following items were chosen to assess the condition of those going through treatment.

---

[5] Since the start of the research several behaviour and symptom measures which were widely used in the US have become more established in the UK: notably the Child Behavior Checklist (Achenbach & Edelbrock, 1983).

Standardised Instruments

1. ADULTS

- self-esteem: Great Ormond Street Self-image Profile
  scoring: 'Very true for me', 'True for me', 'Not very true for me', 'Not at all true for me'
  0,1,2,3 to give a high score indicating high self-esteem

This self-report questionnaire developed by the researchers drew on the Adult Self-perception Profile (Messer & Harter, 1986). It identifies domains, including the care of others, and running the household. These items allow scope for unemployed parents to gain higher scores than on self-esteem questionnaires which only ask about work outside the home, and for women (in particular) to record self-esteem in their traditional home-care roles.

- psychiatric state: 28-item General Health Questionnaire (Goldberg, 1978)
  scoring: 0011 to give a maximum global score of 28, with a cut-off of 4/5 indicating risk of significant psychiatric disability
  Re-scored 0,1,2,3 in four sub-scores for anxiety depression, social dysfunction and somatic symptoms.
  The GHQ was developed to identify psychiatric illness in community (non-clinic) populations.

2. ALL CHILDREN

- Health and behaviour: derived from 20 interview questions to the mother/carer covering somatic symptoms and behaviour disorder (Smith & Jenkins, 1991)
  scoring: 0,1,2,3 a high score indicates greater frequency and severity of problems
- Teacher's behaviour checklist (Rutter, 1967)
  26-item questionnaire designed to indicate behavioural problems in the school setting.
  scoring: 0,1,2—a high score indicates greater problem behaviour or symptoms. A cut-off point of 8/9 differentiates those with significant behaviour problems.

3. CHILDREN aged between 8 & 16 years (inclusive)

- self-esteem: a self-completion Self-Esteem checklist developed at the Institute of Child Health, based in part on domains identified for children (Harter, 1985) and for adolescents (Harter, 1987).

The ICH checklist included the 10 items of the Rosenberg (1965) Self-esteem Inventory, which measures global self-esteem.
Scoring: options of 'Very true for me', 'Quite true for me', 'Not very true for me', and 'Not at all true for me', score 0,1,2,3 to give a high score for high self-esteem.

- Depressive symptoms: Children's Depression Inventory (Kovacs & Beck, 1977)
  A 27-item self-completion questionnaire.
  Scoring: each item presents three options for the child to check. Score: 0,1,2 to give a high score orgreater depression
  A cut-off point of 18/19 differentiates those with significant depressive symptoms.

- Health & behaviour: questions asked within the interview from the child (Smith & Jenkins, 1991) 12 of the 20 items in the mother's interview
  Scoring: 0,1,2,3 a high score indicates greater frequency and/or severity of symptoms.

4. CHILDREN aged 6 and 7 years

- The Pictorial Scale of Perceived Competence and Social Acceptance for Children (Harter & Pike, 1982).
  This pencil-and-paper test is designed to elicit the child's self-perception in a number of domains such as relationship with mother, or physical competence.

- The Family Relations Test (Bene & Anthony, 1978).
  The FRT was designed as a clinical measure for testing the direction and intensity of a child's feelings towards members of the family, and the child's perception of their feelings for him/her.

5. CHILDREN AGED 4 AND 5 YEARS were not interviewed, but information was collected from the mother/carer on their health and behaviour over the previous month and before disclosure; from teachers when it was appropriate and from the family social worker.

Additional research measures

Sexualized Behaviour

The clinical team identified sexualized behaviour as a source of concern, and a common presenting symptom. No standardized measures were available for recording this particular symptom in detail, although six questions on the social workers' Impact Checklist refer to sexualized behaviour and sum into a sub-score on 'sexual problems' (Berliner & Conte, 1986). It was decided, however, to augment this information by asking the mothers/carers to describe any inappropriate or embarrassing sexual or sexualized behaviour which they had observed. This information was in addition to the 20 items on health and behaviour referred to above. The type and frequency of sexualized behaviour reported by the mothers/carers was subsequently rated by the research team. The type of behaviour which qualified for inclusion included inappropriate masturbation (not in private), 'sexual' behaviour towards other children or adults (touching, French kissing, simulating sexual intercourse or buggery), sexual precocity or verbal pre-occupation. Sexualized behaviour was scored on a four-point scale as 'None'; 'Occasional, not very noticeable'; 'Moderate, mother (carer) has to correct'; and 'Persistent and/or very inappropriate'.

Information about the abuse

i) Information was obtained on features of the abuse from the child or mother/carer when that was judged appropriate, but other sources (eg social worker's reports, police reports, clinical assessment) were also used. If the child showed clear distress in this section of the interview they were not pressed to answer the questions, and information was sought from other sources. The features of abuse included the age of the child when abuse began and ceased (therefore the duration), the people who had been involved and the ways they had been involved, the nature of the abuse (both the most frequent type of activity and the 'worst'—see below), the number of incidents, the strategies used by the abuser(s) to ensure the child's compliance, and the events surrounding the ending of the abuse. The children were also asked about whether they had made any previous attempts to disclose, and the mothers/carers were asked whether they recalled any such requests, or any signs or behaviour which might have indicated that abuse was taking place (accepting that they might not have had the knowledge to interpret those signs at the time). The children and the mothers/carers were also asked whether they (the mothers) had believed the child's story of abuse at the time of disclosure or discovery, and at the time of referral. Note was taken of the child's report of the mother's/carer's reaction to the discovery of the abuse. It is important to record that dilemmas associated with choosing between family members which many mothers face after disclosure of abuse was fully recognised by the research team.

ii) Mothers/carers, abused children and abusers were asked what effect they thought the abuse would have had on the child at present and in adult life.

iii) Details were obtained of the abuser's previous (sexual abuse) offenses, and the extent to which s/he admitted responsibility for the current abuse.

iv) The type of abuse which the child had experienced was classified into an approximate hierarchy of severity, taking account of the extent to which the active co-operation of the child was needed and the activities were invasive or involved more than one abuser; the categories ranged from exhibitionism to intercourse, and the use of the child in adult prostitution or pornography. In addition, the abuse was classified as either penetrative (digital/object penetration, penile penetration and pornographic activities) or non-penetrative (the rest).

Additional measures were derived to describe aspects of the child's relationship with the caring parent and the perpetrator. Information was obtained about the child's friendships (Smith & Jenkins, 1991) and school adjustment, a global rating made of the quality of the partnership of the abused child's parents, and the marital and work histories of the carer and the perpetrator.

Clinical Measures

Based on the previous work of the CSA team (Bentovim et al, 1988), a strategy was devised to assess the family at referral in terms of the treatment aims. These aims can be summarised under three main headings: assigning appropriate responsibility for the abuse and reversing the effects of the abuse, working on the family context in which the child lives, and understanding the origins of the abusive behaviour. Twelve areas of family dysfunction or distorted cognition were then identified which typically affect the members of families in which child sexual abuse has been detected. The aims of treatment included reversing the traumatic effects of the abuse on the abused child(ren) and improving those aspects of family functioning which would prevent further abuse. The assessment was completed for each 'caring' (non-abusive) parent, abused child and perpetrator who was attending the hospital. The areas of concern are listed below: each aim was assessed on a four-point scale ('not at all', 'minimally', 'partially', and 'wholly').

FAMILY TREATMENT AIMS

1. FOR ALL SUBJECTS

A. Assigning responsibility for the abuse

1. Subject acknowledges abuse is responsibility of perpetrator, not abused child or adolescent.
2. Subject shows evidence of resolution of conflicted feelings (over-closeness, loyalty, etc) towards
    a) mother
    b) perpetrator
    c) father if not perpetrator

B. Working on the family context in which the child lives

3. Subject is able to express appropriate anger within the group which was 'family ' at the time of abuse
4. Subject is able to share painful issues with
    a) natural mother/maternal figure at time of abuse
    b) perpetrator
    c) substitute carer (when applicable)
    d) child (when applicable)
5. Family (at time of abuse) does not scapegoat abused child
6. Abused child's age-appropriate needs recognised in family (or substitute family)
7. Subject recognises positive qualities in
    a) self
    b) natural mother
    c) perpetrator
    d) other family members
8. Adults in current living context
    a) recognise the damage which might result from CSA
    b) deal with behavioural results of CSA
    c) recognise damage in adulthood arising from abuse in own childhood (adults only)
9. Subject works cooperatively with professionals

NON-ABUSING PARENTS ONLY

10. Subject acknowledges lack of availability to abused child(ren).
11. Subject establishes and maintains appropriate generational boundaries in family/provides adequate protection for child

FOR ADULTS WITH ABUSIVE PARTNER

12. Subject recognises the need for help on individual or couple basis including specific help with sexual difficulties

In addition to the treatment aims outlined above, each individual and family was assessed on the likelihood of a 'good' outcome, on the basis of treatment aims being reached. The outcome was assessed as 'hopeful', 'doubtful with some hope', 'doubtful with little hope', and 'hopeless'. (For details of these derived variables, see Appendix B.)

SUMMARY OF INITIAL RESEARCH INTERVIEWS

1. CARING PARENT
    i) own marital and work status
    ii) abused child's legal status
    iii) positive interaction/confiding with child
    iv) warmth & criticism of child
    v) child's current symptoms
    vi) child's peer relationship
    vii) own mental health—28-item GHQ
    viii) abuse information
    ix) history of care in own childhood
    x) history of abuse in own childhood
    xi) self-esteem questionnaire

2. ABUSED CHILD
    i) school attended; school enjoyment
    ii) relationships with household members at disclosure and referral
    iii) reports of family arguments
    iv) view of parental 'marriage'
    v) confiding and supportive relationships
    vi) health & behaviour in last month
    vii) abuse information
    viii) friendship patterns and adjustment
    ix) self-esteem questionnaire
    x) Children's Depression Inventory
    xi) non-abusive parent's reaction to disclosure

3. PERPETRATOR
    i) living arrangements
    ii) access to abused child(ren)
    iii) relationship with abused child(ren)

     iv) legal status/procedures
     v) quality of current 'marriage'
     vi) employment history
     vii) history of childhood care
     viii) history of childhood abuse
     ix) self-esteem questionnaire★
     x) 28-item General Health Questionnaire★

4. ADDITIONAL INFORMATION
     i) Social worker checklist—child's symptoms★
     ii) Rutter Teachers' behaviour checklist★

★ Questionnaires described in full in section on MEASURES.

Chapter 6

The findings in the descriptive study

The study population

Over the period of 19 months, during which the study was conducted, 299 families with 391 sexually abused children, were referred to the CSA clinical team at The Hospital for Sick Children. Social Services Departments were by far the largest source of referrals, followed by General Practitioners, and child psychiatrists and psychologists from the Child Guidance clinics (Figure 1).

**Figure 1
Source of referral to child sexual abuse clinical team: Hospital for Sick Children—for 299 families**

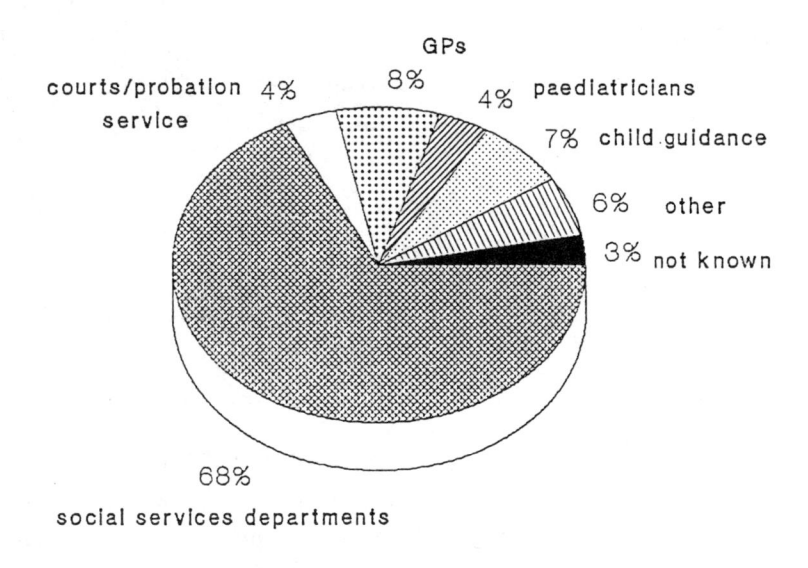

courts/probation service 4%
GPs 8%
4% paediatricians
7% child guidance
6% other
3% not known
68% social services departments

Figure 2 shows how many of the families were eligible and available for inclusion in the descriptive study, and the reasons for exclusion. The major reasons for exclusion were either that the family was not seen in the hospital or that the research criteria were not met. Eighty-three cases which were referred to the CSA team were not, in the end, seen by them. A small number of these referrals were not followed up by the referring agency, but the large majority of the referring agencies were provided with specialist advice on management or local resources for assessment or treatment. When sexual abuse was not confirmed, this was the outcome of a clinical assessment.

A group of 15 families (23 children) were eligible for the research, but were not finally included, usually because they were uncooperative on the few occasions they attended the Hospital. While not openly refusing to join the research, many of these families evaded committing themselves, and regularly postponed research appointments. A further five eligible cases were not contacted by the research team, as a result of administrative errors. The final number included in the descriptive study was 74 families, with 99 abused children; this represented 25% of the referred families and 25% of the referred children.

**Figure 2
Cases included and excluded from study population**

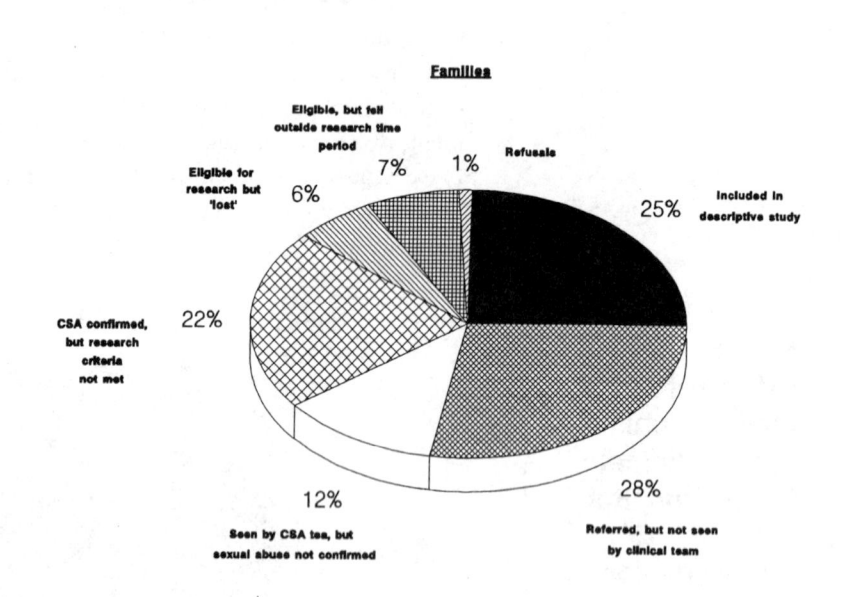

Families

- Included in descriptive study: 25%
- Refusals: 1%
- Eligible, but fell outside research time period: 7%
- Eligible for research but 'lost': 6%
- CSA confirmed, but research criteria not met: 22%
- Seen by CSA tea, but sexual abuse not confirmed: 12%
- Referred, but not seen by clinical team: 28%

Completed interviews

Table 2 indicates the number of successful research interviews which were held as a proportion of the number of people who should have been interviewed. Among the abused children, 12 were aged 6 or 7 years and were given the shortened research interview. Of the remaining 68 children and adolescents who were aged between 8 and 16 years, 54 (79%) were interviewed by the research team, and 14 were not interviewed. The reasons why the researchers were not able to see these 14 abused children tended to be idiosyncratic: eg one child was smuggled out of the UK to live in her country of origin, the severe psychiatric disturbance of another led to her (temporary) elective mutism, one child had a profound hearing impairment. The largest single reason, however, was the difficulty of obtaining interviews with children living in local authority Children's Homes. There seemed to be some reluctance on the part of Children's Home staff to negotiate successfully with the children about their involvement.

Table 2 **Completed interviews compared with informants invited to participate**

	Invited to participate	Interview Achieved	Achieved as a % of expected
Abused children age 4,5	nil	n/a	n/a
age 6,7	12	12	100%
age 8–16	68	54	79%
Parents as primary carer	74	61	82%
Perpetrators: adult*	30	25	83%
: adolescent*	8	6	75%

* Not analyzed in this report

The most common reason why non-abusing adults who had occupied the role of caring parent at disclosure were not seen by the research team was because they had declined to take current responsibility for the abuse or for the abused child. They were not attending the hospital and were not being seen by social services.

It will be seen that only a small number of perpetrators were considered to be available for the research; this is a reflection of the fact that, at the time of the study, very few perpetrators were being seen at the Hospital for Sick Children. A large number denied all responsibility for the abuse, and the evidence was insufficient for the Director of Public Prosecutions to allow a criminal charge to be brought against the perpetrator. These adult perpetrators were thus able to avoid contact with any treatment services. At the other end of the spectrum, a number of perpetrators were in prison, either convicted or awaiting trial. The 25 who were seen represent only a small proportion (21%) of the total number of perpetrators involved with the abused children.

It must be emphasised that the numbers given in the following analyses do not always match the figures given in Table 2. This is because, although an interview was conducted, questionnaires were sometimes incomplete. In the following sections we have chosen to present the findings almost without comment: comparisons with other studies are discussed in Chapter 7.

The personal and family characteristics of the children

Gender

Of the 99 children and adolescents in the descriptive study, 78 were girls and 21 were boys; a ratio of 3.7:1. There were few sex differences and these are noted below in the appropriate sections.

Age at referral to CSA hospital team

The ages of the abused children at the time of the first interview ranged from 4 years 2 months to 15 years 9 months, with a mean age of 9 years 10 months (SD 41.6 months). There was a tendency (non-significant) for the boys to be younger (boys: mean age 8 years 6 months; girls: mean age 10 years 3 months) (Table 3).

Table 3 **Ages of the abused children at the time of referral to the hospital**

	Age of child (years)				
	<7 n (%)	7–9 n (%)	10–12 n (%)	13+ n (%)	n (%)
Girls	17 (22)	18 (23)	18 (23)	25 (32)	78 (100)
Boys	8 (38)	7 (33)	3 (14)	3 (14)	21 (100)
Total	25 (25)	25 (25)	21 (21)	28 (28)	99 (99)

$\chi^2=4.88$, 3df, p .18 [not significant]

Dividing the children into those under ten years and those ten years and over at referral, it was found that the older group reported significantly more symptoms of disturbed health and behaviour than the younger group, and also recorded higher scores on the Children's Depression Inventory (t=−2.10, 25.89df, p<.05, and t=−2.59, 30.40df, p<.02 respectively).

It is worth noting that, the children's own assessment of their attendance at school was related to age; older pupils were significantly more likely to say they had been absent from school more than once a month in the previous year, and those who admitted they were away more than once a week were all over 13 years old at referral (Table 4). Although there was no control group on self-reported attendance, teachers rated the abused children as significantly less good attenders compared with their classroom controls, but they did not rate older children as significantly worse than the younger ones.

Among the children who were interviewed, being 10 years or over at referral appeared to be associated with feeling that the responsibility for the abuse rested with the perpetrator ($\chi^2=7.79$, 2df, p<.03). Further investigation, however, showed that among girls but not boys, being over 10 years old was significantly associated with feeling personally responsible for the abuse having occurred ($\chi^2=10.02$, 3df, p<.02). This finding illustrates the importance of distinguishing the feelings as well as the experiences of boys and girls.

The older children were at referral the greater was the likelihood of their having experienced penetrative abuse. Set against this it was found that the mothers/carers were more likely to believe the younger than the older children's description of the abuse (Table 4).

Table 4 **Summary findings of variables associated with older age of child**

	Analysis of variance		
Older age	F	(df) level	Significance (p)
Penetrative abuse	4.69	(1,92)	<.04
Mother's disbelief in abuse	3.93	(2,44)	<.03
Poor school attendance reported by the children	3.55	(3,50)	<.03
Absence of sexualized behaviour	3.19	(3,92)	<.05
Higher scores on Children's Depression Inventory	5.07	(1,53)	<.03

Figure 3 Distribution by socio-economic group of families of sexually abused sample compared with households in Greater London

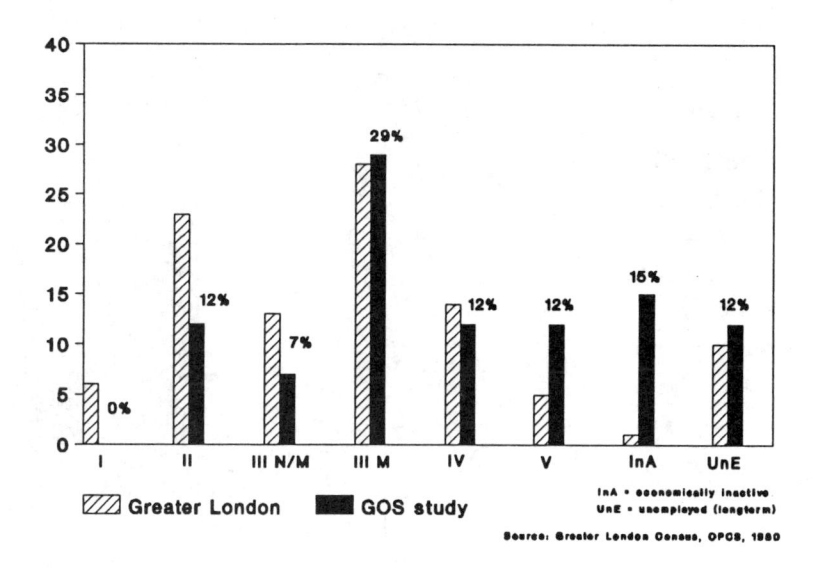

Socio-economic status of parents

The families were assigned to one of the six socio-economic groups used by the Registrar-General (OPCS, 1981) on the basis of the principal wage-earner's occupation in the previous year. Figure 3 shows the distribution by socio-economic group compared with that of Greater London (a few families came from a wider geographical area).

Living arrangements of the abused children at disclosure and at referral to hospital

For a substantial number of children household composition had varied over the (often lengthy) period of the abuse. An impression is gained of a high rate of family dissolution and reconstitution. Using all available sources of information it was clear that 49 (49%) children were living in a household with their own parents at the time of the abuse, while the remaining children lived in reconstituted households, often with step-, foster or adoptive siblings.

At the time of the hospital referral, 45 children were living at home with a parent who had cared for them at disclosure; 8 were living with other relatives, including two natural fathers who took responsibility for their teenage daughters after the latter had been abused by their stepfathers; 22 were living in Children's Homes and special residential communities; and 22 were living with foster parents (Figure 4). Sixty-three children (63%) were not living with their father or any father surrogate. One of the effects of the changes in residence after disclosure was that 37 abused children were parted not just from parents but also from some, though sometimes not all, their siblings and half/step-siblings.

Figure 4 Children's living arrangements at referral ($N=99$)

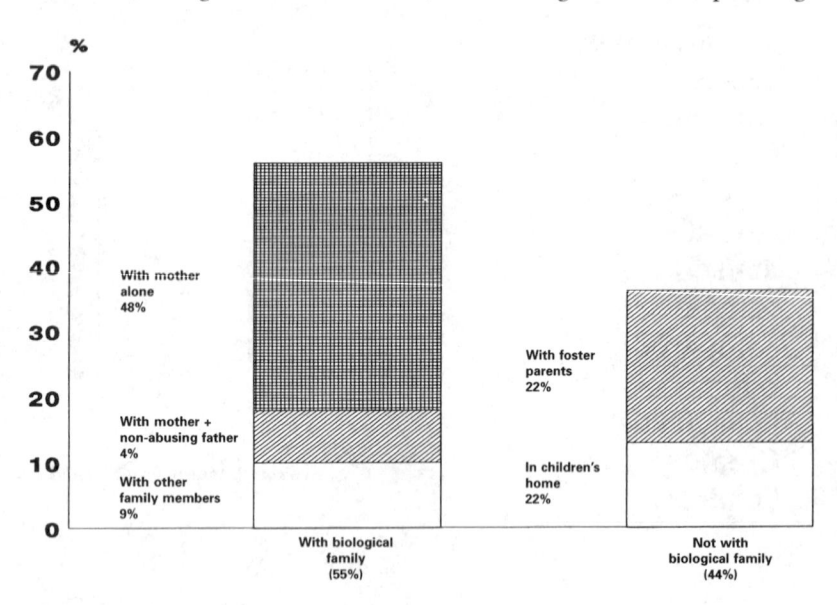

Bars show:
- With biological family (55%): With mother alone 48%, With mother + non-abusing father 4%, With other family members 9%
- Not with biological family (44%): With foster parents 22%, In children's home 22%

The children's legal status at referral

Information on the child's legal status at the time of referral to the hospital showed that some form of legal constraint was in place for 56 (56%) of the children; 69 separate actions had been taken under child protection guidelines, with some children affected by more than one action

(Table 5). The individual circumstances of these orders were extremely diverse. The families for whom there were no legal conditions were those in which the case had been resolved because the perpetrator was in prison, or the family were voluntarily cooperating with social services.

Table 5 **Children's legal status at referral**

	N*	(%)
Safety Order	6	(6)
Ward of Court	11	(11)
In care	24	(24)
Child Protection Register	24	(24)
Supervision order	4	(4)
Total number of children with legal orders	54*	
Total number of children in study	99	(100)

* Some children fell into more than one legal/child care category

Menarche

Among the 78 girls, twenty had reached puberty by the age of referral to the hospital; the age of menarche varied from 10 to 16 years, with a mean of 12 years 8 months.

Abuse Characteristics

Age at onset of abuse

Reliable information about the age at onset of the sexual abuse was available for 66 children. Table 6 shows the distribution of age at onset by gender; the mean age at onset for girls was 7 years 8 months and for boys 6 years 4 months (SD=39, and 32 respectively).

Table 6 **Age at onset of abuse (N=66)**

	Age (years)				
	<7	7–9	10–12	13+	Total
	n (%)	n (%)	n (%)	n (%)	n (%)
Girls	22 (41)	16 (30)	12 (22)	4 (4)	54 (100)
Boys	5 (42)	7 (58)	0 (0)	0 (0)	12 (100)
Total	27 (41)	23 (35)	12 (18)	4 (6)	66 (100)

χ^2=5.88, 3df, p.12 [not significant]

As has been noted the age at referral of the abuse was significantly related, but only for girls, to the perception the child had about responsibility for the abuse. Using the four age-bands shown above, girls who were older at onset were more likely to say that they felt the abuse had been their fault ($\chi=10.02$, 3df, $p <.02$).

An older age at onset was significantly related to the experience of penetrative abuse; this held true for boys and for girls. It will be recalled that there was no significant link between age at referral and the experience of penetrative abuse. This suggests that penetrative abuse of the younger child may be being detected sooner than in the older child.

Age at cessation of abuse

Table 7 shows that, at cessation of abuse, the boys were significantly younger than the girls. Since there was no significant difference in age at referral between boys and girls, it seems that the cases involving boys may have taken longer to be referred to the specialist team.

Table 7 **Age at cessation of abuse**

	Age (years)				
	<7	7–9	10–12	13+	Total
	n (%)	n (%)	n (%)	n (%)	
Girls	10 (16)	17 (26)	20 (31)	17 (27)	64 (100)
Boys	8 (50)	4 (25)	2 (13)	2 (13)	16 (100)
Total	18 (23)	21 (26)	22 (27)	19 (24)	80 (100)

$\chi^2 = 9.44$, 3df, $p <.03$

Duration of abuse

The absence of accurate information about the age at which abuse had started or ended obviously affected the figures on duration. A reliable estimate of duration was only available for 66 children. Table 8 suggests that it was rare for there to have been a short duration of abuse; the mean duration was 32 months, with a range between one month and 96 months (SD 26 months). It should be borne in mind that, to be included in the study, a child had to have experienced more than one sexual contact with the abuser; 'duration' of only one event was not therefore possible.

Table 8 **Duration of abuse**

	Duration (months)				
	<2	2–11	12–35	36–96	Total
	n (%)	n (%)	n (%)	n (%)	n (%)
Girls	3 (6)	9 (16)	22 (41)	20 (37)	54 (100)
Boys	1 (8)	3 (25)	4 (34)	4 (34)	12 (100)
Total	4 (6)	12 (18)	26 (40)	24 (36)	66 (100)

$\chi^2=0.67$, 3df, p.88 [not significant]

The duration of the abuse was greater for children aged over 10 years at referral (t= -2.62, 61.28 df, p<.02). Longer duration was associated with having experienced penetrative abuse ($\chi^2=11.30$, 4df, p <.03). Thus penetrative abuse is related to being older at onset, older at referral, a longer duration, and being older than 10 years at referral was also associated with longer duration. These inter-linking facts about age and duration seem, at first sight, to be counter-intuitive: it might be expected that younger age at onset would be related to longer duration. But this would only hold if the chances of being discovered were the same at all ages. It may be that the younger children's abuse is discovered relatively faster than those whose abuse starts later.

The number of abusers

It was not uncommon for a child to have experienced abuse by more than one perpetrator. Fulfilling one of the criteria of the study, the most recent perpetrator had to have been a close family member or a member of the child's household, but 20% of the children had been abused by other members of their family or by non-family at some stage in their lives.

Relationship of the child to the abuser

The majority of abusers were biological fathers or the father-figure for the child at the time of abuse. In a small percentage of cases (5%) the child's mother, and in one case the child's grandmother, were involved in the abuse, although in only one of these cases was a woman the sole abuser (Table 9). Altogether, only 6% of abusers were female, 94% were males, and males were involved in 98 (99%) of the cases.

Table 9 **Abuser's relationship to the child**

	n	(%)
Natural mother	6	(5)
Natural father	44	(36)
Stepfather	19	(16)
Other 'fathers'	8	(7)
Brothers	18	(15)
Other relatives	19	(16)
Other household members	7	(6)
Total	121*	(100)

* Totals are more than the number of children as several children had been abused by more than one relative/household member

Type of abuse

The type of abuse experienced by the children was classified into the nine categories shown in Table 10. Information was sought on the most frequent abuse, but also on all types of abuse which the child had experienced.

It will be recalled that abusive experiences were also re-classified as penetrative or non-penetrative. Penetrative abuse included digital or object penetration, and penile penetration; in this sample the girl who had been used for pornography was also placed in this category, because of the penetrative nature of her pornographic experiences.

Table 10 shows that, using the definitions given above, some form of penetrative abuse was the most frequent experience for 49 (53%) of the 93 children for whom this information was reliably available, and the most severe ('worst') experience for 59 (62%) of the 95 children for whom there was information. Perhaps unsurprisingly there was a highly significant association between penetrative abuse having been the 'worst' experience and having also been the most frequent experience (χ^2=50.37, 1df, p <.0001).

The child's age at referral (though not, as noted above, their age at onset) was significantly related to the nature of the abuse; children in the older age categories shown in Table 3 were more likely to report that the worst (but not the most frequent) abuse had been penetrative.

As has also been noted above penetrative abuse both as the 'worst' and as the most frequent experience was also significantly associated with a longer duration of abuse. There may be a connection between this longer (undiscovered) penetrative abuse, and the fact that when penetrative abuse had been the most frequent experience there was a significant probability that the mother/carer did not believe the abused child

(Table 11). It must remain a possibility that the non-believing mother also failed to pick up or act on the warning signs of abuse, and thus (unwittingly or wittingly) extended the duration.

Table 10 **The most frequent and the most severe sexual abuse experiences**

Type of abuse	Most frequent		'Worst'	
	N	(%)	N	(%)
Exhibitionism	1	(1)	0	(0)
Fondling/touching	14	(15)	6	(6)
Masturbation (abuser)	8	(9)	8	(8)
Mutual masturbation	4	(4)	4	(4)
Oral-genital contact	13	(14)	11	(12)
Attempted intercourse	4	(4)	7	(7)
Digital/object penetration	11	(12)	13	(14)
Penile penetration	38	(41)	45	(47)
Pornographic activities	0	(0)	1	(1)
Totals	93*	(100)	95*	(100)

* some children were unable to give a clear description of the abuse, or its frequency, either in the study or to other professionals.

It is worth noting at this point that there was an unexpected absence of a direct relationship between more severe (penetrative) abuse and more explicit symptomatology. However, in the current study some of the symptoms of Post Traumatic Stress Disorder (PTSD) were not included in the behaviour ratings, and it is probable that some of the apparently symptom-free children were exhibiting PTSD symptoms.

Table 11 **Summary table on penetrative abuse**

	Analysis of variance		
Association between	F	(df)	significance level (p)
Older age at referral with most severe experience being penetrative	4.69	(1,91)	< .04
Longer duration with i) 'worst' experience being penetrative	3.71	(4,61)	< .01
ii) most frequent experience being penetrative	3.64	(4,61)	< .01
Mother not believing the child with most frequent experience being penetrative	3.40	(2,41)	< .05

Children's feelings about the abuse

The children and teenagers were asked to recall what they had felt about the abuse while it was going on. After being encouraged to describe their feelings spontaneously, they were also asked to agree or disagree with particular adjectives describing the abuse or their feelings. It was possible for the child to check more than one response. Of the 66 children who were interviewed, 31 (47%) said the abuse experiences had been painful, 27 (41%) that they were 'strange', 37 (56%) that they were 'wrong', 35 (53%) that they were 'disgusting', 38 (58%) that they were frightening, 23 (34%) that they felt angry; however, 10 (15%) that the experience had usually been pleasurable, and 8 (12%) that they had believed it was a 'normal' part of family life.

The number of incidents

For a small number of children (17%) disclosure had taken place (or abused ceased) after fewer than 10 incidents. For the majority, however, the number of abusive incidents was at least ten and the total too large for accurate estimation, either by the child or by the non-abusing parent. A larger number of incidents appeared not to be related to a longer duration of abuse.

The perpetrator's strategies for preventing disclosure

The abused children and the mothers/carers were asked in detail about the strategies employed by the perpetrator to ensure that the child did not disclose. Reports from both children and their mothers/carers showed that in more than half (58%) of the 53 cases for which information was available, the abusers had relied on non-violent strategies such as adult 'authority', bribery or non-violent threats to the child or other family members. There was a difference in the reports of children and mothers/carers about how many of the perpetrators used violence or threats of violence towards the child in association with the sexual abuse. Mothers/carers reported that 20 perpetrators (38% of the 53 cases for whom this information was available) used violent threats to the child or to other family members, and that two (4%) of perpetrators had used violence. However, four (8%) of the children gave clear descriptions of actual violence from the perpetrator, and correspondingly fewer children (34%) reported violent threats.

Physical abuse

In addition to the four cases of physical violence associated with the sexual abuse which were reported by the victims, there was evidence that another 20 children also experienced physical abuse in the family from one or both parental figures. In 12 cases the evidence was unconfirmed outside the family, but in 8 cases the evidence was accepted as incontrovertible by the child protection agencies. In 64 cases there was no evidence of physical abuse, and for 11 children the data were missing, but on balance it may be assumed there was no abuse.

For 48 interviewed children who gave valid answers to the questions on the enforcement of discipline in the family before disclosure, 30 (63%) said there were no major physical sanctions. Among 18 children who described major physical sanctions (eg, hard smacking, beating with an object, punching, kicking, or forcing the child to stand outside the house for hours on end after dark), 14 children (78%) said these were administered only by the father, 2 (11%) by both parents and 2 (11%) by the mother alone.

Disclosure characteristics

Choice of whom to disclose to

Some children had attempted to disclose or had finally disclosed, in the sense of speaking about their abuse to another child or an adult. Among the 66 children who were interviewed, 48 (73%) said they had told someone, while the other 18 had been unable to. It became apparent that many of the latter group had attempted to tell, some indeed believing they had adequately described what was happening, but their 'disclosure' fell on deaf ears. Including information from other sources on this aspect of disclosure it appeared that altogether 21 (21%) of the total sample had 'disclosed' unsuccessfully. The remark of one mother is not untypical of a small group who agreed they had been told before, but had not acted: 'She told me two years ago and I watched but never saw anything. I told her to tell me if anything happened but she never did'. The abuse in this case continued intermittently for the whole two years.

Among the 47 children who had successfully disclosed, thirteen (28%) had told their biological mothers, and another 4 (9%) their surrogate mothers; two children (5%) had finally told their grandmothers, three (7%) had told sibs, and 7 (16%) other relatives. Nine children (20%) had told friends, and nine children (19%) disclosed to professionals—teachers, social workers and childcare staff—with whom they already had some relationship. The remaining children were either not interviewed because they were too young or the abuse was discovered, not disclosed.

Circumstances leading to disclosure

The prime reason why the abuse ceased is summarised in Table 12. Even when information from all informants was combined, it was not always clear why the abuse had ended when it did. As we have noted, only 48 of the 66 children disclosed; for the rest the abuse was discovered by others, or the perpetrator left the household; for seven cases it was not possible to establish the reason with any certainty (Table 12).

Table 12 **Reasons for cessation of abuse (n=99)**

	n	(%)
DISCLOSURE		
to biological parents(s)	13	(13)
to other parents	4	(4)
to other family	12	(12)
to teacher/social worker	9	(9)
to adult friend	4	(4)
to child friend	5	(5)
ACTION OF CHILD	4	(4)
DISCOVERY/ACTION BY OTHERS		
by non-abusing parent	14	(14)
by other adult	8	(8)
by sib/peer	12	(12)
Perpetrator left household	2	(2)
other	5	(5)
not known	7	(7)
Total	99	(99)

Since the issue of disclosure is so important in the detection of sexual abuse, and in obtaining the support of others, it is worth outlining the course of events in some families. [Some of the details have been changed to decrease the chances of identification of individuals.]

CASE 26.
The grandparents, especially the grandfather, undertook quite a lot of childcare for the busy parents who had five children and ran their own businesses. He had taken the youngest girl (aged 4) for a car-ride, and a day later, when riding in the car with her mother, the child said 'Can I take off my clothes like I do with grandfather?'. The mother was startled, and asked enough to learn that grandfather often took the child to his garden shed, or out in his van and they took off

their clothes. The next day two of the older children (boy 12 years, girl 10 years) were watching a TV programme about child abuse, during which abuse within the family was described as a not uncommon experience. The mother decided that she would ask these two about their grandfather, using the excuse of the TV programme. She told them that their small sister had described the grandfather taking his clothes off in his van; the boy laughed, but the girl started to cry. Sending her son away, the mother was able to encourage her daughter to disclose, partly because the 10 year old realised she needed to speak to protect her younger sister. Rating: disclosed to own parent.

CASE 31. The girl (aged 14) disclosed to her class teacher. 'I had to find the courage to talk to someone about it. I definitely needed to talk to someone, and I knew something would be done. I couldn't really cope with it, but I went along with it until things got out of hand (escalation of demands). I wanted (the abuse) to stop, but I wasn't aware that I'd have to go through all this (post-disclosure events). I think if I'd known about this (post-disclosure events) I'd have talked to my Mum or my aunt, and tried to keep it in the family'. Rating: disclosed to professional.

CASE 18.
Girl (aged 15 y 4 m): 'I thought about telling my mum, but decided she wouldn't believe me, so in the end I told my mates, and they told the teachers, and they told the police. I didn't want it to happen to my sister.' Rating: disclosed to friends.

CASE 37.
Boy aged 6. Mother found child with his penis in a small girl's mouth when they were in a mother/child Refuge. The boy said 'this is what Daddy does with it'. The mother told the Refuge staff who called in social services; the abuse was finally confirmed several months later at Great Ormond Street. Rating: discovery by non-abusing parent.

CASE 22.
13-year old boy perpetrator: older (foster) sister who was baby-sitting discovered this (fostered) boy and one of the younger fostered boys (aged 8) without their pyjamas on and felt anxious enough to move the 8-year old into a younger sister's bedroom. She did not believe the 13-year old perpetrator's story (of looking at bruises). Later that evening the older girl discussed what had happened with her own parents when they returned. The two boys ceased to share a bedroom. A few days later, the mother's continuing concern led her to ask the 8-year old boy, point blank, whether there had been any sexual activity. The 8-year old was reluctant to talk, and had to be asked direct questions. Next day the mother asked the three younger (fostered) girls. The mother says she was amazed by confirmation of

their abuse by the 13 year old. Parents went to General Practitioner, and counsellor who recommended telling Social Services. Each social worker for each child would have been told anyway. Rating: Discovery/action by sibling.

CASE 57.
Girl and girl friend (both aged 12 years) abused by stepfather. Despite the fact that he had threatened to beat her, and was violent at times towards her mother, she finally took a stand and said 'no'. This was after her mother had thrown him out of the house for other reasons, though he continued to visit. She (and subsequently her mother) were surprised that he gave way, and made no more attempts. Rating: action by child.

Of forty-nine children who answered the question about what their feelings were about having disclosed, 35 (71%) said they were glad, 8 (16%) were ambivalent, and six (12%) said they wished they had not disclosed. All 43 children in the first two groups also said they were relieved the abuse had stopped. The remaining children were either too young to have been asked the question (19) or had never—in any form—spoken of the abuse until it was discovered by other means. If the child reported that their mother/carer currently believed that they had been abused, the child was significantly more likely to say s/he was glad to have disclosed: 87% of those whose mother believed them were glad to have disclosed, compared with 40% of those whose mother did not believe them (Table 13). By contrast, it should be noted that the child's feelings about the abuse having been disclosed or discovered were not related to their depression scores or their self-esteem scores at referral.

Asked about the effects of the abuse, 11 children said they believed there had been no effect on them, two said the effect was positive, 21 said the effect was negative, and fifteen did not know. Their reports of the effects of the abuse bore no relation to their self-esteem scores at the time of referral, but there was a significant tendency for the children who said the effect of the abuse was negative to have higher self-reported depression scores.

Examples: Negative: CASE 031. Girl aged 14. Sexual intercourse with her father for 2 years (between age 11.5 and 13.5 years), up to menarche, when he reduced the severity of his demands. 'I think it is going to affect my relationships with men; at the moment I don't want to let any boy get close to me. I'm very frightened of having a sexual relationship.'
Positive: CASE 066. Girl aged 13.5 years. Sexual intercourse with uncle by marriage for three months when aged 13 years. Always enjoyed it, was very sad it was discovered, and missed him greatly. Said the effect of the 'abuse' (not the intervention) was only positive, and did not perceive it as abuse, but as an affair.

Neutral: CASE 011. Girl aged 9.5 years. Mother's cohabitee (for 1 year) touched her external genitalia, four occasions over 6 months. Previous 'assault' by same aged boy when aged 7. Child unable to describe or agree any positive or negative effects.

Table 13 **Mother's belief at referral in the child's story of abuse and the child's feelings about having disclosed**

Mother's current belief	Child's feelings about disclosing		Total
	Wish had not or ambivalent	Glad	
	n (%)	n (%)	
Does not believe	6 (46)	4 (13)	10
Uncertain or Believes	7 (54)	28 (87)	35
Total	13	32	45

$\chi^2 = 6.06$, 1df, p <.02

Non-abusing parents' or carer's belief that the child was telling the truth about the abuse

The children were asked whether they felt that the parent with primary care responsibility for them at the time of discovery or disclosure (usually mothers) had believed that the abuse had taken place; of the 45 children who answered this question, 27 (60%) said they had felt believed. At the time of referral (often several weeks later) this figure had hardly changed at all; only 64% felt believed. There was a significant relationship between the child's report of being believed at disclosure and being believed at referral, with 86% of the children reporting that their caring parent was expressing the same views. It has to be borne in mind that for some children, the 'carer' at disclosure was not the 'carer' at the time of referral. In these cases data were not available on the attitudes of their original carers because they did not attend the hospital clinic.

As we have noted, it was clear that the younger children were significantly more likely than the older children to be believed by the mother/carer.

There was a significant relationship between being believed and experiencing the *less severe* types of abuse. Among children whose most frequent experience of abuse had been penetrative only 50% were believed, compared with 79% of the children with non-penetrative experiences ($\chi^2 = 6.28$, 2df, p <.05).

The mothers' belief in their children's abuse story was not affected by whether they themselves had experienced either physical or sexual abuse in their own childhood.

Caring parents' reactions to discovery or disclosure

The children reported a range of reactions by their caring parents at the time the abuse first came to light. Among 35 children reporting the reactions of mothers and mother-surrogates, 13 (37%) reported negative, unsupportive reactions, 16 (46%) reported supportive reactions and 7 (20%) reported ambivalent or confused reactions where the child was not the main focus of the mother/carer's concern. Very frequently the ambivalence of her response reflected a mother's concern at the risk of losing her partner/companion, the bread-winner, the second parent for the whole family. Rating the reactions of the mothers to hearing about the abuse, every effort was made by the researchers to make value-free judgements.

Examples: Mother's reaction to discovery of abuse
Supportive: Mother discovered her own boy-friend (not her cohabitee) in compromising situation with her daughter. She was very angry and upset, and 'threw him out' immediately. He had been abusing her child (aged 12) for several years. Court case collapsed on a technicality. Mother believed child, does not want to see boyfriend again, and very concerned about her child.
Ambivalent/confused: With his wife away on holiday with a younger child, a foster father made a sexualized advance to his foster-daughter aged 14 while saying goodnight to her. She told her teacher next day, he made a statement to the police, and agreed he would move out of the home when his daughter returned from a week's holiday. The foster-mother was told when she returned the following week. Very upset that husband had to leave the home, and her considerable concern about their relationship and what would happen to them as a couple dominated her behaviour over the following months. Despite this she remained supportive towards her foster-daughter and wanted her to stay at home; continued to see her several times a week after she left and husband returned.
Hostile: Mother's cohabitee attempted sexual intercourse with 13-year old daughter; medical evidence of penetration. Perpetrator denied even touching the girl, and the mother rejected her story and believed the perpetrator entirely. The girl was removed to foster parents as the court case collapsed and the perpetrator returned to live with mother.

Not surprisingly, there was a significant relationship between the mother/carer reacting supportively at the time of disclosure and her belief that the abuse had happened ($\chi^2=28.57$, 4df, $p<.001$); 77% of those who reacted with hostility did not believe the child, and all those who reacted supportively did believe the child. The significant association between believing the child and reacting supportively was still evident at the time of referral.

Carrying the responsibility for the abuse

The perpetrators' admission of guilt: Information was available for 33 adult and school-age perpetrators on the issue of whether they admitted guilt about the abuse. At the time of referral 25 (75%) of these alleged perpetrators denied any involvement in the abuse; 3 (9%) admitted full responsibility, while the remaining 5 (15%) admitted responsibility for some, but not all, aspects of the alleged abuse. Other abusers were not available for interview.

From 49 valid answers[6], 42 children (83%) were clear that the abuse was not their own responsibility, but 7 (17%) said it had been. Thirty-one children (74%) thought the abuse had been the responsibility of the abusers, while the rest thought it had not been; believing the abuse was the perpetrator's responsibility was significantly associated with an older age at referral ($\chi^2=7.79$, 2df, p <.03). Forty-eight children (98% of the valid answers) thought their mother/carers had not been responsible in any way for what had happened.

The allocation of blame for the abuse was not associated with the children's self-reported depression or self-esteem. Nor was it associated with any of the different strategies used by the perpetrators to ensure the children did not tell anyone.

Psychological impact of the abuse

Child's Depression Inventory (CDI)

Among the 54 children (aged 8 years and over) who completed the CDI at the time of referral, 26% showed scores of 19 and above, indicating significant depression (range from 1–46; mean 15.1, SD 8.8). The children did not report high frequencies of somatic symptoms (disturbance of sleep and appetite, fatigue and general worries about their own health). Two clusters of symptoms were reported as sometimes or frequently present by half or nearly half the children. The first of these reflected anxiety (worries about the future, about their own health or that 'bad things will happen', with problems of decision-making). The second group of symptoms reflected depressive feelings and low self-esteem (crying, suicidal ideation, feelings of low self-worth, hating

---

[6] Some of the interviewed children were unable to answer these questions.

themselves, disliking their own appearance and self-blame). Table 14 shows the proportion of children ticking these items. By comparison with the teachers' perceptions of the children's poor social integration (see below), the children themselves did not report very high rates of loneliness, and perceived themselves to have enough friends and to have fun at school.

Table 14 **Child's Depression Inventory: the ten items most frequently checked (N=54)**

item	percent checking item
Worries about the future	76
Problems with making decisions	74
Anxieties about own appearance	59
Feelings of low self-worth	59
Worries about own health	53
Suicidal ideation	48
Crying a lot	48
Worried that bad things will happen	47
Hating self	45

CDI scores were unrelated either to age or sex, or any of the measures of the mothers/carer's relationship with the child, or to any of the measures of the severity of the child's abuse experiences. CDI scores were also unrelated to age when the latter was treated as a continuous variable, but dividing the children into those above and below the age of ten years showed that the older children had, on average, higher scores. This was not unexpected: there is a rise in depressive symptoms and CDI scores as children they move into adolescence.

Among those who filled in the CDI, scores were significantly lower when the mother/carer indicated at referral that she believed the child's abuse story. For the 52 children who filled in both questionnaires, low CDI scores were also significantly associated with reporting higher self-esteem (Table 15).

Among the seven boys old enough to complete the CDI, higher scores were associated with the mothers'/carer's report of persistent sexualized behaviour before disclosure ($\chi^2 = 6.99$, 1df, p <.03), but not with any of the abuse severity variables. The numbers here are too small to draw any firm conclusion from this isolated finding.

Children's Self-esteem Inventory

On the GOS Self-esteem inventory the children's scores showed a very considerable range from 37 to 128 out of a possible maximum score of 135 (mean 90.8, SD 20.4). Because there are no comparative data

Table 15 **Summary table on Children's Depression Inventory scores**

Low CDI scores associated with	Analysis of variance		
	F	df	significance level (p)
Being believed by mother/carer	4.60	(1,44)	< .04
High self-esteem	2.57	(26,25)	< .01
Being under 10 years	5.07	(1,53)	< .03

from non-abused children, it is not possible to say whether the abused children had higher or lower self-esteem compared with the general population. However, as noted above, there was a significant association between lower self-reported CDI scores and higher GOS Self-Esteem Questionnaire scores (Table 15), following a pattern found in innumerable studies comparing self-esteem and depressive symptoms. It is worth noting that using this scoring system some children had self-esteem scores nearly four times higher than others.

The children's higher total self-esteem score was significantly related to their own report of fewer symptoms but bore no relationship to the mothers'/carers' symptom reports (Table 16). However, the child's higher total Self-Esteem score was significantly related to the mother's/carer's report of greater enjoyment of the child's company at the time of referral, to lower levels of the mother's/carer's criticism of the child, and to the child's report of a better overall relationship with the mother/carer (Table 16).

For the children who completed the self-esteem questionnaire, scores were not related to the duration or severity of abuse. However, a higher score (indicating higher self-esteem) was related to feeling that their mother believed them at the time of the referral interview. The children's self-esteem was also related to their friendship patterns: higher self-esteem was associated with a higher score on the global estimate of good peer relationships at referral (Table 16).

There were no significant differences between boys and girls, nor between different age-groups in the total Self-Esteem scores, bearing in mind that they were all over 8 years old if they completed the questionnaire.

The mean scores in the seven domains included in the self-esteem questionnaire are shown in Table 17. Although there are no normative data for the self-esteem questionnaire, nor for the domains (with the exception of the Global domain derived from the Rosenberg Self-esteem Inventory), the relative mean scores are of interest. Given the frequency of disturbed relationships observed by clinicians in the families of sexually

Table 16 **Summary findings on high total self-esteem: Self-esteem Questionnaire**

High self-esteem associated with	(Analysis of variance)		
	F	df	Significance level (p)
Child's low self-report CDI score	2.57	(25,26)	< .01
Mother/carer's enjoyment of child's company	4.40	(3,31)	< .02
Mother's low criticism of child	3.20	(3,32)	< .04
Mother's good overall relationship with child	3.05	(3,32)	< .05
Mother's belief in abuse history	4.94	(2,42)	< .02
Good peer relationships	4.63	(2,31)	< .02

abused children, it is perhaps surprising to find that the highest means are to be found in the domains of the children's relationship with their mothers, their status within the family and their relationships with their fathers. There may be some confusion about whether the children were rating an ideal or the reality (see below). On the evidence of the adult-recall literature, it is less surprising to find that the mean score for 'body image' domain is the lowest, and the clinicians had predicted low self-esteem in this area.

The global (Rosenberg) self-esteem sub-score was significantly associated with the child reporting a supportive reaction to disclosure from the mother, and the child reporting that the caring parent believed them at the time of the referral interview. There was also a significant association between a high global self-esteem score and the mother's report of greater enjoyment of the child's company, and the child's report

Table 17 **Mean scores of seven domains of self-esteem**

Domain (maximum score)	Mean score	SD	Range
Global (max 30)	21.0	5.9	6–30
Relationship with mother (max 15)	12.1	3.4	1–15
Status in the family (max 15)	10.7	2.9	2–15
Relationship with father (max 15)	10.2	4.4	0–15
Satisfaction at school/college (max 15)	9.8	3.0	2-15
Peer relations (max 15)	9.8	2.7	3–15
Body image (max 15)	8.9	3.7	0–15

of a good overall relationship with the mother/carer before disclosure (Table 18). The direction of the association cannot, of course, be determined from the present data: that is to say, high self-esteem may derive from or contribute to these aspects of a better relationship between mother and child.

The validity of the sub-score on relationships with fathers receives support from the finding that the mean self-esteem score in this domain was significantly lower when the father (or father surrogate) was the perpetrator ($t=-3.84$, $p<.001$). Some support for the validity of the school sub-score can be found in the fact that the school domain mean score was significantly lower for children whose teachers judged them to have more behaviour problems ($t=2.95$, $p<.01$). It had been expected that mean scores in the 'mother' domain would be affected by whether the mother had believed the child, but this was found not to be significant. However, the mean scores on the 'mother' domain were significantly higher when the level of warmth shown by the mother to the child was higher ($t=-2.70$, $p <.03$). The results suggest that some children may be responding to the 'mother' domain with an ideal rather than a real picture of their relationship with their mother.

The finding of the effect on self-esteem in the 'fathers' domain of the father's role as perpetrator led to an investigation of the effect of father-figure as perpetrator on the total self-esteem score. Multi-variate analysis showed no effect on total self-esteem of the father being the perpetrator when either the overall relationship with the mother, or the mother's supportive reaction to the detection of the abuse were taken into account. The effect of the father as perpetrator appeared to be limited to the father domain sub-score.

Table 18 **Summary of findings on global self-esteem (Rosenberg Self-esteem inventory)**

High global self-esteem associated with	Analysis of variance		
	F	(df)	Significance level (p)
Mother enjoying child's company	3.16	(3,30)	< .04
Good overall relationship with mother	3.44	(3,31)	< .03
Mother's belief about abuse	3.24	(2,41)	< .05
Mother's supportive reaction to disclosure	4.51	(2,33)	< .02

Mothers'/carxers' reports of the children's behaviour and symptoms

Mothers (or current carers) in charge of the children were asked about 20 aspects of the children's behaviour and health in the month before referral. The summed scores ranged from 1 to 32 (out of a possible maximum of 60) with a mean of 13.3 (SD 8.3). Table 19 shows the frequencies for each item from the mothers'/carers' reports, ranked by the percentage of children showing marked or very marked presence of the symptom. Irritability, aggression, misery and fears were most frequently noted by mothers. Comparison with non-clinic children in good and bad marriages in the study from which the measure had been derived (Smith & Jenkins, 1991), showed that the current sample were seen by their mothers as significantly more symptomatic.

Analyses of variance showed that the mother's/carer's total symptom scores were not significantly associated with any of the ratings of a positive relationship between the mother/carer and the child (such as warmth), but a high symptom score was significantly related to the carer showing a high level of criticism of the child.

Among the mothers who completed the General Health Questionnaire there was a significant relationship between their report of symptomatic behaviour in the children and both a high GHQ score

Table 19 **Mother/carers' reports of the abused children's health and behaviour symptoms in the month before referral (N=48) (ranked by proportion scoring presence of symptom)**

Item	absent		frequent
		percent scoring	
	0	1	2/3
1. irritability	29	24	47
2. temper tantrums	53	16	30
3. misery	39	33	28
4. fears	33	41	26
5. physical aggression	61	14	24
5. disobedience	49	27	24
5. lying	60	16	24
5. attention seeking	56	20	24
9. sleep problems	69	10	21
9. bedwetting	79	0	21
11. eating problems	69	13	18
11. hyperactivity	71	10	18
11. concentration	50	31	18
14. headaches	56	27	17
15. anxiety	42	42	16
16. stomach aches	52	35	13
17. feeling/being sick	65	25	10
18. nightmares	81	9	10
19. stealing	83	14	2
20. soiling	96	4	0

(indicating their own disturbed mood), and a negative self-rating of their own health (Table 20). Unexpectedly, a *low* total score from the mother/carers' reports of the children's symptoms was significantly related to the child having experienced penetrative sexual abuse—both as the most frequent and as the 'worst' experience. The findings are summarised in Table 20.

There was a significant difference in the mean symptom scores of boys and girls, with mothers/carers reporting higher scores for boys (t=-2.48, p <.03). This parallels the finding from teachers (see below).

Sexualized behaviour of the abused child

In addition to the twenty health and behaviour items on which mothers/carers were asked to report, they were also asked about overt sexualized behaviour observed in the month before disclosure and the month before referral. Just over one quarter of the children (28%) were reported to have shown any sexualized behaviour in the month before disclosure, 16% doing so persistently; and only 22% were reported to be showing such behaviour at referral, 10% persistently. At referral there was no difference between the boys and girls in the frequency with which sexualized behaviour was reported.

Table 20 **Summary of associations with mother/carer's symptom score**

Reports of children's symptoms from mothers	Analysis of variance		
	F	(df)	Significance level (p)
High symptom score & high level of mother's criticism	3.58	(3,53)	<.02
High symptom score & mother's high symptom score on GHQ	6.52	(1,49)	<.02
High symptom score & mother's negative view of own health	8.10	(3,45)	<.01
Low symptom score & penetrative abuse as most frequent experience	10.15	(1,63)	<.002
Low symptom score & penetrative abuse as 'worst' experience	8.10	(1,65)	<.01

For the 45 children for whom information was available at both times, there was a highly significant relationship between the reports of sexualized behaviour at disclosure and referral: in other words the symptoms when present persisted for many months.

Using information from interviews and clinical notes for the group as a whole, sexualized behaviour reported at referral was not related to penetrative abuse having been either the worst or the most frequent abuse experience. This range of behaviour was also not related to duration or age at onset. There thus appears to be no association with the variables most commonly used to define severity of abuse experience. On the other hand, sexualized behaviour at referral was significantly and inversely related to age at referral; all the children in whom sexualized behaviour was either 'persistent' or 'needed correcting' were under the age of 12 years (F=3.19 (3,92) p <.03).

A different pattern was observed among the 49 girls (as opposed to the whole group) for whom relevant information was available. There was a significant association between persistent sexualized behaviour reported at disclosure and the experience of abuse by more than one perpetrator (χ^2=19.66, 6df, p <.004). For the 31 girls for whom onset age was established there were, in addition, significant associations between persistent sexualized behaviour reported at referral and an early age at onset (χ^2=18.29, 9df, p<.04), and a longer duration of abuse (χ^2=13.71, 6df, p <.04).

There was no significant relationship between the mothers'/carers' reports of the children's health and behaviour items and their reports of sexualized behaviour, either at the time of disclosure or the time of referral. There was also no association between the report of sexualized behaviour and the child's report about whether or not the mother/carer believed, at the time of the referral interview, that the abuse had taken place. The findings on sexualized behaviour are summarised in Table 21.

Child's report of own symptoms

The children were asked to report on 12 of the 20 items of health and behaviour which made up the parents' symptom report; a maximum score of 36 could be obtained from the children's symptom checklist. Among the 52 children who gave complete answers to this section of the interview, scores ranged from 0 to 24, with a mean of 10.75 (SD 5.9). The frequencies of individual items are given in Table 22.

For these 52 children a higher self-reported symptom score was significantly related to a higher depression score (F=2.52, (27,24), p <.02); and to low self-esteem (F=2.75, (36,14), p<.03). The child's self-reported symptom score was not, however, related to the length or

Table 21 **Summary findings relating to sexualized behaviour at referral**

	Analysis of variance		
	F	df	Significance levels (p)
Sexualized behaviour & younger age at referral	3.19	(3,92)	<.03
FOR GIRLS ONLY	χ^2	df	Significance levels (p)
More than one abuser	19.66	6	<.004
Early age at onset	18.29	9	<.04
Longer duration	13.71	6	<.04

severity of the abuse or to global rating of the child's friendships. The self-reported symptom score was also not related to any of the variables reflecting the mother/carer's attitudes towards them (such as warmth, criticism or positive recognition), but reporting more or more severe symptoms was related to getting on badly with their caring parent before disclosure (F=3.59, (3,44), p <.03), and at the time of referral (F=4.86, (1,34), p<.04).

Table 22 **Frequencies of children's responses on twelve items of health and behaviour present in the month before referral (ranked by presence of symptom)**

Item	absent		frequent
	\[percent scoring]		
	0	1	2
1. depression/misery	37	19	44
2. stomach aches	40	19	42
3. headaches	25	35	40
4. sleep problems	52	8	40
5. restlessness	58	9	40
6. fears/phobias	31	44	25
7. anxiety	39	41	20
8. feeling sick	67	14	19
9. concentration problems	58	27	16
9. nightmares	69	15	16
11. lying	61	25	14
12. eating problems	71	19	10

For girls, a lower self-reported symptom score was also related to her report of the mother/carer having reacted supportively to the disclosure of abuse (F=3.76, (2,30), p <.04); this association was not found among the small number of interviewed boys.

Teachers' reports of behaviour in school

The teachers' reports of the abused children's behaviour in school were obtained from the Rutter 'B' questionnaires. Not all schools or teachers were able to respond, and only 54 questionnaires were completed. Out of a possible maximum score of 60, the mean score for the abused children was 10.1 (range 0-29, SD: 7.0); 54% scored over the 8/9 cut-off point used to indicate significant behaviour difficulties.

The teachers also provided completed behaviour questionnaires for two same-sex classroom controls for all except two of the abused children and teenagers. Compared with the abused children the controls (called A Controls and B Controls) had mean scores of 4.0 and 4.1, respectively (range control A: 0-16 (SD: 4.3); range control B: 0-20 (SD:4.5)). There were no significant differences on the total scores or item scores of A Controls compared with B Controls.

Table 23 gives the frequencies of the items on the teachers' questionnaire for the abused children and one of the two groups of classroom controls (B). On a number of items teachers were significantly more likely to say it 'applied somewhat' or 'certainly' to the sexually abused children than to the controls.

Comparisons of the scores of abused index children and their classroom controls show that for some individual items on the teachers' rating scale there were no significant differences between the three groups (see Table 23). However, on some items the abused children were significantly differentiated from the controls, and emerged as a distressed and difficult group. The most significant differences showed that the abused children were more fidgety, more likely not to be liked, and to be solitary, unhappy and worried; but they were also significantly more likely to be restless, disobedient, fearful, and to have tics and aches & pains, to wet or soil themselves and be involved in bullying.

Teachers' reported behaviour problems significantly more frequently for the boys than for the girls; the mean score for the 10 boys for whom reports were available was 14.20, compared with a mean score for the 44 girls of 9.11 ($t=-2.38$, $p < .04$).

The teachers' ratings of behaviour were not significantly related to either the mothers'/carers' reports or the children's reports of health and behaviour in the month before referral. A comparison was made of the items on the mothers' reports of behaviour and ten very similar items on the teachers' reports: these were restlessness/hyperactivity, fighting/physical aggression, worrying/anxiety, irritability/irritable mood, disobedience, lack of concentration, fearfulness, lying, stealing, and unhappiness/misery. Teachers and mothers were in agreement in rating markedly disturbed behaviour on lying, disobedience, fearfulness, irritability and unhappiness (McNemar's test: binomial 2-tailed p <.02, p <.001, p <.001, p <.001 and p <.001 respectively).

Among the 54 children who were asked about social problems in school, 29 (54%) described no problems, 18 (33%) described minor problems and 7 (13%) had major problems. Asked about problems with their school work, 33 (61%) said they had no problems, 16 (30%) had minor problems and 5 (9%) had major problems. Only half the children (26) described friendships which were sufficiently close to include confiding, though 12 children (22%) said they confided in teachers. Since this material was derived from interviews, there are no comparable data on the classroom controls.

Table 23 **Frequencies of scores on the Rutter 'B' teachers' ratings of behaviour in school—abused child and same-sex classroom control (group B).**

Item	Abused child		Control B		Significance differances
	percent scoring				
	0	1/2	0	1/2	
1. Restless	64	36	85	15	*
2. Truants	94	6	96	4	
3. Fidgety	64	36	90	10	**
4. Destructive	89	11	96	4	
5. Fights	70	30	85	15	
6. Not liked	59	41	88	12	**
7. Worries	28	72	61	39	**
8. Solitary	43	57	77	23	***
9. Irritable	61	39	82	18	
10. Unhappy	43	57	81	19	***
11. Tics	80	20	98	2	*
12. Sucks thumb	89	11	96	4	
13. Bites nails	67	33	81	19	
14. Absent/trivial	85	15	84	16	
15. Disobedient	56	44	80	20	*
16. Low Concentr'n	37	63	56	44	
17. Fearful	45	55	67	33	*
18. Fussy	82	18	75	25	
19. Lies	65	35	85	15	
20. Steals	82	18	90	10	
21. Wetting/soiling	85	15	98	2	*
22. Aches/pains	71	29	92	8	*
23. Refusal	84	16	96	4	
24. Stutter	96	4	98	2	
25. Speech difficulty	91	9	84	16	
26. Bullies	69	31	87	13	*

Significance p<.05*; p<.01**; p<.001*** McNemar's test

Protective factors

In the literature on child sexual abuse, references may be found to protective factors which it is claimed act to ameliorate the adverse effects of certain abuse variables. Although the present study had not been designed to test the effect of protective factors on the symptoms shown by the abused children at referral, it was felt that an attempt should be made to distinguish protective factors, so far as they could be derived from the data.

Three putative protective factors were chosen. These were (a) the supportive reaction of the mothers/carers or non-abusing parents, (b) the mother's/carer's belief that abuse had occurred, and (c) measures of positive aspects of the relationship between the child and mother/carer. Analyses of variance were undertaken to determine whether these variables were acting to reduce the child's symptom scores in the presence of certain abuse variables such as severity (penetrative/non-penetrative), whether the perpetrator was the father(-figure) or not, or duration which had been selected from the literature. There were no clear effects, and it has to be concluded that despite the overall importance of the three putative protective variables in relation to some other abuse factors, they did not act in a protective manner in the present study. Clearly this aspect of the role of protective factors on the impact of abuse requires specific study.

Characteristics of the primary carers

Mother's marital status

Among the 61 mothers (and surrogate mothers) who had been caring for the child during the abuse period and who had been interviewed, 27 (47%) had been married once, 21 (37%) twice, and eight (14%) more than twice, one mother had never married, and for four mothers there was no reliable information. Their ages at first marriage (or co-habitation) ranged from 15 years to 30 years; and these marriages (co-habitations) had lasted between 3 months and 25 years. In those cases in which the first marriage had ended, 26 (46%) of the women took the decision, in three case (5%) the partner had taken the decision and in four cases it was mutually agreed; (in the rest of the cases the reason was not known). Various reasons were given to explain the failure of the first marriage. The most usual reason (18 cases) was the violence of the woman's partner, followed by his infidelity, the woman feeling neglected, sex problems and the children's unmet needs, money problems, the informant's infidelity and the sexual abuse of the children.

At the time of the referral, only 30 of the 61 interviewed mothers were married or cohabiting.

Mother/carers' childhood experiences

The fifty-five biological mothers who were interviewed were asked to provide some information on their own childhoods. Only 23 (42%) of them had spent all the first 18 years of their life with both biological parents; the rest had spent varying amounts of time with both parents, 6 (11%) had never lived with them and a further 6 (11%) spent less than five years with them. Only 10 (18%) had spent any time at all in reconstituted families; 10 (18%) had spent varying amounts of time (from 1 to 15 years) in institutional care. Nearly a quarter of the biological mothers (12—22%) described their upbringing as providing only poor care, and for 10 (18%) this also included having no happy memories of their childhood or 'teens. Even among the majority (42—77%) who felt they had had adequate or good care, 16 (30%) had no happy memories. This means that more than half (28—51%) of the women had no happy memories of their own childhoods.

It might have been expected that poor care or unhappy memories from their own childhood would affect the quality of their relationships with their own children. No ratings were made of the mothers' relationships with children other than the abused child, but there were no effects of childhood care or unhappy memories on their current relationships with the abused children.

Sexual abuse had been experienced by a large number of these mothers in their own childhood and 'teens. Nearly half (24—43%) of the women reported some sexually abusive experience at some stage before they were 18 years old; for 5% this was limited to exhibitionism (from strangers), for 16% the abuser was a father-figure, for 7% other family members, and for 11% family friends or well-known neighbours. Two women had experienced rape by a stranger. Some of the women had told no-one since the abuse, and it only came to light in the context of the research interview about their child's experiences. Of the 22 who could remember what happened in detail, only one remembered telling someone and being well supported, the rest either told no-one (15), disclosed and were not believed (1), or disclosed and received no support (5). Fourteen of the 24 women (58%) believed the abuse had had major adverse effect on their lives subsequently. These reports of adverse effects were taken at face value and not subjected to any further rating by the interviewers. In addition to the sexual abuse, 10 (19%) of the women who were interviewed described some form of physical abuse from family members as well.

There was no relationship between the mothers' belief in their own child's disclosure of abuse and either the early experiences of childhood abuse or neglect.

Mothers'/carers' psychological state

Sixty-four mother and surrogate mothers completed the 28-item General Health Questionnaire (GHQ) at the time of referral, reporting their feelings in the month before referral; 44 (69%) scored over the cut-off point indicating that they were at risk of affective disorders. The range of scores was from 0—25, with a mean of 9.8 (SD. 7.6). The distribution of scores on the 28 individual items is given in Table 24.

Table 6.10 **Mother/carers' scores on individual items from the 28-item General Health Questionnaire**

item	percent scoring	
	Nil/mild	Moderate/severe
Somatic symptoms		
1. Not feeling well	60	40
2. In need of a tonic	59	41
3. Feeling 'out of sorts'	43	57
4. Feeling ill	81	19
5. Pains in the head	65	35
6. Pressure in head	65	35
7. Hot and cold spells	84	16
Anxiety		
8. Lost sleep through worry	34	66
9. Difficulty staying asleep	42	58
10. Felt under strain	39	61
11. Felt bad-tempered/edgy	48	52
12. Felt scared/panicky	75	25
13. Things getting on top of self	48	52
14. Felt nervous	47	53
Social dysfunction		
15. Not kept self busy/occupied*	80	20
16. Taken longer to do things	71	29
17. Felt not doing things well*	74	26
18. Not satisfied with self*	75	25
19. Not playing a useful part*	68	32
20. Not making decisions well*	68	32
21. Not enjoying normal activities*	47	53
Depression		
22. Thinking self worthless	72	28
23. Finding life is hopeless	73	27
24. Life not worth living	83	17
25. Thought of suicide	88	12
26. 'Nerves' very bad	79	21
27. Wishing was dead	82	18
28. Persistent suicidal thoughts	88	12

* shortened versions of the questions have been rephrased to indicate that score of '0' conveys well-being.

Four sub-scales can be derived from the 28-item GHQ: these cover the symptoms of social dysfunction, anxiety, and depression, and somatic symptoms. Out a possible maximum score of 21 on the four sub-scales the mean scores were social dysfunction—8.50 (range 1-21, SD:4.8); anxiety—10.76 (range 0-21, SD:5.7); depression—4.28 (range 0-19, SD:5.1); and somatic symptoms—7.88 (range 1-21, SD:4.8).

Mothers'/carers' self-esteem

The Great Ormond Street/ICH adult self-esteem questionnaire was filled in by sixty-five mother/carers. A range of scores from 47—134 was recorded (mean 94.66—SD. 19.4): a high score indicated high self-esteem and the maximum possible score was 150. There are no established norms for this questionnaire, so it is not possible to comment on the range or mean, except to point out that the range is wide, with some women's scores nearly three times as high as others'.

The adult self-esteem questionnaires covers ten 'domains' of self-esteem (each maximum score—12) and one global self-esteem sub-score (maximum score—18). The mean scores are given in Table 25, with the ten 'domains' in rank order. There is no clear pattern to the rank order of the mother/carers' responses, but it may be important that relationships with their own peer group, the ability to confide and elicit confiding and their own body image are low down the order. It is of particular interest that the body image domain has the lowest mean scores for both the abused children and their mothers/carers. There was no association for the mothers, of low body-image sub-scores and having been abused themselves when young.

Table 25 **Domains of self-esteem for the mother/carers at the time of referral.**

Domain (Maximum score)	Mean score
Global (max. 18)	10.37
As caring person (max 12)	9.95
Providing for others (max 12)	8.91
Being a 'good' person (max.12)	8.57
Work (max. 12)	8.36
As housewife (max. 12)	8.20
Being fun to be with (max. 12)	7.76
Relationship with own peers (max. 12)	7.69
Having confiding skills (manx. 12)	7.11
Being 'clever' (max. 12)	6.65
Body image (max. 12)	6.37

In the families in which both mother and child had completed self-esteem questionnaires there was a significant association between their own and their child's total self-esteem scores (Table 26). There was also a significant association between the mothers'/carers' total self-esteem scores and their belief in the child's abuse story but with 'unbelieving' mothers showing the highest score, and ambivalent, uncertain mothers showing the lowest (Table 26).

Table 26 **Summary of findings relating to mother/carer's self-esteem**

Mother's/carer's high self-esteem associated with:	**F**	**Analysis of variance** df	**Significance level (p)**
Child's high self-esteem score	4.68	(9,33)	<.01)
High positive interaction with child	2.91	(3,40)	<.05
Not believing the child's abuse has happened	3.69	(2,29)	<.04
Having good overall relationship with child	8.52	(1,23)	<.01

Chapter 7

Discussion of the findings of the descriptive study

Introduction

In any consideration of the findings in this study it is important to re-emphasise that the Department of Psychological Medicine in the Hospital for Sick Children acts as a tertiary referral centre for a wide region covering Greater London, and some parts of the Home Counties. The Child Sexual Abuse team also receive some referrals from more distant locations. The level of experience in the CSA team has led to cases being referred not only for treatment, but also for assessment or advice on management. In 12% of referred cases, sexual abuse was not substantiated by the CSA team. This is a much lower figure than the 60% reported from the United States (American Association for Protecting Children, 1986) suggesting that other agencies had already filtered out most of the unconfirmed cases.

The sample drawn from these referrals may therefore have some unusual characteristics; for example, the family circumstances may be unusually complex, the abuse particularly severe, or the abused child may be presenting a particularly wide range of severe symptoms. However, while this point needs to be borne in mind, at this stage in our knowledge of sexual abuse cases in the UK it is difficult to establish with any exactitude the extent of the differences between the current cases and other abused children. There is some evidence, as we shall see from the points made below, for believing that they may not be unrepresentative of detected cases elsewhere in the UK.

The hospital CSA team follow a policy of not treating cases in which parents who still have some responsibility for the child deny completely that the abuse took place. They also insist that the abused children are allocated to a social worker, which not all local authorities have been able to achieve in recent years. This latter 'rule' ensures the collaboration between hospital and local welfare services which are essential for effective treatment, but it also means that there is some prior selection of well-supported cases.

It does not follow, however, that all the children were (back) with their own families: indeed, many were very isolated. Only 45 (45%) were living in the same household at referral as they had been at disclosure, and 54 (54%) had been removed from their families before referral, thus placing them in the 'double jeopardy' of having been abused and excluded, and for many, rejected as well.

While the refusal rate from families (and their social workers) in the present study was low, the failure rate on individual interviews within the families was high. In a not dissimilar study of children referred to a psychiatric clinic, Conte & Schuerman (1987) reported that social workers frequently ruled out research involvement with the families of sexually abused children, and, largely because of this, their success-rate was only 28%. The current research had not been designed to include interviews with all the parents who had had responsibility for the children but who denied the abuse and no longer had the child living with them; among other issues this would have added to the expense by involving many home visits. It is likely that there would still have been a high failure rate from these parents even if they had been approached for interviews. This point should be borne in mind in designing future research.

Gender ratio

The sex ratio in the present study (3.7:1—girls to boys) is almost identical to that noted in other recent UK and US clinic studies and data on officially reported cases (American Association for Protecting Children, 1986; Bentovim et al 1988; Gomes-Schwartz et al, 1990; The Research Team, 1990), but Hobbs (1990) has reported a higher proportion (one third) of boys in cases referred to a UK hospital paediatric clinic.

Age, onset and the duration of abuse

The children in the present descriptive sample appear to be younger at referral than those described in other studies; for example, 45% of the present sample was under 9 years old, compared with 33% of the sample reported by the NSPCC for 1983/7 (Creighton & Noyes, 1989). This is all the more striking because the role of the Hospital for Sick Children as a tertiary referral centre would lead one to expect that the children would be older than they would be in any sample from primary assessment centres.

Other studies have noted a higher ratio of boys at younger ages, and the absence of a significant difference in the present study requires some explanation. Earlier studies have suggested that, compared with girls, boys are abused more frequently by people outside the family and household, and their abuse is likely to be more quickly detected and reported (eg, Finkelhor, 1984) leading to boys being younger than girls at referral. This difference in detection is well illustrated by the results of the Northern Ireland incidence study, in which the gap between onset and reporting the abuse was three years for the girls and one year for the boys (The Research Team, 1990). When the abuse of boys is within the family/ household their ages conform more closely to the girls' (Finkelhor, 1984, quoting American Humane Association data). Since the population in the present study was drawn from those whose abuser was within the family or household, it was expected that the boys would have shown an age distribution not unlike the girls, and this has been confirmed. The clear bimodal distribution noted in some other studies (DeJong, Hervada & Emmett, 1983; Eckenrode, Munsch, Powers and Doris, 1988; The Research Team, 1990) was not apparent for the present sample, but this may have been an artifact of excluding cases under 4 years of age.

Mrazek et al (1983) found that doctors in the UK reported that only 15% of the sexually abused children they saw were under 5 years old, but that situation is changing rapidly, with a considerable increase in the detection of sexual abuse in younger boys. Hobbs (1990) has noted that the recognition of sexual abuse in younger children is a relatively recent phenomenon; in his own report of cases referred to a hospital paediatric service the average age in 909 cases was about 6 years 6 months.

It was not always possible to establish when the abuse had started: some of the children were too young, and/or had an imperfect grasp of the concept of time. Non-abusing parents were asked to give this information, but frequently said they felt they were unlikely to give reliable information, even with hindsight. Perpetrators were also vague about time, for reasons which can only be guessed at: some perpetrators who had admitted the abuse may have tried to diminish the seriousness of their offence by 'forgetting' how young the child was when it began. Analysing a number of research reports Finkelhor (1979) found that the average age at which children were being abused was between 9.3 and 10.6 years. Fromuth (1983) reported the pre- adolescent period as the age at which adult women recalled peak vulnerability to childhood abuse.

It might be thought that an accurate date of the end of abuse would be easier to obtain than for the onset of abuse; however, for 19 children even the date of cessation was not available. As we have noted, gaining accurate information about the start and end of sexual abuse affects the information about duration, but the present study indicates that abuse continued for considerable periods for a large proportion of the children, (76% for more than 12 months and 36% for more than three years), and frequently involved a large number of incidents. In the present study the duration was considerably longer than that reported in the previous study of Great Ormond Street cases, in which 44% of children for whom information was available were abused for less than 12 months (Bentovim et al, 1988), or the Tufts New England Medical Centre study in which 53% of the children experienced abuse for less than 12 months (Gomes-

Schwartz et al, 1990). In both these surveys, however, cases were included which involved non-family members and strangers, and there is now considerable evidence that children disclose such abuse far more quickly (and thus the abuse has a shorter duration) (Gomes-Schwartz et al, 1990). While the relationship with the abuser is not the only influence on the speed with which a child discloses, it does appear as an important one. The association between believing the child's report of abuse and the relationship of child and abuser deserves further investigation.

It is important to note that, for those children for whom information was available on duration, longer duration was associated both with a greater likelihood of having experienced penetrative abuse and with older age at referral. The combination of these factors may be responsible for the small but clinically significant group of more disturbed young adolescents found in the present study. A disproportionate number of these adolescents (compared with younger children) ended up in Children's Homes. An impression was gained that these difficult and challenging youngsters were often placed in less than optimal conditions. It is suggested that further research is needed to determine the best living arrangements and therapeutic settings for those whose abuse is only detected in their young or mid-teens, and whose families are totally rejecting.

The difficulty of obtaining accurate information about the date of the cessation of abuse which was encountered in this study, has also been noted in other studies (Resick, Calhoun, Atkeson & Ellis, 1981; The Research Team, 1990). Apart from any difficulties abused children may have in recalling such information, the professionals with whom the child first has contact may have little interest in establishing exact dates, unless a legal case against the perpetrator hangs on it. But building an accurate picture of the aetiology of sexual abuse may be affected by this difference of emphasis between researchers and clinical or legal workers. Devising better ways of obtaining this material accurately from abused children should be regarded as an important area of clinical development.

The abuse and the abusers

In the present study the range of relationships of the abuser and the abused child was constrained by the sampling to exclude abuse by distant friends, casual baby-sitters or other child-carers, and strangers. Bentovim & Boston (1988), in a previous study of all referrals to the CSA team at the Hospital for Sick Children, noted that 75% of the perpetrators were household members, and that the majority were in the age-group expected of parents. A comparison with the present sample is complicated, but it appears that the proportion of abusers who are brothers or 'other relatives' may have increased since the first survey (from 9% to an estimated 23%). Hobbs (1990) has reported that, in a sample of 900 cases over 4 years, two-thirds of the abusers are relatives, 95% are known to the children, and 25% are teenagers. In the Tufts New England Medical Centre study 29% of the abusers were aged 18 years or under (Gomes-Schwartz et al, 1990). In an earlier survey of New England college students recalling their childhood experiences, Finkelhor (1979) found that the largest group of abusers had been older male teenagers. There is thus substantial evidence that teenagers are an extremely important group of child abusers; a fact which has considerable implications for preventive work, for treatment, and for criminal sentencing practices.

As with other aspects of the child abuse literature a variety of definitions of 'family', 'household', 'family friends' and 'strangers' have been used, which can hamper comparison. In general, however, it appears that the very large majority of abused children and teenagers know their abuser. For girls, biological fathers are the largest single group of relatives, but, for both girls and boys, abusers are drawn from most parts of the family and from many groups of adults with responsibility for the child (The Research Team, 1990—Table 48).

In addition to the considerable agreement, across time and between countries, about the age-profile of the abusers, there is also widespread agreement about abusers being predominantly male. The findings of the present study replicate many previous studies, both prospective and retrospective (summarised in Finkelhor, 1979). In the present study the women who were involved in the sexual abuse were, with one exception, always associated with a male abuser.

Very little work has been done which looks specifically at the female abuser. Faller (1987) has reported on 30 clinic cases: as with the 6 cases in the present study, the majority (76%) were involved in the abuse with a male partner. Overall, the abusive women represented 10% of Faller's clinic sample, compared with 6% in the present sample. It was not possible in the present study systematically to test Faller's observation that there was a high incidence of mental retardation, psychiatric illness and substance abuse among female abusers, but the researchers' and clinicians' impressions would support her findings (Faller, 1987).

There are references in the literature to the exceptionally severe impact of the abuse if the perpetrator is the father/father-figure (Finkelhor, 1979; Groth, 1978; Russell, 1984), although much of this information comes from adult recall of childhood abuse. The present study did not find higher symptom scores among the children abused by fathers, but there was a significant association with low esteem in the father's 'domain' when the perpetrator had been the father. We have not found references to such a phenomenon in other studies. By contrast the children's global self-esteem (Rosenberg's inventory), the total self-esteem scores and self-reported depression scores were not related to who the perpetrator had been.

In the present study there were no associations between symptoms at referral and the measures of severity (longer duration, penetrative abuse, father-child incest, the use of violence) which had been observed in other studies (eg, Johnson & Kenkel, 1991). It is possible that the absence of significant associations between symptoms and features of severity in the present study arises from the choice of symptom checklists which did not record the full range of possible behaviour in sexually abused children. In particular, information on the symptoms of dissociation and post-traumatic stress were not obtained. Future treatment outcome or descriptive studies of sexually abused children and adolescents should employ measures which tap these symptoms.

Evidence of physical abuse

The rate of excessive physical punishment (24%) reported in the present sample is higher than that found by Mrazek et al (1981). Reporting on children referred for sexual abuse to a hospital clinic, these authors found 15% showed evidence of physical abuse in the recent past. An important paper by Hobbs & Wynne (1990) confirmed the high rate of co-existence of physical and sexual abuse: in a hospital sample of over 1500 cases, one in six of 769 physically abused children (16.9%) and one in seven of 949 sexually abused children (13.6%) had suffered both forms of abuse. The higher rate given in the present study almost certainly arises from the fact that it was based on the children's reports, and not on a physical examination. Hobbs & Wynne make the point that they did not make exhaustive enquiries about previous (recent) physical abuse, and therefore only recorded the abuse for which the physical evidence was still apparent at the time of their physical examination of the child (Hobbs & Wynne, 1990).

The children's feelings about the abuse and the circumstances of disclosure

Nearly half the children disclosed the abuse, and in only a third of cases was the abuse 'discovered' by other people. This suggests that increasing the knowledge and understanding of key professionals with whom the child is in contact (eg, teachers and GPs) may lead to earlier detection. Berliner & Conte (1995) report very similar figures to those in the current study: children told parents in 43% of cases, parents discovered the abuse in about 25% of cases, and in the rest of the families the parents heard about it from others.

Mothers'/carers' reactions and support at disclosure

The reaction of the non-abusing mother or parent(s) has frequently been cited as exerting a major influence on the child's reactions, with many studies finding that a supportive parent ameliorates the effects of the abuse (Adams-Tucker, 1982; Rogers & Terry, 1984; Conte & Schuerman, 1987). The importance for the abused child of being believed by the mother/carer has been noted by other studies (Burgess, Holmstrom & McCausland, 1978; MacFarlane, 1978; Gomes-Schwartz et al, 1990; Johnson and Kenkel, 1991). In a study of father—daughter incest teenage girls with non-believing mothers were reported as exhibiting more problems (Sirles & Smith, 1990). The Tufts study found that while a negative reaction from non-abusing parents increased the effects of the trauma, a positive reaction did not ameliorate the effects: suggesting that low parental support was a risk factor (Gomes-Schwartz et al, 1990). Everill & Waller (1995) suggested that perceived negative responses to the disclosure is associated, in women, with later psychopathology, particularly eating problems and low self-esteem.

The present study showed that being believed, which is the first step towards getting external support, was influenced by the characteristics of the child and her experience, as well as characteristics of the mother or the mother-child relationship. Being believed was significantly associated with being younger and being *less* likely to have experienced penetrative abuse, a point which must have implications for therapeutic work (Hyde, Bentovim & Monck, 1995). Among the children, being believed, was positively associated with higher self-esteem and lower depression scores.

In some families there appears to exist a cluster of positive responses such as support and belief which were related to fewer symptoms in the child, and higher self-esteem in mother and child. If mothers believed the child when the abuse was first revealed this belief persisted and appeared to be associated with positive mother-child relationships. Few mothers or carers came to believe the child if they had disbelieved earlier, and there was a strong sense of regret among those children who had not been believed when they disclosed.

It is, perhaps, appropriate to stress that the non-offending parent is not the only person who has to believe the child. The present study was, necessarily, only concerned with children whose abuse was believed or suspected by agencies sufficiently influential for the case to be referred to Great Ormond Street. We are, of course, unable to judge how many children had disclosed to parents and others (teachers, friends, etc) and not been believed. These case fall, as it were, at the first hurdle. The difficulty of ascertaining the 'truth' about sexual abuse has been vividly illustrated in recent years in the UK.

The children's symptoms

Many research reports fail to distinguish exactly the point at which information about symptoms has been collected, with the consequence that studies often imply that the symptoms observed and reported after abuse are caused solely by the abuse. There are two reasons why this is a doubtful assumption. First, the symptoms shown at the time of detection or referral to an assessment or treatment centre reflect the child's combined experiences of the abuse and the detection and its immediate consequences. And while some symptoms may diminish in severity because the abuse has ended (eg physical trauma, or sexually transmitted diseases), others may remain or even increase as a result of the events following detection (eg depression, anxiety or sleeplessness) (Goldstein, Freud & Solnit, 1979; Summit & Kryso, 1978). Second, the child's symptoms after detection will also reflect other (and perhaps more persistent) aspects of his/her environment and temperament. These will include features which have acted and continue to act to exacerbate or diminish the effects of the abuse and the events following detection (for a discussion of the problems of separating the effects of early and later events, see Rutter, 1983). Thus it is not claimed in the present study that the symptoms observed in the children were necessarily caused only by the abuse, or by the detection process or both. The design of the study did not enable us to determine which characteristics of the child's behaviour or health preceded the abuse, and which had their onset after the abuse began. It is anyway unlikely that symptoms have a simple 'on-off' relationship with sexual abuse. The abuse experiences frequently vary over time in frequency or severity. For some children behaviour may change more noticeably as abuse experiences change, rather than between the pre-abuse period and the abuse period.

The range of symptoms reported in the present study by teachers, mothers/carers and the children themselves suggested levels of pathology and disturbance well above what would be found in 'normal' populations of the same age. For example, the teachers' scores showed 53% of the children with significant difficulties in school, a figure identical to that found by Rutter (1967) in a group of children referred to a psychiatric clinic. This figure was considerably higher than the figure of 25% given for 'school difficulties' in the Northern Ireland study (The Research Team, 1990), and far higher than the 10% reported by Rutter (1967) from a non-clinic population. Other studies have noted higher levels of symptoms than 'normal' control groups (Tong, Oates & McDowell, 1987), or have found higher proportions of behaviourally deviant children when standardized behaviour scales are employed (eg Freidrich, Urquiza & Beilke, 1986). However, although the literature on sexually abused children are found to have significantly more emotional and behavioural disturbance than non-abused children, this is not a universal finding. By contrast, Mannarino & Cohen (1986) studying 24 sexually abused girls referred to a child psychiatrist, found no evidence of rates of anxiety or depression or lower self-esteem when scores were compared to standardized norms for self-report questionnaires (including the Children's Depression Inventory). This highlights one of the numerous methodological difficulties associated with research in this field: relatively few studies provide large enough samples for researchers to be in a position to make firm conclusions; thus the issue of increased symptomatology cannot be regarded as proven one way or the other.

A comparison of the mothers'/carers' ratings and the children's self-ratings of the children's health and behaviour with scores from a community population of primary school-aged children showed that the abused children had higher scores on a number of items (cf, Smith & Jenkins, 1991). This pattern is familiar from other studies. For example, over half the detected cases in the Northern Ireland incidence study were reported by parents and teachers as showing significant emotional disorder at the time of detection (rated on a 7-point scale) (The Research Team, 1990). In nearly a dozen studies reviewed by Berliner (1991) parents reported sexually abused children as more distressed and symptomatic than non-abused children. Salzinger, Kaplan, Pelcovitz, Samit & Krieger (1984) found that both teachers and non-abusive parents reported maltreated children showing significantly more problems than children from a non-psychiatric hospital out-patient clinic. However, it has sometimes been noted that sexually abused children and adolescents do not exhibit as much psychiatric and behavioural disturbance as children referred to psychiatric clinics (Gomes-Schwartz et al, 1990).

More importantly, it is clear that the symptoms shown by many sexually abused children are non-specific. Beitchman, Zucker, Hood, da Costa & Akman (1991) have pointed out that, with the exception of sexualized behaviour, symptoms shown in the short-term by many sexually abused children are typical of those referred to child psychiatric clinics. Adolescents differ more clearly in showing significantly raised rates of depression, suicidal ideation and suicidal actions than non-abused and non-clinic adolescents. This is important for those with responsibility for protective and preventive services for children. There is no easy way of detecting that a child has been abused from the behaviour s/he exhibits. When help is provided on the basis of the child's disturbed behaviour, the causes may remain obscure. It is seldom that the cause will itself be the trigger to professional intervention.

In the current study, the absence of agreement on the children's symptom level between teachers and mother/carers may have been a product of using two different scales, but the point has also been made that parents in dysfunctional families may not be reliable witnesses of their children's behaviour (although not all the families could have been described as 'dysfunctional'). Salzinger and her colleagues found that teachers differentiated the behaviour of referred and non-referred children within the maltreating families more accurately than the parents did (Salzinger et al, 1984). The clinical importance of teachers' reports of adjustment and symptomatic behaviour is considerable if only to indicate how the child is functioning in a setting outside the family.

Most studies using parents' and teachers' reports of symptoms have found these differentiate successfully between sexually abused and non-abused children, while studies using children's reports of their own symptoms have found these do not differentiate. However, in the present study as in others, many children were reported either by parents, teachers or themselves as having few symptoms or none. It is claimed that the absence of symptoms may be due to dissociative thought processes, and that as the child 'gets in touch with' the abuse, symptoms will increase. This is no doubt the case for some children; it is possible, however, that some children may have or may acquire coping skills which are, in themselves, generalisable to other stressful events and other stages of life. For example, the ability to choose and use good confidants is a skill which can carry someone through their whole life.

The association between severity of abuse experience and negative circumstances in the child's past and a raised level of symptoms at the time of referral found in the present study has been noted in several previous studies of clinic populations (MacVicar, 1979; Finkelhor, 1987; Wozencraft, Wagner & Pellegrini, 1991), although the evidence of a consistent pattern is lacking (eg, Rimsza, Berg and Locke, 1988). As Hartman & Burgess (1989) point out, the impact of individual variables such as the nature of the abuse masks the interaction with other variables such as duration or relationship with abuser. These interactions, for example, may affect the findings of some studies (Finkelhor, 1979; Fromuth, 1983; Russell, 1984) that the symptoms of child victims are greater when the abuser is male and older: two variables which are inevitably associated with being a father(figure).

The present study has found a group of adolescents who were older at referral, and who presented a particular challenge to the therapist: for these young people the duration had been longer, the abuse was more likely to have been penetrative (severe), the mother was less likely to believe the child, and absence from school had become an established pattern.

The mothers' negative outlook

The discovery, in the present study, that the mother's/carer's estimate of a worse score on child's symptoms was significantly associated with her report of her own worse overall psychiatric health (on the General Health Questionnaire) finds echoes in other studies. Conte and Schuerman (1987) found a significant association between the parent's 'negative outlook' and a higher (worse) score on the checklist of the child's symptoms, although they also found that, compared with parents of non-abused children, the parents of sexually abused children were more likely to have a negative outlook in general. As they point out, this 'negative outlook' for the child may also have formed part of the less favourable social environment, which may leave the child at risk of sexual abuse by others. Together with the child's having previously experienced stressful life events, large family size and the parents' low educational attainment, parents' negative outlook explained 22% of the variance between the abused and non-abused sample. These authors emphasise the critical importance of support networks for the abused child (Conte & Schuerman, 1987). Madonna, van Scoyk & Jones (1991) have also reported negative and dysfunctional family interaction patterns in sexually abusive families. Rating the interaction within incestuous and non-incestuous families of sexually abused children, they found significantly more parental neglect and emotional non-availability in the former than the latter.

Sexualized behaviour in the abused children

The presence of sexualized behaviour was noted in a minority of children in the present sample both at the time of disclosure and the time of referral. In the Northern Ireland study more than a third of the children were reported to have shown 'over-sexualized behaviour' in the previous six months, defined by the presence of one or more items on a nine-point checklist. The proportion was higher among boys than girls, though the difference did not reach significance (The Research Team, 1990). Bentovim et al (1988) found that, at referral to a clinic, 35% of victims showed this type of behaviour, compared with 5% of non-abused siblings. In a sample of sexually abused children under 7 years seen at a community mental health centre 41% were recorded as showing sexualized behaviour, in contrast to 5% of non-abused controls (Gale, Thomson, Moran and Sack, 1988). Despite some agreement between studies on the proportion of children and adolescents showing this type of behaviour, there must be some doubt about the reliability of data. 'Age-inappropriate' sexualized behaviour is likely to be a subject on which adults may have very different perceptions of what qualifies for inclusion. The Research Team (1990) and Hobbs (1990) give lists of behaviour which might be included, and standardised checklists have more recently been developed (Friedrich et al, 1986; White et al, 1988), but this is an area which requires more research and clinical attention.

The mothers/carers

The disclosure period is almost inevitably a distressing time for mothers of sexually abused children, but the mothers reported far higher levels of anxiety than depression. This was not predicted by the clinicians, but it does make good sense. For most families and mothers clinical assessment was a time of great uncertainty, carrying the possibility of separation from their children, or partners, with a high level of concern about the future.

Many of the mothers had experienced physical or sexual abuse in childhood and had few happy memories of family life. Perhaps in compensation, being a parent and looking after others were revealed on the mothers' self-esteem questionnaire as areas in which they felt most effective. Being confronted with the serious abuse of their children challenges exactly those areas; it is not surprising, therefore, that anxiety is a predominant response. However, this should elicit different professional responses from an assumption that depression is most commonly experienced by mothers. Treatment and assessment centres should consider how they can reassure parents and children about the length or content of intervention work, and the specific expectations the centre workers have for the family. Setting clear targets and checking that clients really understand their part in the therapeutic process would go some way to reduce anxiety.

There is a strong relationship between the self-esteem of the mother and the child's self-esteem and more positive interactions with the child. Conversely there was an association between the mother's low self-esteem, the child's low self-esteem and negative ratings on their relationship. Children felt more positive about themselves, and less depressed when their mother enjoyed their company, and was not critical of them and when she believed their description of the abuse. Other studies have noted the importance of initial belief and subsequent support in reducing children's and adolescents' symptoms (Johnson and Kenkel, 1991). Encouraging mothers (primary carers) and non-abusive fathers to understand and believe their abused child must therefore be an important part of post-disclosure family work. Specific work on reducing the parents' criticism of their abused child, and increasing the opportunities for relaxed enjoyment of their child's company should also be included. Many programmes for improving parenting skills include these elements and might be a useful component of post-disclosure therapies.

When mothers were highly critical of the child then the child was perceived by them as more symptomatic. Mothers/carers reported a lower level of symptoms in the children who had described penetration as the most frequent and the worst abuse experience. At one level this is a surprising finding, as the majority of other studies have found a significant association between severe abuse and high symptom level. However, mothers were also found to believe less frequently in the children's description of abuse when it had been penetrative, and this may have mediated their perception of symptoms. The children themselves reported more symptoms when they had experienced penetrative abuse. In the present study it was clear that the mother's perceptions and her own needs are an extremely influential component of the child's self-esteem, behaviour and psychological status when they both enter treatment.

Beitchman et al (1991) have drawn attention to the very high rates of marital breakdown and psychopathology among the parents of sexually abused children and teenagers, which makes if difficult to separate the specific effects of sexual abuse on behaviour from the effects of adverse family environments. This is not a universal finding, however. Mannarino & Cohen (1986) reported that 82% of the parents of their sample of 24 abused children were not involved in marital separations; however, this study does not report on the important issue of the quality of the parental relationship, which is known to affect the behaviour and emotional well-being of children in the family. Specific marital work should form part of the planned provision for families in which intra-familial sexual abuse has been disclosed.

Among the children who were interviewed the very large majority had found the abuse frightening, disgusting or painful. A small but significant group had believed it was part of 'normal' family life. This highlights the importance of effective sex education about what is appropriate and inappropriate behaviour between adults and children and teenagers, and the importance of children knowing about Childline and similar advice lines which enable them to check whether what is happening to them is right or wrong.

Chapter 8

Conclusions

Many of the findings from the present study have been similar to those from other descriptive studies of child and adolescent victims of sexual abuse (largely in the United States and Canada). In the event, the size of the sample and some aspects of the design of the study made it difficult to make major advances in our knowledge of the sequelae of sexual abuse and disclosure. For example, the difficulty of obtaining co-operation from all those who had responsibility for the children in the period before detection meant that it was difficult to answer some of the most important questions about the protective factors in the children's lives.

In this study the children's and adolescents' coping responses were not investigated and there is now some evidence that the ability to cope may be an important influence on response to therapy (Johnson and Kenkel, 1991). Future studies should address the issue of coping strategies and social support for the abused child and the non-abusing parent, and monitor the occurrence of stressful events or ongoing difficulties which child and parent(s) have to face. It is difficult to avoid the feeling that many researchers and clinicians in this field regard the abuse as the only likely cause of the child's symptoms or the non-abusing parent's distress. In reality, other stressful events (connected or unconnected with the abuse) may adversely affect their progress through treatment. Conversely, some positive experiences may, independently of treatment, exert a beneficial effect (Brown et al, 1992).

Part II

The Treatment Outcome Study

Chapter 9

Introduction

There have been remarkably few studies of the outcome of treatment of the sexually abused child, despite the fact—noted in the introduction to this Report—that the number of treatment programmes has increased considerably in recent years.

In addition to an absence of prospective studies of treated children it is clear from the literature that there is still considerable confusion about the most suitable ways of measuring the outcome of treatment following sexual abuse. In studies of abused children the rate of re-victimisation is an important consideration, but it may be equally important to assess the outcome of treatment in terms of family relationships, psychiatric symptoms, school performance or peer relationships, and the internal psychological adjustment of the abused child. As we have noted in the main introduction, early retrospective studies suggested a wide range of maladaptive and anti-social behaviour in the later lives of childhood victims, for example, prostitution (Silbert & Pines, 1981) and sexual offending (Groth, 1979). There is also evidence of considerable intrapsychic disturbance and pain (Courtois & Watts, 1982), unsatisfactory marriages or hetero-sexual partnerships (McGuire & Wagner, 1978).

It is likely that the design of many of the earlier retrospective studies produced a biased picture of the abused child as an adult; for example, some of the samples were drawn from deviant adult populations such as incarcerated offenders or prostitutes. More recently, better designed studies have suggested links with anorexia and bulimia nervosa (Hall et al, 1989) and somatization disorders (Morrison, 1989). But later studies have found no link between childhood sexual abuse and adult women's sexual dysfunction (Fromuth, 1986; Greenwald, Leitenberg, Cado & Tarran, 1990).

Despite these difficulties of interpretation, the results from retrospective studies can indicate some of the areas of disturbed and deviant behaviour which need to be investigated in prospective outcome studies of treated populations (see Chapter 7, Gomez-Schwartz et al, 1990).

Most of the studies evaluating the treatment of sexually abused children have focused on relatively short-term outcomes. In an 18-month follow-up of 156 abused children and adolescents referred to the Family Crisis Program, at the New England Medical Center, Massachusetts three measures were used to assess progress: the Louisville Behavior Checklist, two self-concept /self-esteem scales, and an assessment of inappropriate sexual behaviour (Gomes-Schwartz et al with Sauzier, 1990). At the conclusion of the follow-up period the treated children, as a group, showed significant reductions in psychopathology, and an increase in self-esteem. Comparisons of individual assessments showed that the majority had 'improved', but some had not changed and a minority had worse problems at follow-up than at the start of treatment. Neither age, sex, race or socioeconomic status influenced outcome; similarly no association was found between relationship with the offender, or aspects of the severity of the abuse and an improved outcome. The authors concluded that there was great variability in progress, with some children developing symptoms during the 18-month follow-up who had not shown them at referral, while the majority showed reductions in the symptoms reflecting anxiety in particular (eg, sleep problems, worry, nervousness) and in sexually inappropriate behaviour.

In the UK Bentovim, van Elburg & Boston (1988) followed up 120 sexually abused children treated at The Hospital for Sick Children, London between 1981 and 1984; the follow-up period varied from two to five years, which makes the interpretation of results rather difficult. Social workers in the community rated the 'overall' situation of the child as improved for 61%, the same for 24% and worse for 10% (the rest being unknown). Only 23% of the children were in the same family setting as at referral, and only 14% were living with both their own parents. The proportions of children showing emotional problems had declined from 69% to 49%, and those showing sexualized behaviour from 36% to 21% . There had been re-abuse in 16% of the cases and a suspicion of re-abuse in a further 15% (Bentovim, van Elburg & Boston, 1988).

At the start of the work reported here few prospective studies of the effects of treatment had been completed, and the present study was necessarily explorative. Studies undertaken since the present study began are reviewed in the Discussion section.

The current treatment study was based on the clinical work of an unusually experienced team. A recent survey has shown that only 27% of the treatment for abused (not just sexually abused) children in the UK takes place in hospital-based clinics (NCH, 1990). Unlike the majority of other treatment centres in the UK at the time, the CSA clinical team in the present study had worked with sexually abused children for over a decade.

Chapter 10

Aims, Hypotheses and Procedures

It will be recalled that, overall, the research had two aims. The first was to provide a description of the personal and family characteristics of the children and the results of that part of the study have formed the first part of this report. The second aim was to assess the outcome of two treatment programmes offered to the sexually abused children and their families or carers, and this forms the focus of the second part of this report.

The main hypothesis to be tested (noted on page 8) was that those children and family members receiving group treatment in addition to family network treatment would have an improved outcome compared with those receiving family network treatment only. All families in the programme were also appropriately supported by community agencies, who had responsibility for child protection, and these professionals were actively involved throughout the treatment.

The procedure followed for the treated sample has been described on page 12. All the measures which were used to describe the parents' or the childrens' mood and feelings were repeated at the second end-of-year interview. Thus, parents filled in questionnaires on mood and self esteem and on the behaviour and symptoms of their child. The children completed the Children's Depression Inventory and the self esteem questionnaire.

The families which entered treatment were randomly allocated to one of the two treatment programmes. The families and individuals seen by researchers at the start of treatment were not seen again by researchers until approximately twelve months later, and the second, post-treatment research interviews were usually held in the setting in which the child lived. The second interview was held after a programme of treatment had been completed, but clinicians sometimes judged that more treatment was required. This further treatment was not randomly assigned, as the research monitoring had by then been completed. The second research interview was therefore not always held at the end of treatment, but only at the end of a particular programme. Throughout this section of the Report, measures taken at the first interview are referred to as Time 1 data and at the second research interview are referred to as Time 2 data.

Chapter 11 The Treated Sample

Among the 99 children from 74 families who were included in the descriptive study, 47 children from 37 families were accepted into treatment and randomly allocated to one of the two treatment programmes. Three boys who had been referred to the hospital as teenage abusers, had also been abused and met the research criteria as abused children. They were, however, excluded from the following analysis because their own treatment had focused on their needs as abusers not as abused children. The 47 children, with their parents or carers, became the population for the treatment outcome study. Table 27 shows the main characteristics of the 47 children in the treatment study.

Table 27 **Characteristics of the children included in the treatment outcome study (n=47)**

	n	(%)
Sex		
boys	7	(15)
girls	40	(85)
Ages		
4–7 years	10	(21)
7–9 years	11	(23)
10–12 years	13	(28)
13+ years	13	(28)
Severity of abuse (worst experience)		
not penetrative	18	(38)
penetrative	29	(62)
Length of abuse		
up to 2 months	2	(6)
2—18 months	11	(31)
19—36 months	11	(31)
37—60 months	4	(8)
61 months +	7	(15)
(not known	13)	
Relationship of abuser to abused child		
biological parent	30	(46)
surrogate parent	12	(18)
brother	7	(11)
other	16	(25)
Living arrangements		
with parent who was in charge at abuse	21	(45)
with foster parents	10	(21)
Children's Home	6	(13)
other family	9	(19)
Special residential school	2	(4)

Chapter 12

The Treatment Programmes

The content of the two treatment programmes is described in full in the Appendix. A brief description of the treatment received by the two treated populations is given here.

Family network treatment

This programme consisted of regular meetings at four to six week intervals with family members and support professionals from the community. Initial tasks include ensuring adequate protection for the child, assessing the degree of responsibility taken for the abuse by the perpetrator, and ensuring family support for the child. The format of family work lends itself, in particular, to addressing the aims of empowering the non-abusive parent(s), improving communication between members, and clarifying roles and boundaries within the family. Family network treatment may work towards rehabilitation within the family, or towards separating the child from the family and establishing them in substitute care. Overall contact with the family was maintained for between one and two years.

Family with the addition of group work

Families allocated to this category received, in addition to the family network meetings described above, a series of group meetings appropriate to each member's age and developmental stage. Mixed groups were held for younger children of six, eight and ten years, and single sex groups for three age ranges of teenage girls. Other groups were held for teenage perpetrators (all boys), mothers, adult perpetrators, couples and caretakers. The groups were held weekly, and the duration varied with age and needs from between 6 and 8 weeks for the youngest children and up to 20 weeks for older children and adults. The group work format lends itself in particular to issues of de-stigmatisation, to raising self-esteem and assertiveness, and to future relationships and self-protection.

Collaboration with community-based professionals

The professionals in the community (probation officers, social workers, residential child-care staff) were closely involved with all aspects of the work—attending family network meetings and caretakers' groups. The work they conducted with the families in the community formed an important part of the treatment, and included case-work with the families, individual children and adults.

Chapter 13

The Findings

Introductory note

The findings are presented in two sections. First, results are given of the effect of treatment for the whole population, looking initially at the data from the research interviews and then the data from the clinical assessments. Second, results are presented which test the original hypothesis that the addition of group treatment would lead to a better outcome. In this section the data from the two treatment groups will be compared, looking first at the evidence from the research interviews and second the evidence from the clinicians.

Comparison of the treated and untreated populations

In order to establish the extent to which the findings on the treated populations could be extrapolated to all cases in the descriptive study, it was important to identify whether the children and mother/carers who had been accepted for treatment differed in any respects from those referrals not receiving treatment. On all except one measure the children and mothers entering treatment did not differ in any significant way from the children and mothers who did not have treatment. Comparison of the treated and untreated groups [using 2-tailed t-tests] showed no significant differences in age or sex balance, or in the distribution of scores on the children's own report of symptoms, the teachers' reports of symptoms, or the children's reports of self-esteem or depression. In terms of the duration or severity of abuse, the characteristics of the perpetrators or the age at which abuse was initiated there were also no differences between the treated and untreated children. There was, however, a significant tendency for the mothers to report a higher symptom score at the time of referral for those children who entered treatment at Great Ormond Street.

Numbers involved in the following analyses

The size of the treated sample was dictated by the intake to the clinical treatment programme during the 19 months of the research (N=47). On some of the measures the numbers of completed schedules or questionnaires available for comparison of scores at Time 1 (before treatment) and Time 2 (after treatment) was smaller than this total. This was because not all those who were interviewed at Time 1 were available for re-interview at Time 2, and for a few cases children or mothers/carers agreed to be interviewed at Time 2 when they had not been at Time 1. For example, one child elected not to speak in the first interview but was willing to talk at the second. The child's age also influenced the availability of data. Among the 47 treated children there were five 4 and 5 year odds who were too young to complete self-report questionnaires, for one of whom there was no teacher's report because she did not attend school. There were eight 6 and 7 year odds who completed different schedules on self-esteem and family relationships from those completed by the older children.

Since the size of the sample has had an inevitable effect on the power of the analyses we have undertaken, and on the confidence we can place in the results, we have drawn attention to the numbers involved in each analysis. Overall, we must warn against treating the results as being anything other than indicative: much larger samples are need to draw firm conclusions on treatment efficacy.

Chapter 14

Assessment of the Whole Treated Population—Non-Clinical Measures

A comparison was first made between research measures at Time 1 and Time 2 for the treated group as a whole: in other words, we first investigated the effects of treatment rather than looking separately at the two treatment programmes.

Reports of the child's symptoms by teachers, by mothers and by the children themselves at the start treatment (Time 1) were compared with scores 12 months later (Time 2). This procedure was repeated for the child's self-esteem and depression scores at Time 1 and Time 2. Table 28 shows that, for the children for whom there was information at both times, there were significant improvements in the 8-16 year olds self-reported depression scores and the mothers'/carers' reports of the children's health and behaviour. There were, however, no significant changes reported on the other three items (the 8-16 year old children's self-esteem and reports of their own health and behaviour, and the teachers' reports of behaviour). Given the significant inverse association between the children's report of their depression and their report of their self-esteem at the start of treatment, it was interesting to note that this association no longer existed after 12 months. While depression had reduced there had not been a commensurate significant rise in self-esteem, despite this being a stated target of treatment. More detailed analyses of the measures is given below.

Among the younger children, aged 6 and 7 years, self-esteem scores showed little change, but Family Relations Test scores showed some improvement: the numbers of children were too small for conclusions to be drawn.

Table 28 **Comparison of abused children's individual scores before and after treatment: all treated children**

		(Wilcoxon's 2-tailed test)	
	(N)	Z	significance level (p)
Mother/carer's report of symptoms	(26)	−2.93	<.004*
Teachers' reports of symptoms	(21)	−1.48	NS
Child's report of symptoms	(23)	−1.61	NS
Child's CDI score (8–16 year olds)	(28)	−3.44	<.001*
Child's self-esteem (8–16 year olds)	(27)	−1.62	NS

Children's Depression Inventory

Analysis of the individual items of the CDI showed that the large majority of interviewed children recorded almost the same scores on the individual items at the start and end of treatment. Nevertheless, as a group, significant improvement was recorded on the items identifying sadness, worries about the future, having fun, thinking bad things would happen, bad things being the child's own fault, suicidal ideation, anxiety, and ease of decision-making (Table 29). On the remaining items while some individuals improved their scores there was no significant improvement for the group as a whole. Very noticeably the children did not record significant improvement on somatic symptoms. Only four of the items which had appeared at the start of treatment among the ten most frequently recorded by the children were among those which had significantly reduced. These items were: worries about the future, problems with making decisions, worries that bad things will happen, and suicidal ideation. It must be obvious that, when children had reported problems infrequently at Time 1 there was less scope for marked improvement.

Table 29 **Items on the Child Depression Inventory on which there was significant improvement from Time 1 to Time 2** (N=28)

Item	Z	Wilcoxon 2-tailed test Significance level (p)
Worries about the future	−2.66	<.01
Has fun	−2.66	<.01
Worries that bad things will happen	−2.78	<.01
Thinks bad things are own fault	−2.57	<.01
Feeling sad	−2.20	<.03
Anxiety	−2.19	<.03
Problems making decisions	−2.09	<.04
Suicidal ideation	−2.02	<.05

The mothers'/carers' reports of the children's symptoms

While the total scores at Time 2 showed a significant reduction (improvement) compared with Time 1, the only individual symptoms for which mothers reported a significant reduction were headaches, misery and temper outbursts (Table 30). Of these it will be recalled (page 48) that misery and temper outbursts were among the three most commonly reported symptoms at the start of treatment, while headaches were reported by relatively few mothers. As with the CDI items, when problems had only infrequently been reported at Time 1 there was less opportunity for children to show 'improvement' by reducing scores which were low ('normal') anyway.

Table 30 **Individual items on the mother/carers' health and behaviour checklist showing significant improvement after treatment**

Item	Z	(Wilcoxon 2-tailed test) level of significance (p)
Temper tantrums	−2.81	<.005
Misery	−2.31	<.02
Headaches	−2.10	<.04

Teachers' reports of children's behaviour in school

The total scores on the teachers' behaviour scales (Rutter 'B') showed no significant improvement the treatment period, and only one of the 28 items (the presence of tics or mannerisms) showed a significant reduction. Nevertheless, teachers rated some improvement for 52% (11/21) children for whom reports at both times were available, but the numbers of reports available for school-aged children at Time 1 and Time 2 was disappointingly low.

Children's self-reported symptoms

For the group as a whole, the total self-reported symptom scores showed no significant reduction over the treatment period, and no individual item on the child's report of 12 health and behaviour inventory showed a significant reduction. However, half the children who were old enough to provide this information (14/28) rated themselves as slightly improved. The absence of large improvements was particularly unexpected as several items (misery, aggression, anxiety and fears) had been targets of the children's treatment programmes.

Children's self-esteem inventory

There was no significant improvement in the children's self-esteem scores, although 18 (64% of those who had completed questionnaires) showed some improvements in their self-esteem. Analysis of individual items on the self-esteem questionnaire revealed that there was significant improvement on only one global item ('I don't think much of myself'). The absence of change on other items was accompanied by the fact that as many children had lower (worse) scores as had higher (better) scores. There were also no significant differences in the scores of any of the seven self-esteem domains (relating to mother, father, family, peers, body-image, school performance or global self-esteem) between Time 1 and Time 2.

Among the younger children who had been given the Harter/Pike Pictorial Scale of Perceived Competence, three had lower scores at Time 2 (indicating lower self-image) and four had higher scores. Among these 6 and 7 year olds lower scores were associated with mothers reporting more sexualized behaviour.

Set against these figures for children who improved during the year of treatment, some children developed more, or worse symptoms. Six children (21% of those filling in the CDI) had higher depression scores at the end of the year. The comparable figures for other measures were 39% (11/28) showing lower self-esteem, 33% (7/21) rated as showing more disturbed behaviour by teachers, 31% (8/26) rated as showing more disturbed behaviour by mother/carers, and 36% (10/28) of the children rating themselves as having more symptoms.

Relationships between mother/carers and abused children before and after treatment

Against prediction, there were no significant changes in the relationships between mothers/carers and the children before and after treatment (as measured by the research team), with the exception of the fact that the children were *more likely* to have become a target of maternal criticism (Wilcoxon Z= -2.06, 2-tailed p=<.04). There is no obvious explanation for this individual finding. It is possible that, as the children and the household returned to 'normal', some mothers may have reverted to higher levels of criticism, which they had been withholding in deference to the child's state of 'shock' during the disclosure processes. Alternatively, some children may genuinely have become more difficult to care for as the year passed; but this seems a less likely explanation in view of the significant improvement in health and behaviour recorded by the mothers/carers themselves. Among the mothers and their children who were seen at Time 1 and Time 2 the absence of significant change in hostility and warmth is striking. Whatever the origins of their attitudes towards one another, these appear markedly stable throughout treatment.

Mothers and carers

It is important to bear in mind that 'treatment' for the adult carers was available in the form of support and instruction to those foster parents who had charge of the abused children, as well as to the step- and biological parents. For this reason, we have considered the changes in their self-reports alongside that of the biological parents. The progress of the mothers/carers through treatment produced a more positive outcome on non-clinical measures than that shown by the children. Mothers recorded significantly lower depression scores and higher self-esteem scores at the end than at the start of the year of treatment. This may provide some explanation for the mothers'/carers' reports of improvement in the health and behaviour of their children, but it makes the absence of marked improvements in the mother-child relationships rather more difficult to explain.

The mothers' General Health Questionnaire and self-esteem scores

For the group as a whole there were significant improvements in the scores (Wilcoxon $Z= -3.38$; 2-tailed $p <.001$) (Table 31). Seventeen mothers (65%) reported reductions in symptoms (improvement), though six mothers (23%) reported feeling worse; the remaining three (12) had identical scores. There was strikingly significant improvement in the reports of somatic symptoms (Wilcoxon $Z= -3.56$; 2-tailed $p <.001$), and of anxiety (Wilcoxon $Z= -4.12$; 2-tailed $p<.0001$), a less noticeable, but still significant reduction in symptoms of social dysfunction (Wilcoxon$= -2.43$; 2-tailed $p<.02$), but no significant change in reports of depression, on which 50% of the mothers/carers said they felt the same or worse than they had at the start of treatment.

Twenty-five mothers/carers completed the self-esteem questionnaire before and after treatment. As a group their scores showed significant improvements (Table 31). Sixteen women (64%) reported higher scores, though nine (36%) reported lower (a lower score indicates lower self-esteem). More detailed assessment of the 'domains' of the adult self-esteem questionnaire showed that improvement was not confined to any particular area.

Table 31 **Comparison of mother/carers' scores on the self-esteem questionnaire & the GHQ before and after treatment**

	(N)	Z	(Wilcoxon's 2-tailed test) Significance level (p)
Mother/carers' self-esteem	(26)	−2.44	<.02
Mother/carers' GHQ scores	(25)	−3.38	<.001

External sources of stress for mother/carers

Mothers/carers were asked to comment on the amount of stress they felt they had experienced during the year on a range of issues arising from the abuse and the after-effects of detection such as losing a partner, financial worries, or loss of friends. On each of nine items, mothers were coded consensually by the research team as having experienced no difficulties, or mild, moderate or severe difficulties. The question of whether the item was applicable to the mother or not, was left to her to define. Thus if she had no financial difficulties the question was not applicable; if she had difficulties but described them as unimportant, and the research interview established the evidence for her view, she would be rated as 'mild' (Table 32).

Table 32 **Levels of difficulty experienced by mothers during the year of treatment.**

	None or mild	Moderate or severe	(N)
	%	%	
Single parent-hood	22	78	(18)
Loss of children	29	71	(7)
Loss of relatives	50	50	(8)
Financial	50	50	(22)
Loss of friends	62	38	(21)
Employment difficulties	64	36	(14)
Dispute over abuse	64	36	(11)
Moving house	66	33	(9)
Loss of partner	71	29	(14)

While it might have been expected that this group of mothers would find the loss of their children hard to bear, it is important to note that a large number of mothers (78%) also found the new task of single parenthood presented moderate or severe difficulties. (These were the women who had not been single parents immediately before the detection of the abuse.) These two results have to be set against the fact that only 29% said they found the loss of their partner (in these cases the abuser) presented moderate or severe difficulties. If they wished to have a partner at all, the majority of these mothers did wish to have the same partner back again. Not surprisingly, half of those with financial difficulties found these to be moderately or severely stressful. All the women reporting this level of difficulty were single parents, a group which it is well-known are at risk of receiving very low incomes in the UK. Overall almost none of the women who had experienced the nine different circumstances had survived without seeing them as difficulties.

Chapter 15

Clinical Assessment of the Whole Treated Population

In this section we consider the effects of treatment on the whole treated group as it was assessed by the clinical team.

Clinical assessments were made of the abused children and the mothers/carers who had been responsible for the child during the abuse, when these latter were going through treatment. The clinician with responsibility for the case made the ratings at the start and end of treatment. For a small number of cases which were still involved with final sessions of group or family work 12 months after the treatment had begun, the ratings were made at this point.

Clinical expectations of improvement in the child

At the start of treatment, clinicians were asked to give an opinion about the likelihood of a good outcome for the child themselves and for their relationship with their mother/carer. At this stage, before any treatment had been received, the clinical team were significantly more likely to predict a 'hopeful' outcome for the child if he or she had been allocated to the treatment which included additional group work, rather than family network only. Unfortunately these ratings were not made blind to the programme to which the family had been allocated.

Family treatment aims

An assessment was made by clinicians of the extent to which individuals had achieved each of the family treatment aims at Time 1 and Time 2, and these were compared (using the Wilcoxon matched pairs signed ranks test). Results showed that clinicians rated a significant improvement over a number of items (Table 33):

a) in the relationship between the mother/carer and the abused child;

b) in the children's capacity to recognise positive qualities in themselves;

c) in the children's understanding that the perpetrator was responsible for the abuse;

d) in the resolution of their feelings (including a reduction in their positive feelings) towards the perpetrator.

e) in the extent to which the rest of the abused child's family could understand his or her age-appropriate needs.

On the other hand, clinicians judged that there had been no significant changes in the following areas:

a) the child's ability to express appropriate anger in the family in which the abuse had occurred;

b) the child's ability to function in that family without being made a scape-goat;

c) the ability of the adults in the child's end-of-year family to recognise the possible damage which sexual abuse might do to children;

d) the ability of those adults to deal effectively with the behavioural consequences of the abuse.

Table 33 **Aspects of clinical improvement (the family treatment aims) in relation to the abused child: both treatment groups (Wilcoxon matched-pairs signed-ranks test)**

	(N)	Z	level of significance (2-tailed p)
Abused child resolves feelings towards mother	(40)	−3.54	<.0001
Abused child sees positive qualities in self	(39)	−4.32	<.0001
Child agrees that abuse is the fault of perpetrator	(41)	−3.15	<.002
Child resolves feelings towards perpetrator	(41)	−3.76	<.001
Child less likely to see positive features in perpetrator	(38)	−2.54	<.02
Family perceives abused child's needs appropriately	(36)	−2.06	<.04

Chapter 16

Assessment of Treatment : a Comparison of the two Treatment Populations

Comparison of the initial status of mothers and children in each of the two treatment programmes

In order to establish whether the addition of group treatment led to a better outcome an essential first step was to make sure that those receiving the additional group treatment were not significantly different before treatment began from those receiving family network treatment only. In terms of the children's age distribution, sexes, level of symptoms reported by mother/carers, teachers, or the children themselves at the time of referral, the severity of abuse and socio-economic status there were no significant differences between the children allocated to the two treatment groups. Assessment by clinicians about the degree of outside support in terms of intensity and type did not reveal any difference between those offered family work only, and those offered family network plus group work. However, the children following combined family and groupwork were judged by clinicians at Time 1 to have significantly more positive attitudes towards their mothers/carers and towards their perpetrators than the children following family network treatment only.

Chapter 17

Assessment of treatment: progress within each treatment group

Family treatment only

Initially a comparison was undertaken of the extent of improvement within, rather than between, the two treatment groups. Considering first the children who received family network treatment only, we found that there was a significant improvement in the mothers'/carers' reports of children's symptoms and in children's self-reported depression scores (Table 34). In each case a high proportion of the children showed improvement (9/11 on the mother's/carer's symptom report, and 12/13 on the Children's Depression Inventory). It will be recalled that when the whole treated population was investigated together these were also the only two standardised measures on which significant improvement was recorded for the children (Table 28). The numbers of reports which could be compared on each measure varied: the reasons for this were mentioned in Chapter 13.

Outcomes for the abused children: non-clinical standardised measures

Among the children following family treatment only there were no significant improvements in the teachers' reports of behavioural disturbance at school, nor in the children's reports of their own symptoms or self-esteem (Table 34). However, despite the absence of significant changes for the group as a whole on all three measures some individuals showed improvement: teachers rated 47% (8/17) with improved scores, but 41% with worse scores; 61% of the children (8/13) reported higher self-esteem but the remaining 39% (5/13) reported lower self-esteem; and 45% (5/11) children reported the same or worse symptom scores.

Outcomes for the mothers/carers: non-clinical standardised measures

Among the very small number of mothers who followed family treatment only, self-esteem showed marked improvement. On the GHQ three women reported improved scores and three the same scores at the start of treatment, but these results did not reach significance (Table 35).

Table 34 **Comparison of abused children's scores at Time 1 and Time 2: Family treatment only**

	(N)	Z	(Wilcoxon matched pairs) p
Mother/carer's report of symptoms	(11)	−2.09	<0.04
Teachers' reports of symptoms	(17)	−0.17	0.63 (ns)
Child's report of symptoms	(12)	−1.16	0.25 (ns)
Child's depression score (aged 8–16)	(13)	−3.01	<0.01
Child's self-esteem (aged 8–16)	(13)	−1.08	0.28 (ns)

Table 35 **Comparison of mothers' self-esteem and General Health Questionnaire scores at Time 1 and Time 2: family treatment only**

	N	Z	Wilcoxon matched pairs Significance (p)
Mothers' self-esteem	6	−2.20	< .05
Mothers' General Health Questionnaire	6	−1.60	1.08 (ns)

Clinical outcomes following family treatment only

Clinicians judged that 50% of the children and 50% of the mothers had a good or moderate outcome after family treatment. In 43% of mother-child pairs good or moderate improvement was observed in their relationship. On other Family Treatment Aims children and mothers/carers did not make significant improvement.

Family and additional group treatment

Outcomes for the abused children

Among the children receiving the additional group treatment significant changes from Time 1 to Time 2 scores were found on two of the five standardised child measures (Table 36). One of these—the mother's rating of the child's symptoms—had also shown improvement when the children followed the family treatment only. Teachers' scores indicated that the majority of the children had fewer symptoms at the end of treatment. The three scores that relied on the children as reporters showed no significant improvements (Table 36).

Detailed investigation of the direction of change on these standardised measures revealed that mothers reported that 62% (13/21) had fewer symptoms, and teachers reported that 69% had fewer symptoms. By contrast the children gave less optimistic reports: 75% (18/24) either had lower or equal self-esteem scores compared with Time 1; 60% (15/25) had higher (worse) depression scores; and 58% (7/12) had higher or equal scores on their own symptoms.

Table 36 **Comparison of children's scores at time 1 and time 2: Family treatment and additional groupwork.**

	(N)	Z	(Wilcoxon matched pairs) p
Mother's/carer's report of symptoms	(21)	−2.36	0.02
Teachers' report of symptoms	(13)	−2.27	0.02
Child's report of own symptoms	(12)	−0.65	0.51 (ns)
Child's depression score	(14)	−1.60	0.99 (ns)
Child's self-esteem score	(24)	−0.71	0.48 (ns)

Outcomes for the mothers/carers following additional groupwork

Among the mother/carers who received family network treatment and the additional group treatment improvement showed on both the standardized, self-report measures (Table 37). Improved GHQ scores were reported by 94% (15/16), and higher self-esteem by 69% (9/13) of the mothers who filled in questionnaires on both occasions. The change was highly significant on the GHQ scores and nearly reached significance on the self-esteem scores.

Table 37 **Comparison of mothers'/carers' self-esteem and General Health Questionnaire scores at Time 1 and Time 2: Family treatment with additional group treatment.**

	N	Z	Wilcoxon matched pairs level of significance
Self-esteem	(13)	−1.92	<0.06
General Health Questionnaire	(16)	−3.28	<0.001

Clinical outcomes following the additional groupwork

Clinicians rated 80% of the children and 89% of the mothers/carers as having made good or moderate overall improvement after the additional groupwork. In 89% of child–mother pairs, clinicians rated good or moderate improvement in their relationship. Clinical ratings of Family Treatment Aims before and after treatment showed that the children had made significant progress in three areas: understanding that the abuse was the responsibility of the perpetrator, not themselves; resolving their feelings towards the perpetrator; and learning to share their pain with the perpetrator. Family members were judged to have made significant progress in recognising the abused child's needs.

Chapter 18

Assessment of Treatment: Comparison of the two Treatment Programmes

In order to test the original hypothesis that the addition of group work would produce a better outcome, the changes on the standardised measures were analyzed by the type of treatment the mothers/carers and children received. Account was taken, in these analyses, of the small numbers in each treatment programme.

Non-clinical standardised measures of outcome

The abused children

There were no significant differences in the levels of improvement on any of the five standardised non-clinical measures during the year of treatment between the children attending family treatment and those attending the additional groupwork (Table 38).

Table 38 **Comparison of the children's scores at Time 1 and Time 2: the effect of the two treatment programmes. (Mann-Whitney U-Tests)**

	(N*N**)	Z 2-tailed	Significance
Mothers' reports of child's symptoms	(11;15)	-0.81	0.42 (ns)
Teachers' reports of symptoms	(10;10)	-0.83	0.40 (ns)
Children's reports of own symptoms	(12;14)	-0.59	0.55 (ns)
Children's self-report depression	(13;15)	-1.11	0.26 (ns)
Children's self-esteem	(13;14)	-0.33	0.73 (ns)

N*—number in Family network treatment only
N**—number in Family network and group treatment

The mothers/carers

Although there had been a significant rise in self-esteem scores and an improvement in GHQ scores for the treated mothers as a whole, there was no significant difference between the effectiveness of the two treatment programmes (Table 39). The addition of group treatment had led to a greater reduction of GHQ scores, and this difference nearly reached statistical significance. This group of women started with significantly higher GHQ scores than the women following the programme of family network treatment only, and the reduced scores was undoubtedly an achievement. But with higher initial scores there was also a greater chance of achieving significantly reduced scores.

Table 39 **Comparison of the mother/carers' scores at Time 1 and Time 2: the effect of the two treatment programmes. (Mann-Whitney U-tests)**

	(N*N**)	Z 2-tailed	Significance (p)
Mothers' GHQ scores	(6;16)	-1.89	0.06
Mothers' self-esteem	(6;14)	-0.05	0.96

N*—number in Family network treatment only
N**—number in Family network and group treatment

Clinical measures of outcome

Unlike the five standardized measures or the interview derived measures of family or peer relationships, clinicians' ratings of the progress made by the children in the Family Treatment Aims showed some significant differences according to which treatment programme they followed.

The mother-child relationship

Clinicians' judged that improvement was significantly more frequent in the overall mother-child relationship when the family had received the additional group treatment: 65% of those having group treatment had made 'good' clinical progress in this area, compared with only 19% of those receiving family treatment only ($\chi^2=5.94$, 1df, p <.02). We have already noted that 43% of the mother-child pairs who received family treatment only showed good or moderate improvement, compared with 89% of the pairs receiving the additional groupwork: this difference reached statistical significance. For the 25 cases in which the perpetrator was still in communication with the abused child, the clinicians were significantly more likely to rate an improvement in the relationship between perpetrator and child when the family had received group treatment ($\chi^2=7.03$, 1df, p<.01).

The Family Treatment Aims

The clinicians noted that the successful achievement of the six Family Treatment Aims listed below was significantly related to having followed the additional groupwork (Table 40).

Table 40 **The effect of type of treatment on the clinical ratings of improvement of the children (Chi-squares)**

Family Treatment Aim	Type of treatment	% improved	(N/N)	significance (p)
Child shares pain with mother at time of abuse	family network family+group	0 38	(0/17) (6/16)	<.02
Child is able to speak without being scapegoated	family network family+group	5 39	(1/19) (7/18)	<.04
Child's needs are understood by family at time of abuse	family network family+group	11 58	(2/18) (11/19)	<.01
Child's needs understood by current carer	family network family+group	36 91	(5/14) (10/11)	<.02
Child sees positive features of self	family network family+group	15 52	(3/20) (12/23)	<.03
Adults agree that sexual abuse has damaged child	family network family+group	32 78	(6/19) (18/23)	<.01

Clinical ratings of improvement in the remaining Family Treatment Aims did not show any statistically significant differences between the two treatment programmes. However, it can be seen in Table 41 that attending the additional group treatment led to a consistently higher percentage of children showing improvement on each remaining Family Treatment Aim. Overall, the addition of group treatment was associated either with significantly higher improvement or with a trend to higher improvement on all the Family Treatment Aims.

Table 41 **The effect of type of treatment on the clinical ratings of improvement of the children (Chi-squares)**

Family treatment aim	Type of treatment	% improved	N/N	significance (p)
Child resolves feelings towards at-abuse mother	family network family+group	15 36	(3/20) (8/22)	ns
Child resolves feelings to perpetrator	family network family+group	25 48	(5/20) (11/23)	ns
Child shows appropriate anger in own family	family network family+group	6 29	(1/17) (4/14)	ns
Child shows appropriate anger in current family	family network family+group	31 55	(4/13) (6/11)	ns
Child understands abuse was fault of perpetrator	family network family+group	65 78	(13/20) (18/23)	ns
Child sees positive features in at-abuse mother	family network family+group	5 35	(1/19) (8/23)	ns
Child sees positive features of perpetrator	family network family+group	0 9	(0/19) (2/22)	ns
Current adults deal effectively with behaviour	family network family+group	32 52	(6/19) (12/23)	ns
Child works effectively with professionals	family network family+group	45 57	(9/20) (13/23)	ns

Mothers/carers in treatment

The clinicians' ratings of the mothers' own improvement was significantly associated with the treatment programme she had received: 88% of the mothers following family and group treatment had a 'good' outcome, compared with 44% of those receiving family treatment only ($\chi^2=5.45$, 1df, p<.02). A significant association was observed between receiving family and group treatment and improvement in the mother–child relationship ($\chi^2=5.31$, 1df, p<.02) and improvement in total family functioning ($\chi^2=7.09$, 1df, p<.01), but there was no significant association between the treatment programme and clinical ratings of improvement in the mothers' marital relationships (Table 42).

Table 42 **The effect of type of treatment on the clinical ratings of improvement in relation to the mothers. (Chi-squares)**

Clinicians' overall aim	Type of treatment	% improved	(N/N)	significance (p)
Mother's own improvement	family network	36	(5/14)	
	family+group	88	(15/17)	.01
Mother-child relationship	family network	29	(4/14)	
	family+group	83	(15/18)	.01
Total family functioning	family network	21	(3/14)	
	family+group	89	(16/18)	.001

Clinicians' predicted outcome compared with final clinical ratings of children's improvement.

At the end of treatment clinicians made a global rating of improvement in relation to three areas of the child's functioning: the subject themselves, their relationship with their mother/carer and their relationship with the perpetrator. Clinicians judged that 31 children (69%) had made good progress in their own functioning, and 14 (31%) had made poor progress. In their relationship with their (Time 1) mother 18 (47%) had made good progress, and 20 (53%) had made poor progress, including 6 children whose relationship with their (Time 1) mother had deteriorated during the year. Nine children were rated as having shown improvement in their relationship with the perpetrator, while 18 (67%) had not improved (among whom two were worse).

Clinical ratings of outcome showed that the children receiving the additional group treatment were more likely to be rated as 'the same as expected', while 56% of those without the groupwork had done worse than expected (Table 43).

Table 43 **The type of treatment and clinicians' expectations of outcome compared with actual outcome for the child.**

Actual outcome compared with predicted outcome	family network only	Group & family treatment
	n (%)	n (%)
worse	10 (56)	2 (8)
same	6 (33)	17 (68)
better	2 (11)	6 (24)
Totals	18 (100)	25 (100)

χ^2=11.76, 2df, p<.003

A similar pattern was found when clinicians' predictions of improvement in the mother-child relationship was compared with their final ratings. Among the mother-child pairs following the additional group treatment, 21/23 (91%) had an outcome as expected, and 2/23 (9%) had an outcome worse than expected, (no mother-child dyad did better than expected); whereas among those following family treatment only, 7/16 (44%) had an outcome as expected, 8/16 (50%) had an outcome worse than expected, and 1/16 (6%) (χ^2=10.69. 2df; p <.001).

By contrast, there were no significant associations between the type of treatment the family received and clinical ratings of outcome for the mothers alone, or for the perpetrator-child relationship.

The influence of the mother/carers' belief that abuse happened on clinical ratings of improvement

Because of the importance attached by the clinical team to the value of the mother believing her child' abuse, it had been expected that clinicians' ratings of improvement of the mothers, and more particularly of the children, would be associated with the mother's belief, but this was not borne out by the data. In the same way, acknowledging the disruptive effects of children changing households during treatment, it had been expected that clinical improvement would be associated with whether the child had moved, but this also appeared relatively unimportant compared with other influences on improvement.

Chapter 19

Children's and Mothers'/Carers' Attitudes to Treatment

At the end of their treatment the children and their mothers/carers were asked by the research team both about their feelings about the treatment they had received at the Hospital, and the extent to which they felt it had been helpful in addressing particular issues associated with the abuse. The issues on which they were asked about their feelings had emerged as important in the pilot phase of the research study. The issues on which they were asked to comment about the helpfulness of treatment had been identified as important features of treatment by the clinical team during the planning stage of the study.

The children's attitudes to the treatment

The children's responses to the initial questions on their feelings are given in Table 44. The number of children answering each question varied: some of the children were too young to be asked, and for some children an individual question might not be applicable. For example, only the children who were parted from their mother between meetings could answer the question about their feelings on meeting her, and only a few children met the perpetrator during treatment.

The children finished their treatment in the hospital with positive feelings about most, but not all, of these items. For example, while the children were very likely to feel positively about the therapists' gender and therapeutic style, they were more likely to have mixed or negative feelings about some other issues like 'talking about the abuse' (50% had negative or mixed feelings) or 'meeting others with similar experiences' (31% had negative feelings about this). For those children who were not living with their mothers for some part of treatment, it seems that meeting their mother was frequently a negative experience.

It is of interest to note that, for the six children who had a chance of talking to the abuser, this was a positive experience for four; this probably reflects the fact that children not infrequently have very ambivalent feelings, mixing love and aversion or fear, about the abuser. In the hospital setting, in the presence of other people, it is likely that the fear will be less, and either hostility or warmth can safely be expressed.

Table 44 **The abused child's feelings about aspects of the treatment offer (ranked in order of proportion of those with positive response)**

	percentages				
	No feelings	Negative feelings	Mixed	Positive feelings	(N)
Therapists' interview style	3	6	9	81	(32)
Therapist's gender	9	3	6	81	(32)
Feeling understood	0	7	17	77	(30)
Meeting others with similar experiences	0	31	0	69	(13)
Talking to abuser	0	17	17	66	(6)
Talking to mother	0	42	5	53	(19)
Talking to family	0	31	23	46	(13)
Talking about the abuse	6	34	16	44	(32)
Coming to Hospital	16	28	13	44	(32)

Children's views of helpfulness of the treatment

The children were asked about what parts of their treatment they had found particularly helpful. As rather few of the children could recall specific aspects of the treatment, a checklist prompt was used, focusing on those aspects which the clinical team had identified as important components of the programmes. Answering the general question about how helpful the hospital treatment had been to them, the majority of the children again produced positive responses. However, there were two issues (understanding the origins of the abuse, and their relationship with the abuser(s)) about which less than half the children felt they had been helped, and a further three issues (facilitating further disclosures, relationships within the current family and planning for the future) about which only about half the children reported they had been helped (Table 45). The two issues which the children were most positive about (raising self-esteem and preventing further abuse) still left about a fifth of the children feeling they had not had as much help as they would have liked. It is interesting to note the contrast with the children's reports of "feeling better about themselves", which might reasonably be taken to indicate raised self-esteem and the absence of statistically significant improvement on the self-esteem questionnaire. This illustrates one of the difficulties of measuring this particular aspect of children's and adolescents' well-being.

Table 45 **Helpfulness of treatment reported by children (ranked in order of proportion reporting treatment had been helpful)**

	percentages				
	unhelpful	no effect	mixed effect	helpful	(N)
Preventing further abuse	3	14	3	79	(29)
Raising self-esteem	10	17	0	72	(29)
Offering understanding	10	10	16	68	(31)
Resolving guilt	7	18	7	67	(27)
Relations with current family	11	29	7	54	(28)
Enabling further disclosures	14	19	14	52	(21)
Planning for future	7	33	7	52	(27)
Relations with original family	14	32	5	50	(22)
Understanding origins of abuse	13	20	27	40	(30)
Relations with abuser	33	25	17	25	(12)

The type of treatment the child had received made no difference to their descriptions of how helpful or unhelpful the treatment had been, but numbers were small in such analyses casting doubt on the reliability of results. There were also no differences between boys and girls, nor between the younger and older children (either above and below 10 years old, or above and below 13 years old), but we must enter the same cavent about small numbers.

Among the 14 children who attended the additional groups, and were old enough to be interviewed, there was no effect of their commitment to treatment (rated by clinicians) on their view of the helpfulness of the treatment on any of the items above.

Mothers'/carers' attitudes to their treatment

The mothers/carers also reported a range of feelings about the treatment, which are reflected in their reactions to the global question about coming to the hospital. Since much of the burden of organising a family so that some members can attend the hospital falls on mothers, this question was designed to probe their reactions to the effort, time and family disruption of attending a regional hospital as opposed to their local clinic. Inevitably, however, the mothers reported a number of other feelings at the same time. For example, among those who had a negative reaction, some simply said 'It was a waste of time', implying that the return for their effort was negligible. This may have implications for the development of more local treatment resources. Other women reported the extreme difficulties associated with bringing one child out of several, or bringing several small children to the hospital and feeding and entertaining them for several hours usually single-handed.

The numbers of women shown responding to the questions (Table 46) is sometimes quite small: this was usually because particular questions did not apply to all the women. For instance only three mothers were parted from their families, and only six had faced talking to a perpetrator they did not otherwise see.

Table 46 **Mother/carers' feelings about aspects of the treatment programmes**

	(percentages)			
	Negative	Mixed	Positive	(N)
Talking to others with similar experiences	0	0	100	(16)
Talking to child	0	0	100	(8)
Therapist's gender	7	0	93	(14)
Feeling understood (by therapist)	10	10	80	(20)
Therapist's style	6	23	71	(17)
Talking to family	0	33	67	(3)
Talking to abuser	0	33	67	(6)
Talking about the abuse	19	19	62	(21)
Coming to the Hospital	40	13	47	(15)

One of the more striking results of asking the mothers for their feelings was their report of how they felt about talking to the child. Since all the mothers/carers lived with these children, or were expecting to return to living with them, the fact that all of them felt positive about the child is important; even when the relationship had been disrupted by the disclosure, the opportunity to talk together in the therapeutic setting was highly valued. The sixteen mothers/carers who reported positive feelings about talking to others with similar experiences were, of course, attending group sessions. It is important to note that none of them found this produced even mixed feelings, despite the length of the groups (20 weeks) and the often exacting nature of the therapeutic work in this type of treatment. Like their children, the mothers/carers had a very positive response to the therapists' style, and gender, and almost all felt they had been understood.

There were no differences in their feelings about the treatment between those women who had followed family network treatment only, or family network and group treatment, except that the latter were significantly more likely to 'feel understood'. It was not, unfortunately, clear from their responses whether they were reporting feeling understood by the clinicians or by other mothers in the groups. The numbers are small, but the overall impression was of mothers finding much of the treatment a positive experience.

Mothers'/carers' reports of the helpfulness of treatment

Mothers'/carers' comments on how helpful they had found the treatment produced a less positive picture. Bearing in mind that the number of respondents who answered the questions was quite small, it can be seen that the issue on which the largest proportion of women felt helped was over preventing further abuse, followed by feeling that they had been 'supported' by the therapy, and having their own role in the abuse clarified (Table 47). For all other topics less than half the women reported that they had been helped on the issues which clinicians had originally identified as part of the treatment programme.

As with the questions about their feelings, not all the women who were interviewed at Time 2 could answer all the questions. A number of women said that issues had not been addressed in their treatment. Since the issues had been identified by clinicians as integral to the treatment programmes, and since it was unlikely the issues had been avoided, we could only conclude that some mothers simply forgot an issue had been discussed. This does suggest the importance of providing constant re-enforcement for non-abusive parents on key therapeutic themes.

Table 47 **Helpfulness of treatment reported by mothers/carers (ranked in order of the proportion reporting that treatment had been helpful)**

	(percentages)				
	unhelpful	no effect	mixed effect	helpful	(N)
Preventing further abuse	17	11	17	62	(10)
Offering support	21	16	5	58	(19)
Clarifying own role vis-a-vis abuse	22	22	6	50	(18)
Accepting the abuse has happened	22	22	11	45	(18)
Discovery of future abuse	16	42	5	37	(19)
Resolve own guilt	33	22	11	33	(18)
Help relating to family	13	40	13	33	(15)
Planning for the future	17	39	11	33	(18)
Relate to perpetrator	15	7	31	31	(15)
Understanding origins of abuse	33	22	17	28	(18)
Help relating to managing abused child	22	33	17	28	(18)

There were no differences in their views on the helpfulness of treatment between those women who had followed family network treatment only, or family network and group treatment, except that the latter were significantly more likely to 'feel understood'.

The effect of commitment to treatment

The commitment the mothers had shown towards their treatment was rated by clinicians at the end of the year. Out of 21 mothers on whom this rating was made, 18 were rated as showing a high level of commitment, and three a low level. A high degree of commitment to the treatment programmes was significantly related to a lower GHQ score at the start of treatment, but not to maternal self-esteem. The level of commitment to treatment was not related to the amount of improvement in the treatment period.

The three women who had been rated by clinicians as showing only a low level of commitment t treatment reported that they believed the treatment had had "no effect" in seven of the 11 areas shown in Table 44. In this very small group only one woman reported any positive responses: she said she thought the treatment might have helped her plan for the future. This 'unhelped' group are clinically significant, but whether there are long-term implications for their children cannot be established from the present study.

Chapter 20

Discussion—The Treatment Outcome Study

The treatment study was undertaken to test the hypothesis that the addition of group work to the standard family network treatment would improve the outcome for sexually abused children and their mothers (carers) after approximately 12 months treatment. [Too few perpetrators went into treatment to be able to test the same hypothesis for them.] The small numbers who entered treatment and the varying numbers for whom there were completed questionnaires and ratings derived from interviews at Time 1 and Time 2 means that results must be approached with care.

Perhaps the most important conclusion is that the results indicated a considerable difference between the effects recorded on the one hand by the use of standardised measures with children and mothers, and measures derived from a specifically designed research protocol, and on the other the ratings by the clinicians in charge of treatment. A comparison of the two treatment programmes showed that there were no significant differences in the outcome of the treated children either on the research team's measures of symptoms (rated by mothers/carers, teachers, and children), or on the children's self-reported depression and self-esteem. None of the changes in the measures of parent-child relationships derived from the research interviews was significantly related to following one treatment programme rather than the other. Among the interviewed mothers there were no significant differences in the levels of improvement on the two standardised measures (self-esteem and GHQ) when the two treatment programmes were compared, although the GHQ scores for mothers following the additional groupwork tended to show greater improvement.

Clinicians' ratings, however, recorded significantly greater improvement on several of the Family Treatment Aims when mothers and children had received the additional group treatment.

The pattern of these results requires some exploration. It seems reasonably clear that with this particular group of sexually abused children and their mothers/carers, the form of additional group treatment provided by the specialist team did not carry significant advantages for the psychiatric state of the mothers/carers or their children compared with the family network treatment available to the whole sample: at least, this was true at a 12-month follow-up using standardised measures. This lack of group-treatment effect on these measures is unlikely simply to be due to the small numbers as, if this had been a significant factor, trends in the expected would have been consistently apparent, but these were absent. It is possible that the measures used may have been relatively insensitive to improvement in these children, but again if there had been any appreciable effect trends would have been discernible. The results are likely therefore to be valid for this sample.

Nevertheless, in generalising from them, it is important to remember that these findings can only be applied to similar groups of children and mothers, and very probably not all sexually abused children. In addition, it is only the form of groupwork which was used by the Great Ormond Street specialist team which was tested against the family network treatment developed by the same team. The use of other forms of groupwork might have produced a more marked difference between treatment groups.

It should also be borne in mind that there was a wide span of ages among the abused children, and that this had implications for the length of treatment. On the other hand, the varying lengths of treatment had been planned fully to meet the needs of each age group. If the groupwork had been a significantly effective addition this would have been apparent despite the variation in programme length. There is an urgent need to find settings in which larger numbers of treated children and varying treatment programmes can be used to test treatment efficacy.

In contrast to the researchers' measures and ratings, the clinicians' ratings did suggest that the additional groupwork contributed significantly to a positive result. Clinicians, however, were largely rating different factors from the research team. The measures clinicians used related particularly to the child's attitudes to abusive episodes, and to his/her parents, rather than to the child's psychiatric state. The fact that the research and clinical teams differed in their measurement of the child's progress does not necessarily reflect inconsistency. It is quite possible, and indeed not improbable that, for example, group therapy makes a more effective contribution to the child gaining greater understanding of who was responsible for the abuse, but not to the child becoming less depressed.

The positive results in favour of group therapy obtained by using the clinical ratings, highly consistent though they are, cannot be accepted uncritically. The ratings were not independent of the clinicians' knowledge of the type of treatment which the child and parent/carer had received, and clinicians were rating their own cases. The measures used were unstandardised and of unknown reliability. The positive results obtained by the clinicians therefore provide only tentative evidence for the superior effectiveness of group treatment in altering children's attitudes to the abuse and their parents.

Assuming, for the sake of argument, that the findings of both the research and clinical teams are valid, it is possible to draw some conclusions about the processes leading to change in the child and mother/carer. It seems that changes in the child's attitudes do not necessarily lead to short-term changes (observable at 12 months) in the mental state of the child or parent/carer or in the behaviour of the abused child. Nor does the absence of a significant change in psychiatric state or behaviour prevent change in attitudes to the abuse and to each other. It might well be, however, that changes in the child's attitudes to the abuse and to parents do lead to change in mental state, but over longer periods of time than our study enable us to investigate.

The absence of significant differences between the two treatment programmes in the improvement of children or mother/carers on certain Family Treatment Aims and the standardised measures may have arisen from a relatively larger than expected effect of the family network treatment, or from a relatively smaller than expected effect of the group work. It is important that the right conclusions are drawn from this finding. There is no suggestion that the addition of group treatment is not useful, only that it is not strikingly more effective than the family network treatment alone. For families following family network treatment in the hospital setting, professional support was also routinely available in the community, possibly providing a level of support not dissimilar from the weekly groupwork. The children in the present study all followed specialised therapeutic programmes: it may indeed matter less what the content of the programme is than the mere fact of involvement. Gomes-Schwartz et al (1990) found that sexually abused children following a specialist programme showed more improvement after 18 months than children who had only received crisis intervention work with the same team for a shorter time.

No predictions had been made of a general treatment effect (regardless of treatment programme followed). In the event, the standardised research measures recorded significant improvements for the children by the end of treatment in their own self-reported depression and the mother/carers' ratings of the children's behaviour. The mothers also reported higher self-esteem and fewer psychiatric symptoms for themselves, particularly in the areas of somatic symptoms and anxiety. On clinical measures the whole treated population also made significant improvements on a wide range of issues. Clearly the important question is whether these improvements are the result of treatment (with the inevitable passage of time) or simply the passage of time (or both). Mannarino et al (1991) have drawn attention to the fact that improvement after treatment may be the result of passing time, which has the effect of giving greater distance from the abuse and disclosure experiences, and also usually means that most of the criminal proceedings against the perpetrator are resolved one way or another.

One aspect of the results requires further comment. The treatment effect was most significant in those standardised measures for which the mothers/carers were the informants (ie, their own GHQ and self-esteem scores and their estimates of the children symptoms). It appears that, at the 12 month follow-up, mothers/carers made (in some sense) better progress that their abused children. Love (1989) also found that mothers reported substantial benefits over a 12-month treatment programme compared with the lesser progress of their children. Is this because the children inevitably take longer to recover from the multiple traumas of the abuse and the consequences of the disclosure, and if this is so, how long may we expect this recovery to take? Is it because the approach to recovery in the treatment programmes is couched in a manner more immediately 'available' to adults than children? Is it because the non-abusive adults are better at absorbing and acting upon the messages contained in treatment? These are all questions which need systematic assessment through prospective studies of treated samples, but which also have significant implications for the organisation of treatment.

Improvement recorded on the Children's Depression Inventory has been noted in other studies which have used this measure. For example, Deblinger, McLeer & Henry (1990) noted that CDI scores were significantly improved in 10 out of 19 girls following 12 structured sessions of cognitive behavioural therapy. Mannarino et al (1991) found that sexually abused girls, psychiatric clinic controls and normal controls all improved their scores on the CDI at 12-month follow-up, in a study in which not all the abused girls received therapy. This illustrates the difficulties of interpreting material gathered at only one follow-up point: there is a tendency for all subjects to regress towards the mean, which on unidimensional measures such as the CDI can be interpreted as 'improvement'.

Evidence from the Children's Depression Inventory showed particular patterns of change which link with the observations of Terr (1991) on trauma. She has proposed a classification of trauma which distinguishes the single event (Type I) from repeated events (Type II). Most sexual abuse would fall into the second category. Responses to Type I trauma include 'omens', detailed memories, and misperceptions, while responses to Type II trauma include rage, numbing and denial, and dissociation. In particular, she suggests that the second type is characteristically followed by a sense of hopelessness, and "futurelessness". With hindsight, the clinical team considered that many of the children fell into this category at the start of treatment, but they had not rated them on these symptoms at the time. When individual items of the Children's Depression Inventory are considered there seems to be some indication that progress was made in the treatment period on some characteristic Type II phenomena. However, since the children were not systematically assessed on Terr's schema, these points must remain as suggestions only.

The absence of 'improvement' on the children's self-esteem may reflect the fact that this measure taps a trait rather than a state phenomenon, and is less amenable to change through therapy or over time. However, nearly all the children experienced considerable disruption in the year of treatment. Negative experiences like separation from the family, growing awareness of permanent rejection and failures of rehabilitation, mothers maintaining their disbelief, abusers maintaining denial may add up to continuous undermining of the children's sense of self-worth. Indeed, these uncertainties about the present and the consequent uncertainties about the future may, for some children, be a more potent influence on them than the treatment. Berliner (1991) suggests that the reason why so many studies of sexually abused children do not find lowered self-esteem may arise from the existence of a group of children who appear to 'cope', and deny the impact of the abuse. Further research is required on the relationship between coping strategies and self-esteem in these vulnerable children.

It is interesting to note that mothers/carers and children took consistently different views on the children's progress. Children following family treatment only reported significantly less depression at the end of the year, but on the other self-reported measures such improvement was absent. That mothers perceived improvement while children did not may illustrate the difference between observable behaviour and the child's internal world in which the effects of trauma continue to dominate. However our findings are in contrast to those of Mannarino et al (1991) where parents of abused children continued to report serious behaviour problems and psychiatric symptoms at a 12-month follow-up.

Mothers/carers also reported major change and stress in the year, including learning to manage as single parents on reduced incomes, sometimes with fewer friends or relatives as support. At the level of anecdote, several mothers reported their satisfaction with the groups they attended in terms of providing them with a ready-made group of women to relate to in a positive way: not an experience many of them had elsewhere in their lives. The area of the General Health Questionnaire on which mothers recorded least improvement was depression, and it may well be that, for some women at least, the sense of loss associated with some aspects of the post-disclosure events persists even when anxiety and somatic symptoms associated with the early (crisis) stages of intervention have resolved (Hooper, 1992).

Overall, clinicians noted a number of improvements in the children and the mother/carers. However, in the expression of appropriate anger, functioning in the family without becoming a scapegoat, adults recognising the damage child sexual abuse can do, and dealing appropriately with the behavioural consequences little change had occurred. These are important issues and the absence of continuous therapeutic work with siblings may have played a part. More resources would only be a part of the solution, as many families were reluctant to allow siblings into the therapeutic process or to attend the hospital. The researchers were also unable to see the siblings, and there is a need for more systematic evaluation of the relationship of the non-abused to abused siblings, and both to the non-abusing parent who has post-disclosure charge of the family.

It should also be borne in mind that in the present study, as in the Tufts New England Medical Center study, quite a large proportion of the children appeared to have more rather than fewer symptoms at the end of the treatment period (Gomes-Schwartz et al, 1990). Longer term follow-up studies will be needed to answer the question of whether this deterioration is permanent or temporary.

One of the interesting and relatively innovative aspects of the outcome research was the systematic recording of what the mothers and children felt about the treatment they had received, and the extent to which they reported the treatment as helpful. Most of the responses on feelings about and helpfulness of treatment were positive. The children, however, had quite negative feelings about coming to the hospital, which was often associated with having to leave school early, arriving home late, with many hours of travel. It is not surprising either that the children found talking about the abuse or talking to other family members to be a negative experience. Talking to family involved the children in sharing painful memories, not with others who had had the same experience, but with the family who may have complex, even hostile, feelings for the abused child—as the source of current family disruption.

Most of the children felt that the treatment had helped to resolve their feelings of guilt about the abuse, which is a major focus of the treatment programmes. The children reported feeling helped in their self-esteem, although the standardised questionnaire recorded no significant change. They also reported feeling helped with preventing further abuse, which reflects the work done on self-protection. But there was clearly a failure to help many children to understand the origins of the abuse, or feel helped in their relationships with their current family or the abuser. In the absence of full admission of guilt by most of the abusers this latter therapeutic work is difficult to carry out.

The mothers'/carers' opinions about treatment reflected many of the same concerns. There was a very small group of women who held that the treatment had been a waste of time. These mothers might respond to different treatment techniques, but were probably under severe pressure to maintain their relationships with the abusive partner and/or other family members who were encouraging them to reject the child. These pressures on mothers and conflicts of loyalty have been graphically described by several writers (Meiselman, 1978; Dietz & Craft, 1980; Gomes-Schwartz et al, 1990; Hooper, 1992), and clearly form an extremely difficult area of any therapy. When in group settings, the women who wished to maintain a relationship with abusive partners found the angry, confrontational attitudes of women who did not difficult to manage.

Some aspects of the treatment programmes were reported to be very helpful by most mothers. Feeling understood by the therapist and being given the scope to talk to their abused child in the family meetings were seen to be helpful. Like the children they had rather more negative and mixed feelings about coming to the hospital, and this suggests that the development of local treatment centres would be appreciated by mothers with family and work responsibilities.

In a very small study of female incest victims Woodworth (1991) found that one third of the mothers and victims expressed disappointment with the treatment they had received, despite the fact that it had helped them personally. Combined with our own more detailed results, this suggests the need for much clearer communication by therapists of the aims and expectations they have of the treatment, and more opportunities for mothers and children to feed back their own expectations during the course of the treatment. This was, of course, undertaken: the results merely highlight that it does not always work as well as it should.

The discovery that many of the mothers believed the treatment had not been helpful in a number of areas has important implications. Like the results of the measurement of outcome, the views of the children and mothers on their treatment should contribute to the evaluation and review of any treatment programme. For example, the issues which mothers identified as less helpful may in future need to be dealt with in more depth and/or more persistently in their treatment, or even approached in completely new ways: most importantly these parts of the treatment cannot remain unchallenged by the therapists themselves.

Chapter 21

Conclusions

The study sought to test the hypothesis that the addition of group treatment to family network treatment would produce a better outcome for the children and primary carers. While the standardised research instruments showed no effects of the type of treatment, some of the clinicians' measures of improvement confirmed the hypothesis that mothers and children would do better with group treatment. In the treated population as a whole, mothers and children made progress in a number of areas of personal and inter-personal functioning, though not always on the issues which had been the focus of treatment. It is accepted that the passage of time will almost certainly have contributed to reducing pathological symptom levels. The absence of an untreated control or comparison group makes it impossible to determine how much improvement was dependent on time passing and how much on treatment. Barrnett, Docherty & Frommelt (1991) and Finkelhor & Berliner (1995) have pointed out that almost no outcome studies of treated sexually abused children have incorporated control groups, thus diminishing the value of the conclusions and this criticism is fully accepted in relation to this study.

The children and parents in treatment had major emotional problems at the time of referral, and many of the families had a history of dissolution, change and reconstitution. Many of the parents themselves recalled a history of poor early care and few good memories of childhood; many had developed ways of relating to each other, and to their children which were aggressive and abusive. In this context, it may well be that the helpfulness of what is learnt through family and group work may not show itself to the fullest extent until more time has passed, and some of the major changes in the lives of children and their parents have been resolved. The clinical team identified certain areas of work which even after 12 months continued to need monitoring. In particular, children with post-traumatic symptoms, persisting negative mood swings, and self-destructive behaviour, men whose sexual orientation towards children is deeply embedded, or mothers who find the stress of their situation contribute to their depressed feelings, all require further work. Further contact with a treatment agency may be a useful facility to offer abused children and their families, maybe long after their formal treatment programme has ended.

The treatment programmes for sexually abused children and their families at Great Ormond Street can only be available for a small proportion of the children in the region who might benefit from treatment. It is important to develop treatment programmes locally so that time-consuming and expensive travel can be avoided. The opportunities for more integrated services for families or individuals may be also greater when the treatment is organised locally.

Chapter 22

Directions for Future Work

The final stages of a lengthy treatment outcome study which has involved both researchers and clinicians was a suitable time to consider how the work can be fed back into further research and improved clinical practice. We considered these areas separately, although there are considerably overlapping interests in setting up systems to assess treatment outcome.

Adopting agreed definitions

Perhaps the most important development in CSA research and practice would be the adoption of some agreed definitions which can be used by researchers, by those involved in referral and assessment work and those involved in treatment. It is important that agreement should be reached as soon as possible on the 'best' way in which the circumstances surrounding the abused child, the nature of the abuse and other aspects of each case can be recorded. This would greatly improve the knowledge about sexual abuse by reducing the disparities between reports from different researchers and different treatment centres.

This may not be easily achieved, because there are likely to be differences of opinion both about what needs measuring and about what measures to adopt. It is our opinion, however, that while there may be some difficulty in agreeing particular measures, there is widespread support for the idea of adopting some shared definitions and some shared procedures to record information about sexual abuse cases.

However, it is not only those involved in research who need to address these issues; it affects also those who are involved in clinical work. It is our view that there are three ways in which the work of researchers and clinicians can prove mutually beneficial. The first is in the evaluation of treatment for audit, the second the evaluation of treatment for further development of therapy, and the third the further development of systems for recording treatment in this field.

Standardising assessments of sexually abused children

In this context there is a need for some standardisation of the assessment of children before and after treatment, and it is suggested that this should include a psychiatric assessment whenever possible. This would enable researchers and clinicians to observe systematically whether the sexually abused children fall into specific diagnostic categories (eg, using DSM IV), or show a higher rate of psychiatric disorders or largely suffer from a disparate and variable range of symptoms[7].

The example of assessing and recording sexualized behaviour in abused children may be used to illustrate the way in which co-ordinated clinical and research procedures can be established. There are many reports of the differences between abused and non-abused children in sexual behaviour, but it is only recently that attempts have been made to provide standardised systems for recording sexualized behaviour (Friedrich et al, 1986; Gale et al, 1988; White et al, 1988). Measures for recording sexualized behaviour are now available (Friedrich, 1988; The Research Team, 1990). These measures now need to be systematically validated with UK samples of abused and non-abused children and adolescents. We have already drawn attention to the fact that this range of symptoms may be unusually difficult to obtain agreement on, because the behaviour encroaches on taboo areas, and the practice and beliefs of adults as to what is allowable and appropriate in children varies widely (Smith & Grocke, 1995).

There is scope also for defining more clearly on a case-by-case basis, for treatment programmes and for the sexually abused population as a whole, what is meant by "success". The variation is considerable: as Maddock, Larson & Lally (1991) have pointed out, even within intra-familial abuse families some therapists regard "success" as separating the family, some regard "success" as re-forming the family. There is an urgent need to formulate a core of agreed targets and aims for treatment which can be appropriately assessed before, during and after treatment.

Evaluation of treatment programmes

Dubowitz (1990) commented that 'considering the importance of preventing child maltreatment, it is remarkable that relatively few programmes have been rigorously evaluated as to the short-term and long-term outcomes'. This point has been stressed by Finkelhor & Berliner (1995) in their review of treatment outcome studies of sexually abused children. Evaluation is becoming a widely accepted task, both as part of medical and social services audit, and as a systematic way of assessing the effectiveness of clinical work. There is also greater acceptance that effectiveness of treatment cannot be assumed because symptoms reduce over any treatment period: as we have admitted, control or

---

[7] A first step in the direction of standardising procedures in the USA has been taken recently in the adoption and publication of Guidelines for the Clinical Evaluation of Child and Adolescent Sexual Abuse (American Academy of Child and Adolescent Psychiatry 1988).

comparison groups are essential before treatment rather than other factors is identified as the reason for the improvement. Inevitably, in child abuse cases this is a particularly difficult issue, when for legal and ethical reasons child protection procedures have to be completed which may in themselves be therapeutic.

This is an appropriate place to emphasise that the random allocation of clients and their families was accepted and the purpose understood by all those entering the present study. A small number of cases (5) were blocked by family social workers who had asked for groupwork, when they found the random allocation was to family network treatment only. Although the small numbers entering the study have made firm conclusions difficult, it is clear that treatment of either type was largely beneficial. This suggests that clients and referral agencies will probably find random allocation acceptable in future studies.

Record-keeping in child sexual abuse cases

These points are relevant also to the ways in which the records on cases are kept by social services and other organisations who are involved in the disclosure and assessment stages of child sexual abuse work. It is important for those involved in this work to adopt some standardised record-keeping, greater agreement about what information should be collected and shared procedures for acquiring it.

To enable the efficacy of treatment to be more readily assessed, and to lay the foundation for better long-term follow-up of sexually abused children, clinical records also need to be carefully organised. Many clinical records do not report exactly what the aims of the treatment are, nor the criteria used to decide why a case is thought sufficiently 'better' to terminate treatment. There may be scope for improving the ways in which the content of treatment can be recorded so that the process of improvement can be more easily tracked through time. In a recent paper on clinical work with sexual abuse victims Berliner (1991) has given an interesting example of this approach by detailing the reasoning behind each component of treatment.

Such an approach enables a therapeutic team to establish clear treatment targets and the means for reaching those targets. If when they are assessed at the end of treatment the abused child or family have failed to achieve the specific target (eg their self-esteem is as low as or lower than it was at the start of treatment), it may be easier to amend certain aspects of the treatment programme if it is already known how and when the issues of self-esteem are addressed. Alternatively, the therapists may already have recorded the reasons why, despite the treatment, the individual is likely still to feel badly about themselves (for example, because they have faced redundancy failure in a public examination during the treatment period).

Recording the content of treatment programmes

In a related point, there are probably many successful treatment programmes for sexually abused children and their families for which there is little record of the content. This, of course, limits the chances of that successful system being passed on to other therapists. It is increasingly important that the content of treatment programmes for sexually abused children and their families are systematically recorded, so that neither changes in team members nor the wide variation of professional backgrounds found on most teams reduce the value of efficacious therapy.

Uncovering the long-term consequences of childhood sexual abuse

A great deal has been written about the long-term consequences of childhood abuse, but almost all our knowledge is based on the recall of those adults prepared to cooperate in research and able to remember traumatic events. Many of the samples of these adults have been drawn from socially atypical groups like incarcerated offenders or prostitutes. There is a need for long-term follow-up studies of children on whom information was gathered as close to the point of abuse as possible. It will be important also to establish a large enough population of these children and adolescents for conclusions to be drawn about the larger abused population. An opportunity for this might be gained from collaborative studies involving several agencies or voluntary organisations with responsibility in this field of child protection.

Many of the families who are referred for specialist treatment after the discovery of sexual abuse of the children have multiple problems, frequently unrelated in any clear, causal way to the abuse. Assessing the impact of treatment for a few problems which are assumed to derive directly from the abusive situation may leave the child and family still exposed to several adverse social and psychological pressures. For example, in many families the parents had extremely poor current relationships and the children were exposed to considerable inter-personal violence: even specialist sexual abuse programmes need to attend to such issues or risk failing to achieve significant improvement.

We do not wish to give the impression that these suggestions for future work will suddenly change the extent or accuracy of our knowledge about abused children, or improve the effectiveness of treatment, but we offer them in the hope that they will be seen as essential first steps to providing more responsive and effective services for the sexually abused children and their families.

Chapter 23

Clinical Implications of Study Findings

Being involved in a treatment outcome study provides a clinical team with an opportunity to look carefully at the effectiveness of the work they do. It is unusual for the outcome of psychiatric treatment programmes to be beneficial to all those who take part. Some patients show marked improvement; some show small improvement; others show no improvement or even appear more symptomatic than at the start of their treatment. The evaluation of treatment will identify those who have or have not benefited. The content, timing and length of treatment can then be considered in relation to the successes and failures. In the following section we consider the implications for clinical work of the results of the treatment outcome study.

The treatment context

Any consideration of the clinical implications of a study of outcome of treatment must start by addressing the context in which treatment is taking place. In the present study the fact that all abuse as perpetrated by a close family or household member meant that family disruption was an inevitable consequence of disclosure. It also became clear that, for many of the children and families, treatment was taking place against a background of ongoing stress and instability.

Nearly half the children were living apart from their original family at the time of referral and had suffered separation not only from their parents but also in many cases from their brothers and sisters. At follow-up a third (37%) were still living apart from family members. While some of these had undoubtedly benefited by removal from abusing and rejecting families into stable and supportive foster placements, others were still in children's homes or other temporary settings, some of which could not be described as providing optimal childcare.

Many families were in a state of flux and over 40% of the children moved residence at least once during their treatment. Three-quarters of the perpetrators could not be included in the treatment either because they were in prison, or had disappeared, or took no responsibility for their abusive actions. As a result, the abused children were denied a chance of working directly on their relationship with the abuser. Mothers who had chosen to support their abused child frequently had to adjust to considerable change, in addition to accepting that the abuse had taken place. Some mothers had to face the loss of their partner or another child (when these had been the abuser), and the emotional and/or financial implications of such changes. During treatment some families and abused children still faced the stress and uncertainty of impending court proceedings. Finally, and significantly, over a third of the children in treatment were disbelieved by their mothers in their disclosures of abuse.

These rejected and uprooted children attended treatment alongside others who enjoyed the absolute support and belief of their mothers (and sometimes both parents), and who had been spared the trauma of removal into substitute care. The high levels of continuing disruption and change in some children's lives also meant that they received a higher level of involvement from the child protection professional network in the community

In the context of such continuing experiences of rejection, stress and family disruption, attempts to separate out the impact of treatment are inevitably problematic. The clinical implications of the wide variation in circumstances of the treated children emphasises the importance of detailed attention to the context of each child and family in planning therapy. In particular, it raises questions about the appropriateness and difficulties of treating together in groups individuals whose past and current experiences vary so widely.

The effect of treatment for the whole group

Clinically, it is important to note that the children showed a wide spread of symptoms at the start of treatment, and that this was evident also at the 12-month follow-up. As we have noted, other studies of sexually abused children and adolescents have reported similar results (e.g. Gomez-Schwartz et al., 1990). Mannarino et al. (1991) found that sexually abused girls between 6 and 12 years and a matched control group drawn from a child psychiatric clinic population showed group differences on behaviour checklists when compared with non-clinic normal girls at the initial assessment. Self-report measures were used to assess depression, self-esteem and anxiety in the girls, while parents completed the Child Behaviour Checklist (CBCL). Initially, the groups showed no differences in self-report measures except in the field of current anxiety. The parents, however, rated the abused girls and the clinic controls as showing more problems than the normal controls. Differences in parental ratings remained at 6-month and 12-month follow-up, suggesting that observable behaviour problems persist over time. However, the picture was more complicated by the end of the 12-month follow-up. All three groups made considerable progress on the self-report questionnaires, and the abused and clinical control groups showed fewer problems compared with their scores at the start of the year. At the 6-month and 12-month follow-up: sexually abused girls and clinic controls did not differ from one another, but did show significantly more severe symptoms than the normal controls.

The variability in children's scores on the standardised measures at follow-up are likely to reflect not only different reaction to treatment but wide variation in the children's original abuse experiences, the degree of support in the current living context, the stressful events of the previous year, and the children's coping strategies and cognitive processing. Many writers have noted the variability of children's reactions to stressful events (Rutter, 1985; Freidrich, 1988). In the present study coping and cognitive processing were not measured, but will have influenced not only the children's capacity to deal with the original abuse trauma, but also to process the treatment effectively. From an analysis of the children's personal history no clear prognostic factors emerged in the present study, but better initial scores on the standardised measures tended to be found among the younger children who were believed and supported and had better relationships with their mothers and peers. In these families therapists have natural allies in the family or other living context in which the child can openly share their abusive experiences and restructure their negative self-image.

The tendency for a poor outcome, at least on the standardised measures, for the older children who were disbelieved and criticised by their mothers with whom they have poor relationships, gives cause for particular clinical concern. It is also clinically relevant that this group of teenagers had, without exception, experienced penetrative abuse. Mannarino et al. (1991) noted that, at 12 month follow-up, girls who had experienced intercourse reported significantly more depression, anxiety and self-esteem problems, than those who had experienced fondling. It appears that family network treatment together with group work did not meet their needs sufficiently well to lead to improved scores on self-reported psychopathology or adults' reports of their behaviour. The results suggest that groups might usefully be selected on the basis of similar experiences of family rejection, disbelief or disruption. On the other hand, such issues might be best met or consolidated in an individual setting. Glaser (1991) has commented that uniform treatment programmes (like group work) are unlikely to meet the needs of each individual. In the present sample clinical work continued for some children after the 12-month research follow-up, and this frequently took the form of individual work.

It could be argued that treatment should not be initiated until uncertainties about the child's living context are resolved. These questions about the modality and timing of treatment must be explored further in order to meet the needs of this vulnerable and highly symptomatic group of children and adolescents more effectively.

Scrutiny of the individual items on the Children's Depression Inventory showed that the children's suicidal ideation, worries and fears about the future, sadness and anger had improved. The similarity between several of these items and some features of post-traumatic stress disorder (PTSD) suggests that a measure of PTSD would have been clinically useful in this study. PTSD appears to be common in sexually abused children. McLeer et al. (1992) found that over 40% of the sexually abused children referred to a specialised clinic met DSM-III-R criteria for PTSD. In addition, the numbers of children meeting partial criteria was very high (e.g. 87% showed symptoms of re-experiencing behaviour).

One of the prominent features Terr (1991) describes is dissociation, which was not recorded systematically in the current study. However, both mothers and teachers were asked to report the children's levels of concentration; poor concentration may be an indicator of dissociation. At the start of treatment mothers had identified that half the abused children showed poor concentration (compared with 11% of the non-clinic sample reported by Smith & Jenkins, 1991). By contrast, in the school setting teachers reported that 63% of the children showed poor concentration at the start of treatment, but this was not significantly different from the proportions among classroom controls. On this item, neither mothers nor teachers reported the children making significant improvement during treatment. However, it is possible that a decrease in dissociation manifests itself as an increase in depression, as the young people face their feelings about the abuse. If this were true it might help to explain the wide range of CDI scores at follow-up, some of which will have reflected a reaction to 're-discovering' the abuse.

The fact that, in the present study, few of the symptoms which were probably reflecting post-traumatic stress showed improvement after treatment emphasises the importance of identifying these symptoms more precisely at referral and (possibly) of providing more specific therapy.

Among the mothers/carers, progress recorded on the standardised self-report measures was more favourable than similar measures for the children, with significant improvement in GHQ and self-esteem scores at the end of treatment. On the GHQ the items showing most improvement were those for somatic symptoms, anxiety and social dysfunction, while only half the women reported improvement in depressive symptoms. It is possible that, among the smaller number of women with high depression scores, more specific psychiatric treatment was needed.

Some of these women may have experienced long-term depressive symptoms, and it is important to remember that there were no measures of symptomatology before the women entered the treatment programme. It is conceivable that their depressive feelings were not the result of the abuse situation at all, but arose from previous experiences on vulnerabilities.

Treatment over a 12-month period had a beneficial impact on a range of symptoms shown and reported by mothers and abused children, but failed to produce improvements on a number of other symptoms. What then are the clinical implications of the poor outcome for children's behaviour and self-esteem, and for mothers' depression, bearing in mind that the latter two were explicit and important treatment goals? In this respect it is interesting to note that mothers whose children had been more severely abused tended to view their children as less disturbed, and to have fewer concerns about their own health and self-image. Despite these mothers' views of their children, the children reported themselves as having low self-esteem and high symptom levels. This group of children may reflect in their own negative image of themselves some lack of concern by their mothers, which leads the mothers to under-report their child's continuing needs. Clinically, it may therefore be important to identify and work on ways in which all mothers should act supportively and empathetically towards the abused child. This issue is frequently addressed in the course of treatment by joint work with mothers and children and by encouraging the mother to accept the abused child's story. However, this may not be sufficient to create a truly understanding climate for the restoration of the child's self-esteem.

Among mothers who do accept the child's allegations and who choose to support their children their own continuing depressive symptoms may arise from unresolved self-blame and guilt. Many mothers also lost their partner (when he was the abuser) and faced the financial hardship and sometimes the emotional difficulties associated with single parenthood; it will be recalled that 78% of these mothers rated adjustment to single-parent status as moderately or severely stressful. The adult woman's continuing depression may also relate to the stressful life events and long-term difficulties which continued in many of their lives. Treatment programmes may need to be more sensitive to the impact to such events, and future evaluation studies should take proper account of life events and ongoing difficulties.

Differences between the two treatment groups

While the standardised measures showed few differences between those receiving family network treatment alone compared with those receiving the additional group treatment, clinicians ratings showed several advantages for those in groups, who were significantly more likely to show improvement in the mother-child relationship and overall family functioning, and to meet or exceed the clinical predictions of outcome.

The fact that the clinicians' ratings were not blind to treatment type has already been noted, as has the lack of reliability or validation of the clinical measures of outcome. The weight that can be given to the more favourable clinical findings for those with the additional group treatment therefore has to be open to question. However, although discrepant results were obtained from the clinical and standardised measures, they are not necessarily incompatible. The standardised measures sought to establish the individual functioning of mothers and children, while the clinical ratings were designed to explore changes in family interaction. Research interviews provided some similar but not identical ratings to those chosen by the clinical team. The overlap between areas targeted by clinical and research teams was less than had been planned, despite the fact that the research measures had been selected after consultation with the clinical team on the main targets of their work.

The clinical emphasis was on enabling those family members in treatment to allocate responsibility for abuse appropriately with the perpetrator, to generate a protective and supportive family atmosphere where feelings could be articulated and accepted, and to work towards understanding the origins and effects of abuse.

The research measures which were used to classify the appropriate allocation of blame, and understanding about the origins of the abuse did not show significant change; but it needs to be borne in mind that although these were rated by researchers' consensus, they were not formally validated or tested for reliability except between raters. There is considerable scope for improving clinical or research assessments of 'blame', 'guilt', 'mother-child communication' and several other issues which lie at the heart of much clinical work with sexually abused children.

Accepting the limitations of the clinical ratings, the implication is that the clinicians' treatment aims—with their focus on communication and articulation of feelings within the family—were more successfully addressed when group treatment was available. However, successful achievement of these treatment aims was not always reflected in improvement in individual symptoms, and some important family treatment aims rated by clinicians were not met. For example, the continued scapegoating of the abused child and the difficulty expressing appropriate anger within the family may reflect the fact that it was seldom possible to include the perpetrators in family treatment. In addition, these aims are difficult to address successfully when the abused child is living in substitute care.

Other treatment aims not adequately achieved included the capacity of caretaking adults to recognise damage resulting from the abuse or to deal effectively with the behavioural consequences. Although avoidance and denial may play some part, this apparent failure of treatment to achieve these essentially educative aims is disappointing, and warrants attention to the way in which these issues are addressed and to factors which might inhibit their assimilation. The use of written materials, reinforcement across different treatment settings and regular reviews of progress and topics are among possible improvements.

Allowing, sometimes encouraging, the clients and patients the opportunity of commenting on their treatment is becoming more common (see for example: De Luca, Boyes, Grayston & Romano, 1995). The responses of mothers and children to the treatment provides important feedback to clinical staff. Inevitably, clinicians feel reassured by the positive feedback, but the negative or confused responses are equally important. For example, while 69% of the children attending group sessions reported that it had been a positive experience, there must be concern that a third of the children did not rate it positively. While half the children felt positively about talking to their mothers and family, the important detail is that the other half were unhappy with this role. Reluctance with having to discuss the abuse is understandable, and a typical feature among traumatised children (Terr, 1991), but for some children such 'emotional processing' of the abusive events is thought to be an important therapeutic task. Group treatment settings, or large family meetings may not always be conducive to such work, and alternative settings like separate meetings for mothers and children may be preferable and should be assessed by clinical teams.

Clinical work may need to pay greater attention to improving ways in which children can feel comfortable making any necessary further disclosures, or reaching an understanding of the origins of their abuse. Wattam (1992) has discussed the difficulties inherent in asking children to talk about sexual matters with strangers, even when the latter are professionals. She points out that 'Talking about sex allusively (is) the proper way in which sex is talked about . . . It is not assumed that people who talk this way are avoiding the real topic of sex; rather they are pointing to it in the way that talking about sex is socially organised.' (Wattam, 1992; page 95). Children, and sometimes non-abusing carers, have to go through a learning process in talking to professionals about the abuse—a process which may, in itself, be somewhat shocking. Indeed, Jones (1988) has suggested that this aspect (among others) of therapeutic work with sexually abused children should be offered as individual therapy. The restrictions normally placed on talking about sex in public are likely to need particular attention when children and carers are meeting in groups. Group work requires that the usual rules are broken, and new rules adopted, at least within the group setting, allowing the children to be open and free to discuss this normally taboo subject. However, while some will find that learning to talk openly is hard, others will gain support from being able to share this information in a group setting. For example, the group setting has been presented as helpful in reducing the isolation felt by many abused children (Furniss 1983) and providing the support needed for children and non-abusing carers to cease denying the abuse trauma (Glaser, 1991). It has to be admitted that neither of these statements has yet been systematically tested. It is possible that the greater progress made by the mothers in their groups than the children made in theirs, reflects the greater understanding adults have of the purpose of the groups, which allows them to use the experience constructively. Children and teenagers, by contrast, may need more extensive and more sensitive introduction to the agenda of the treatment presented in this setting.

For the very small group of children who were able to talk to the abuser in a therapeutic setting, most reported this was a positive experience. Help in resolving the commonly conflicted feelings towards the abuser was an important therapeutic goal, yet for the majority of families this work had to be conducted by proxy, because of the absence of the perpetrators. The criminal justice system is, at present, very unlikely to encourage admission of responsibility by perpetrators, until and unless they are convicted. The proportion of perpetrators who are convicted is very small indeed. It has been claimed that clinicians in countries which operate confidential reporting systems (e.g. Belgium) may be able to work more effectively with the whole family, including any familial perpetrator. It remains to be tested whether this provides clear benefits for the abused child.

Finally, it is important to recognise that systematic assessment of clients in treatment can co-exist with individualised treatment programmes. The study by Friedrich, Luecke, Beilke & Place (1992) of sexually abused boys specifically states that the treatment varied from individual to individual: 'a standard approach with all boys could neither be developed or justified.' (Freidrich et al, 1992; page 397). Nevertheless, within this individualised model of treatment, information on the functioning of the boys and their families was systematically collected, and a positive outcome for many could be demonstrated. Although the present study offered a chance to compare two different treatment modalities, the same principles of assessment can be achieved without such constraints.

Summary

Treating the abused child and helping the family effectively are not simple tasks. Several external factors affect the context in which clinicians work. Sexual abuse is widespread, and affects families with a wide range of characteristics, some of which will continue to work against the interests of the child, despite professional support. In addition, the legal system frequently affects the work that can be done with perpetrators, the majority of whom are either not charged or not convicted, and very few of whom accept responsibility for the abuse.

Against this background, we may summarise the possible modifications to treatment in the light of the study findings.

1 Give more attention to the context and ongoing life stresses of each individual and family. This would include working with the community-based professionals to maximise stability and support before treatment, and to minimise disruptions and stress during treatment.

2 Adopt the use of systematic assessments of behaviour and symptoms in the child, and in the mother, to be repeated periodically during treatment and specifically at the end of treatment. Use standardised assessments of post-traumatic stress disorder in the child.

3 Ensure that post-traumatic symptoms are specifically addressed in treatment, either in group or individual work.

4 Give consideration to the appropriateness of treating together in groups, individuals with widely differing abuse experiences.

5 Consider development of separate treatment programmes to meet specific needs of children and adolescents who have been rejected and disbelieved by their families.

6 Allow the opportunity to family members in treatment to comment on their own progress, and contribute to setting new targets where these are necessary.

7 Provide the opportunity for those adults or abused with specific psychiatric needs, such as depressive or anxiety disorders, or PTSD, to receive specialised therapy.

Appendix

The Sexual Abuse Treatment Programmes

Introduction
Arnon Bentovim, Caroline Hyde & Anne Elton
The sexual abuse treatment programme was initiated in the Department of Psychological Medicine (DPM) at the Hospitals for Sick Children [Great Ormond Street] in 1980. This followed a survey of children being seen by a variety of professionals in the UK (Mrazek et al, 1983), which had indicated that the usual response to an abusive episode was prosecution of the perpetrator and placement in care of the child. Only 11% of the children identified as sexually abused was receiving any treatment, and the professional response at that time was to avoid speaking of the abuse to the child. It was expected that the children would recover from the abuse without specific intervention, and there was no research available which any other conclusion.

A combination of concerns from both feminists and from child care professionals began to articulate a different perspective about the long-term harm of sexually abusive experiences on children and young people; these have been discussed in the introduction to the main study. Research also started to indicate that there were short-term effects on emotional functioning and behaviour, also reviewed above.

The early approaches to treatment by the specialist team in the DPM were based on two major elements which still remain as part of the treatment programmes. These were a group work approach, and a family-systems approach linked with statutory child care and criminal justice agencies.

1. Group work approach

The first major element in the treatment programme was the establishment of groups for family member to work with their peers. Giaretto (1981) introduced this approach to family work, by not so much working with the family conjointly, but by offering each family member the opportunity of working in separate groups.

From early in the work of the specialist DPM team, group work was introduced for children of various ages, for the mothers of abused children, for perpetrators and for couples when, although one was the abuser, there seemed some chance of rehabilitation.

Groups for younger children had a 'psycho-educational' approach, with similar components for older children, but with more opportunity of sharing feelings and responses. Group work, like family work pursues a number of goals, which are detailed in the following pages. As well as the goals shared with family work, group work offers the opportunity for sharing and emotional support from peers who have had similar experiences: raising self-esteem, developing assertiveness and self-protective approaches, and developing understanding concerning sexual issues, the nature of abuse and its effects.

Work achieved in groups was integrated at regular intervals with family work, with professionals in the community carrying out appropriate tasks following discussion with the DPM team in network meetings and in separate professionals' meetings.

Our earlier research indicated that this approach for sexually abused children and their families was helpful in reducing the rate of re-abuse, and in improving situations emotionally for children (Bentovim et al, 1988). Maintaining the protective context, however, meant that only 14% of the children were living at follow-up with both their parents, a further third with their mothers, about a quarter in alternative family or care context, while about a fifth were living independently.

Recent concerns about the nature and extensiveness of offending behaviour have been incorporated into the treatment through the introduction of more intensive groups for parents who had offended and young people who had begun to offend. There is a growing consensus about the need for group work for offending behaviour to be focused on detailed consideration of the abusive cycle which involves the perpetrator with victims both within and outside the family, patterns of arousal to children and their origins, and the way these are maintained. There is a need for the perpetrators to understand the effects of abuse and take responsibility for this. Their abusive behaviour needs to be linked to their own experiences of abuse, sexual, physical and emotional when this is relevant.

2. Family Systems Approach

The second major component of the specialist team's work is the family systems approach. This involves attempting to understanding sexual abuse in the family as arising from a cyclical set of experiences: the childhood experiences of the parent who subsequently abuses, with current abusive actions being aspects of the current family context and the vulnerability of the children, which together maintain the sexual of violent actions of some members.

Furniss (1983) explored the different family patterns which were seen including those with a high degree of secrecy who often functioned reasonably well in the community, in contrast with families where there was a far higher level of violence and more obvious poor child care.

The major therapeutic steps which flowed from this model included the necessity for absolute open-ness between the family members and the professionals about the nature of the abuse. There was also a necessity to give the protection of the children the highest priority involving working with the child care professionals, the criminal justice system and the probation officers in charge of the perpetrator. This gave a mandate to ensure ongoing protection of the child and for the therapeutic agency to work with the protective agencies to deliver a therapeutic service.

The first task was to ensure that the relationship between the mother and the abused children, which has often broken down, was strengthened and built up, to explore the origins of the abusive behaviour, and to test whether the relationship between the parents was sufficiently protective and to reverse abusive patterns of parenting. In cases where there had been satisfactory earlier work it was possible to consider the restoration of the relationship between the abuser and the abused child.

In such an approach it was essential to pay proper attention to the inter-professional working relationships of the therapeutic agency professionals, and the care and probation professionals, to assign tasks and gather information about progress to ensure that family members were confronted with issues which needed change both within and between individuals.

Assessment and Aims of Treatment

Caroline Hyde

For most of the children and families in the current study the initial 'diagnosis' had been completed in the community before referral to the specialist hospital team. In a few cases it was necessary for the clinical team to carry out further disclosure work in conjunction with community professionals, and this was completed before treatment by the hospital team began. This work ensured the safety of the children.

In the majority of cases, therefore, the initial contact of the family with the clinical team occurred some time after the acute phase of disclosure. The task in assessment was to estimate the positions reached art referral by each family member and by the family overall, and to judge the prognosis for treatment.

Where possible the whole family was seen for assessment. In cases where the perpetrator appeared unable to take any responsibility for the abuse he (nearly all were males) was seen separately. Children in foster care or living in children's homes were seen with their substitute carers, and the biological family was seen separately when this was appropriate, for example where return to the family seemed likely. In all cases the key professionals from the community (social workers, probation officers, care workers, etc) were included in the assessment.

The framework for this initial assessment and for monitoring progress during treatment was a series of twelve Family Treatment Aims on which families, mothers and the abused children were rated by the clinicians. These twelve aims have been given in detail in the section on Measures in the main Report. These aims centre on reducing the risk of re-abuse, and focused in particular on appropriate allocation of responsibility for the abuse (located with the perpetrators not the children) by all family members, on the protective capacity of the non-abusing parents, on parental willingness to prioritise the needs of the abused children, and on open communication between all family members. While these treatment aims were rated from the point of view of the key individuals (children, mothers and perpetrators), they incorporated measures of individual and dyadic functioning as well as functioning of the family system as whole.

A. ASSIGNING RESPONSIBILITY FOR THE ABUSE

1. Father/mother/children acknowledge that abuse is the responsibility of the perpetrator not the child
2. Mother/father acknowledge own responsibility for abuse, or for lack of availability to child (when applicable).
3. Evidence of the child's resolution of conflicted feelings to perpetrator and parents (eg, over-closeness, loyalty)

B. TREATMENT FOCUSED ON FAMILY RELATIONSHIPS

4. Family members able to express reasonable anger to each other in an appropriate, non-destructive and non-scapegoating manner
5. Development of relationships sufficiently open for painful issues to be confided and shared (a) between child and parents and (b) between parents

6. Family allow all members space to speak and listen to each other without scapegoating
7. Recognition of individual needs appropriate to age, etc
8. Recognition of positive qualities in all family members, including each member recognising positive aspects of self
9. i) Establishment and maintenance of appropriate generational boundaries
ii) Provision of adequate protection from further abuse to children
C. TREATMENT FOCUSED ON ORIGINS AND EFFECTS OF ABUSE
10. i) Recognition of potential and actual damage which might affect child as a result of sexual abuse
ii) Parents show capacity to help child with actual behavioural effects of abuse
iii) Recognition of damage caused to adults by abusive experiences in their own youth
11. Recognition by adults of needs for help on an individual and couple bases, including specific help with sexual difficulties
12. Ability of families and professionals to work together co-operatively

In addition to rating these twelve treatment aims for each child, mother and perpetrator, an estimate was made of the capacity for change of individuals and the family system overall. An important factor in this estimation was the strength of the community professional network and the family's capacity to work co-operatively with professionals.

Estimate of treatability

On the basis of the above factors a clinical prediction of prognosis was made for each family at the time of referral to the specialist team. Families were assigned to one of the following categories.

Hopeful

This category covered those cases where family members had already achieved a number of treatment aims, in particular those concerned with appropriate allocation of responsibility for the abuse, recognition of the child's need for protection, some awareness of the need for change and flexibility in family members, and co-operation with professionals.

Doubtful

This category covered the cases where there was either insufficient recognition of appropriate responsibility for abuse, need for protection, flexibility and co-operation to be hopeful, or there was insufficient reason not to feel hopeless. This category was subsequently divided into 'Doubtful with some hope' and 'Doubtful with little hope'.

Hopeless

This category covered those cases where there was absolute denial of the abuse by the adults responsible, where the child was disbelieved and sacrificed in favour of the abuser buy the non-abusing parent, where the child was scapegoated and their need for protection overlooked in favour of the needs of the adults in the family, and where there was no reason to anticipate change. In cases where no hope was held for the family, children could still be rated as 'Hopeful' if there was a clear plan to allow them to settle in a supportive substitute family and good support by social services and other community professionals, and the children themselves were thought to have the potential to accept and use the treatment.

An estimate of treatability in one of those categories was made for each of the following:
* each abused child
* mothers
* perpetrators
* mother-child relationships
* perpetrator-child relationships
* marital relationship
* family overall

The following examples from the current study illustrate the differences between the categories.

Family rated as 'Hopeful': example 1

The abusing father in the M1 family was on probation and living in hostel accommodation at the time of the assessment. From the time of disclosure he had acknowledged full responsibility for the abuse and accepted the need for separation from his family, although he hoped for rehabilitation ultimately, and expressed a strong wish for treatment. The mother was protective and believed her daughter's allegation from the start. She was able to accept that her passivity and rather peripheral position in the family before disclosure together with the couple's current anxiety about infertility may have contributed to the occurrence of the abuse. Both mother and daughter were keen to receive treatment, and to work towards the rehabilitation of the father. The high quality of input to the family by the community-based professional network (probation service and social services), and the good co-operation between the family and the professionals was felt to increase still further the likelihood of a good outcome for this family.

Family rated as 'Hopeful': example 2.

The family unit in the S1 family consisted only of mother and her 8-year old daughter. The abusing father (who was seen for assessment) denied the allegation entirely, but the mother was allied with the child, had already initiated divorce proceedings, and there was no contact between the child and her father. At the time of assessment the mother was depressed and remorseful at her own failure to protect her daughter. The child was quite withdrawn, and exhibited a number of post-traumatic stress phenomena. Mother and daughter were both anxious to protect each other from further distress and this restricted emotional communication between them, but both were keen to receive treatment, were able separately to begin to articulate their distress, and were well supported by community professionals whom they trusted. They were felt to stand a good chance of benefiting considerably from treatment.

Family rated as 'Doubtful'—1. Some hope

The C family consisted of mother, step-father (perpetrator), a 13-year old abused step-daughter, and a 9-year old daughter of both parents. The step-father was still completing a prison sentence but was granted release in order to attend an assessment interview. In this interview he demonstrated only limited acceptance of responsibility, and tended to minimise the events of the abuse. The mother had oscillated in her allegiance between the children and her husband in the early stages after disclosure, but at the time of assessment had consolidated her priorities in support of the children. However, there was concern over her lack of trust in herself in maintaining separation from the step-father which, at the time of assessment, manifested itself in extremes of over-protective behaviour towards the children. The mother had taken several overdoses around the time of disclosure and the abused girl had also taken an overdose some time before the referral to the specialist team. Rehabilitation of the perpetrator to the family was not contemplated, but hope was held for the mother and children if the mother could be supported in separating fully from her partner, and the child helped to overcome both the effects of the abuse itself and her mother's initial ambivalence.

Family rated as 'Doubtful with little hope—2'

The three children in the W family were abused by their natural father who, with his second wife, had been given care of the children following the break-down of his first marriage. The natural mother neglected the children while she had the care of them, but at the time of the assessment by the specialist team she had established a new relationship and wanted to resume care of them (the children were currently in foster-care).

The natural father denied the abuse allegations absolutely. The mother believed the allegations, but appeared to have limited understanding of the effects on the children, and had difficulty in working co-operatively with professionals with whom she was often in conflict. The attitudes of the children to both parents were confused and conflict-ridden. Although treatment was planned to include the natural mother, little hope was held of rehabilitation of the children to her care, and the outlook of the team for the children was guarded.

Family rated as 'Hopeless': example 1.

In the M2 family the mother was firmly allied with the abusing step-father who demonstrated minimal acceptance of responsibility for the abuse and refused to attend for treatment. The 16-year old abused step-daughter was in foster care at the time of assessment. The motivation of the mother and the step-daughter for the treatment was only slightly better than that of the step-father, and family pathology was so entrenched that the prognosis for any rehabilitation was felt to be nil. Treatment was planned to focus on separation of the abused child from her natural family and re-establishment in substitute care.

Family rated as 'Hopeless': example 2.

The 15-year old abused daughter in the S2 family had attempted to disclose long-standing abuse by her step-father on three occasions before action was taken. The first two 'disclosures' (to a teacher and later her GP) were ignored. The step-father, who was separated from the family at the time of the assessment, denied the abuse allegation vociferously. The mother, with whom the child was living, appeared precarious in her support of her daughter and was pre-occupied with her own needs, saying that she was not sure she could live without her husband and had contemplated suicide. The child was struggling to maintain loyalty to her mother. She was depressed and withdrawn, failing in school, had thought of suicide and was self-mutilating. At the time of assessment it was felt that it might prove impossible for the child to remain in her mother's care, but it was agreed to offer treatment to both with close supervision by social services. The outlook for both mother-child relationship and for each individually was felt to be poor.

At the end of treatment the clinical staff used the same twelve areas to rate progress, they also rated the extent of progress made by each treated individual, and the relationships as described above. The amount of improvement observed at the end of treatment was measured against the clinical expectations recorded at the start by the clinicians.

Treatment Programmes

I. Offenders: Young People and Adults
Arnon Bentovim

Group work for young people and adults who are offenders is predominantly focused on offending behaviour, particularly on the cycle of abusing and what maintains it.

Only when these issues have been addressed is it appropriate to look in detail at any problems which may arise from either the individual's own experience of victimisation or any other early experiences which contributed to the development of the abusive, offending behaviour. Finklehor (1984) in his exploration of the factors which lead to an abusive orientation towards children also noted that victimisation, along with other major stresses in childhood, could result in an abusive orientation.

Research indicates that both sexual and physical abuse in childhood may be important aetiological factors in the development of Sexually abusive behaviour. Early privation and failure of care in boys and young men may lead to a sexualisation of intimacy since they find that the closeness which they lacked early on from their carers may be apparently achieved during later childhood and adolescence through sexual contact. At the same time they discover that it is possible to become powerful through using their own sexuality abusively; this reverses early experiences of powerlessness.

Boys and men who have had a lack of warmth, intimacy and good memories in childhood may develop a compulsively sexualised relationship to children. Those who have been actively victimised themselves may compulsively need to reverse the experience of helplessness, pain and terror.

The only mother in the current study who had abused alone, and not in the company of her male partner, was a woman who had been extensively abused herself in childhood; her caretaking of children was also pervasively sexualised.

Specific issues need to be addressed in groupwork with offenders, these are:—

(1) *Sexual Orientation and Distorted Beliefs*

Those who abuse have major distortions of sexual arousal and interest. These are relieved and maintained both by abusive actions and by masturbation to distorted fantasies, which may be followed by a period of self disgust and a then a variety of justifications of their activities. Their 'grooming' of the child, bribes and threats are all part of this self justification. This may lead to total denial of their actions, and a level of pleasure which can be at an addictive level, leading to there being no desire to take any control. They interpret their victims' silence and lack of any protest as consent rather than frozen anxiety, perceive any protest as lack of affection towards themselves. Any failure of sexual response by their partner is seen as rejection, further humiliation and justification for their perverse activities, rather than arising from their partner's immediate personal state or from their own failures in loving and appropriate attentiveness.

(2) *Distorted Sense of Self*

Offenders can be extremely self-critical. When asked what they feel would happen if they truly and fully acknowledged their abusive acts and orientation, the common response is that violent suicidal action would occur. Often they feel their abusive orientation to be so powerful as to be beyond control. Therefore denial and maintenance of a self-righteous stance is essential to avoid grievance. Emphasising their victim experiences and neediness is used as a way of reducing their own guilt and self-punitiveness, and not facing the addictive and pleasurable nature of their actions and taking responsibility for them. A fundamental lack of honesty makes it extraordinarily difficult for the therapist to know what is the real self and what a defensive mask to maintain a position of power and triumph.

(3) *Sexualisation of Relationships*

The major confusion shown by offenders is that powerful primary emotions, both positive and negative, may only be expressed through sexuality. Both closeness, intimacy, positive caring and also anger and grievance about powerlessness become sexualised. Mr M, a man whom we felt showed some real capacity to change, revealed during the group his idealised attempt to give everything to his daughter in distinction to his own experience of privation. Yet at the same time he could feel a sense of grievance with a child who had more than he did as a child. He thus had a powerful sense of grievance towards women, particularly his mother, wife and daughter, which had an important role in triggering and maintaining his abusive behaviour.

Aims and Focus of Group Work

(1) Sharing

Meeting others in a similar situation has a powerful effect in reducing stigmatisation and so helping towards genuine openess and honesty. Deviant sexual arousal patterns may then be explored in a non-punitive context, but with appropriate confrontation and care.

(2) Work on Sexual Constructs

This work focuses on the origins of deviant arousal in childhood and adolescence, and traces the cycle of arousal, abuse, responses and reinforcement. It is essential that this is done in considerable detail and is examined in current and future relationships.

(3) Distorted beliefs and feelings

Examining, sharing and confrontation of rationalisations by others may gradually help self-examination.

(4) Victim Awareness

Developing a more realistic appreciation of the trauma experienced by victims of sexual abuse and the likely or possible long-term results.

(5) Exploration of Offenders Victimisation Experiences

Work on painful experiences which have shaped the abusive self, their sense of powerlessness, and exploration of ways of having emotional needs met through abusive actions.

Group Activities

A variety of group activities are needed to explore these themes in a number of different ways. Adults find it easier to talk than do young people, who need the medium of questionnaires, constructing responses through group discussion and facilitation by the leaders.

A variety of other approaches, use of videos, reading texts, completing questionnaires and discussing them are all helpful. The programme can take a theme, use a task to explore it, and then set follow up tasks and check responses at the next meeting.

(1) Group Rules

It is essential to establish rules about attendance, confidentiality limits, and language use. It is helpful to have a male and female co-therapist so that inappropriate attitudes to women can be confronted during the life of the group.

(2) Work on Sexual Constructs

(a) Developing a detailed picture of their sexuality and patterns through the use of questionnaires.
   (i) Asking group members to fill in questionnaires which focus on hostility, attitudes to women and rape, sexual fantasies and orientation, views about sexual activities; all these help to put group members in touch with their own patterns and act as a focus for discussion on a variety of themes.
   (ii) Developing their own questionnaires and patterns.
   Developing their own questionnaires can be particularly helpful for young people to do together. For example they can be asked to identify the elements in a normal dating cycles or in abusive activities so that everyone contributes to the total. Members can then be asked to indicate which statements apply to them.
(b) Use of victim statements.
   It is important for victim statements made to the police or Social Services Department to be compared with what the offender has identified and described. This allows for a close examination and taking apart of the self-justifications, rationalisations, minimisations and denials, and opens the way to constructing a more accurate picture of the offending behaviour.

(3) Examination of Cycle of Abuse

After doing this it is now appropriate to look in detail at each member's cycle of abuse. They have to acknowledge the origins of the abusive action, the way in which control was acquired by grooming and power maintained by threat. Patterns of arousal and their maintenance by masturbatory fantasy are examined so that the individual can see himself and his actions fully in a context of support rather than attack.

(4) Response and Prevention.

It is essential to develop preventative strategies for each individual so that they now themselves own their own pattern sufficiently to identify ways of taking avoiding action. There also has to be a deconstruction of previous patterns and construction of new ways of relating which are not abusive.

Attitudes to Self and Others

To understand what has motivated them and led to their own pattern it is essential for each member to do some work on his own experiences.

(1) Family Tree

Constructing a full picture of their families, carers, life experiences and all partners can help members understand how their own behaviour and general ways of relating have developed.

(2) Parenting

It is important to discuss how to manage both being partner and parent in terms of protection. If the offender is going to have continued contact with his family it means identifying and adopting appropriate rules to ensure the safety of any children.

(3) Self Assertion and development of intimacy and emotional sharing

(a) Accepting anger.

One of the most difficult things for adult and young offenders is to face and accept the anger and distress of others. The powerful sense of grievance and anxiety about punitiveness frequently lead to desperate attempts to reduce other's anger and avoid conflict. In the group it is important to be able to express and hear anger appropriately. Inappropriate expressions of grievance, hostility and

attempts to control need to be confronted. Role plays can help define other ways of accepting appropriate anger other than by continued self-justification or attack.
(b) Confiding worries and problems
This is an important dimension offered by the group. It can challenge the habitual pattern of secrecy and silence.

Victim Awareness

Awareness of how victims feel and suffer is an important stage for offenders to reach before being able to speak of any abusive experiences of their own.

(1) Videos

This is a powerful medium in helping offenders begin to appreciate the long-lasting impact of their actions. There are a number of videos of survivors, both adults and young people which can usefully be shown.

(2) Texts

Finding letters published from children and young people who have been abused, and the long term effects on adults is again a powerful way of confronting such experiences to help individual abusers become aware of the effect of their abuse. Dealing with their own abuse and victimisation require many of the same approaches as are necessary with those presenting as victims.

Case example:
Initially Mr M was extremely unforthcoming in the group. He had the ability to sit like a statue without movement and with no or minimal verbal response. He was able to say briefly that he had been abused throughout his childhood, rejected by his parents and brought up in a children's home where he was sexually abused by a woman care worker. This was all described without affect. Similarly he talked flatly of the death (in a road traffic accident) of his first wife. He was convinced that his abusive relationship with his daughter was mutually loving. It was after he saw the video where victims spoke about the lifelong suffering and distress caused by their abuse that he had a breakthrough and was able to ask his daughter if indeed she was afraid of him. She confirmed this which shocked him. He then began to speak in more detail of his own abuse, and the way in which he had developed masturbatory fantasies and so eroticised the abuse that it became an important factor in the

development of his own sexuality. He also spoke of the way in which he idealised speed, dangerousness, which had played a part in the death of his first wife. He began to be rather more understanding his second wife's distress at his lack of emotional response. He began to realise what shell he had grown around him and he developed an increasing preoccupation with his sexual arousal to children. Nevertheless, he continued to take flight either into outbursts of violent anger or non-attendance.

Practical Considerations

(1) Resources

It is essential to consider the way in which young people and adults who have offended are received. If there is a parallel group where partners or victims are being treated in the same building, it is essential that the abusive young people and adults come at somewhat different time so that there is no overlap in waiting facilities. Careful reception, the availability of comfortable rooms, refreshments are all helpful.

(2) Membership

Abusive young people need to be in a size of group which is manageable by co-workers. It is essential that careful pre-screening work is carried through to exclude individuals who, for instance, have used considerable levels of violence, since it is preferable to have a membership who are a reasonably consonant one with another. Individuals who have developed raping patterns of behaviour should be treated separately to individuals who are abusing within family contexts or known contexts, even if there is the threat of violence.

(2) Co-therapy

It is helpful in working with young people and adults who are offending to have a man and woman in a co-therapy team. Issues of gender are important in looking at offending behaviour. There is an advantage in having a woman who can directly confront issues in a safe context and will not take a compliant role in relation to threats or description of violent behaviour. Direct live supervision of work or time to supervise and discuss both before and after groups is essential for work to prevent splitting and separation of the workers.

(4) Duration

Groups for young people need to be at least 16–18 sessions to be effective, whilst with adults, at least 20–30 sessions or more are essential to work through the full programme. Groups need to last an hour and quarter to be effective.

(5) Links with Professional Network

These need to be established and maintained throughout the life of the group. With young people who have offended it was found helpful to ensure that a professional, either a residential or fieldworker accompany the young person to the group meeting, and themselves attend a group for care workers. This close contact with professionals means that it is possible to set 'homework' which can be completed with community professionals between weekly group meetings. It is essential for community workers to have a clear awareness of the programme of work, to be able to feedback on responses to particular groups, and to be able to receive consultation on their work.

Adults need a close link with a community professional, preferably a Probation Officer, who meets regularly with them throughout the life of the group. Again a close link with the individual in the community can be maintained, the rules for contact with children and living contexts can all be monitored, and regularly shared at 4–6 weekly family meetings with appropriate professionals. All this ensures that the work in groups is fed into the family work.

Suggested Programme for Sexually Abusive Young People

(1) Introductions, establishing rules, hopes and fears for the group, development of appropriate sexual language using body map charts. Developing an agreed language for the group of acceptable and unacceptable sexual words.

(2) A group exercise ranking sexual acts, using cards, stating as many different acts, both appropriate and inappropriate that members and leaders that could list. Acts are ranked for abusive content.

(3) Normal dating cycle. A group exercise defining the steps towards making a sexual contact in an appropriate consensual way. Appropriate ways of meeting, talking, fantasising, making dates in a group, non-sexual touches, finding appropriate ways of giving permission and seeking consent to sexual activities.

(4) A number of sessions describing cycles of offending, using the group to build up both pre, during, post responses and each young person having to define what relates to them in detail, using community caretakers to check out and bring back homework after each session.

(5) Comparing with victims' statements and group session exploring victim responses. Seeking list of responses from group members.

(6) Session on their own victim experiences, anxieties, fears about homosexuality, having flashbacks, connection between victim responses and offender activities, how they link.

(7) To prevent reoffending introducing principles of thought blocking, masturbatory satiation, introduction of more appropriate fantasies and masturbation.

(8) There can be additional sessions of development of social skills, assertion, training and sharing, perspective taking, anger management.

Programme for Adults who have Sexually Offended

(1) Hopes and expectations for the group, exploration of attitudes and views, particularly of Social Services Department, Probation Departments, the Courts, and what group members have shared in terms of experience up to the time of coming to the group. Their wish to understand why they abused, and to be able to take control rather than be controlled—the need to work together.

(2) Attitudes towards women and children, e.g. through exploration of rape attitude inventory, and exploration of their own background through genogram.

(3) Exploration of early memories, fantasies and sexual experiences, what they shared, what was different, attitudes to sexuality in their own families, their own abuse, and the place of sexuality in their social lives as they were growing up.

(4) What is left over from their childhood experiences of sexuality and growing up, anxiety, embarrassment and discomfort about sexual matters, earlier sexual experiences and current ones in their own with partners.

(5) Sessions focusing on the development of abusive orientation, introduction of material about sex offending cycles. Detailed accounts of how abusive actions began, how they were maintained, the effect on them, arousal masturbatory patterns, the experience of the power of their own abusive impulses and actions getting each member to describe his own cycle of abuse, and share this in the group.

(6) Sessions on rationalisations. How did they justify themselves, their actions, their attributions to their children, partners, and the need to find ways of controlling rather than being controlled, facing an awareness that they were always at risk and have to take precautions and find ways of controlling themselves in the future.

(7) Awareness of victim effects, through watching videos, texts, comparing the effect of abuse on their children and beginning to look at their own experience of abuse.

(8) Looking at their own future relationships with partners, rules for living, emotional communication, and continuing to think about effects on their children, and ways of more appropriate relating as by this time there may be more contact with children, partners, other family members.

(9) Use of various cognitive techniques introduced, e.g. masturbatory satiation, impulse control, introducing different endings to their fantasies as a means of blocking deviant fantasies. Such work needs to be linked to probation officers' work.

Programme for Couples Group (when one is abusive and one non-abusive

(1) Each couple needs to share with the others what has happened, who has been abused, how abuse was revealed, and who to, the responses of mother's, and the issue of denial or acceptance of abuse. Aims for each family to be compared and contrasted and goals to be reached and how and with whom.

(2) Beginning to explore feelings within group—abuser in relationship to themselves, non-abuser to be able to share how they feel, the dilemma between being a parent and a partner, and being in touch with their own anger and hurt.

(3) The use of sculpting and role play to explore relationships within the family before and whilst abuse was continuing and defining future patterns and how they can be achieved.

(4) Discussion of how to meet children's needs, rules necessary for future safety and protection of children, the need for non-abuser to have a more key role in relationship to protection of children, and to be the person who controls the relationship between abusers and children.

(5) The issue of their own sexual relationships, problems and difficulties, development of tasks for couples to be able to share in different ways and to report back, the future and relationships with professionals, couples issues and parenting issues.

II. Treatment Processes with sexually abused children and adolescents and their families
Anne Elton

1. Family Treatment

The aims of treatment have been detailed. In addition we clearly wished to try and help the victims overcome dysfunctional consequences of the abuse, such as behavioural difficulties, sleeping problems, and post-traumatic symptoms and thus become survivors rather than victims. As we were seeing them early on in this process we will refer to them as victims throughout this section.

Although treatment of the family was seen as a central part of the therapeutic programme it must be emphasised that the family treatment sessions were far more often held with parts of the family. The most frequently seen sub-system was the mother/victim dyad; increasingly it was likely that some time would be given to them individually on occasion. In the few families where the offender has admitted responsibility for the abuse and where the mother wished to have some continuing relationship with him there would be a couple session or sessions. This could occur in cases where mother did not wish the offender ever to live with her or the children again, but where she felt that it was in the interest of the children (or at least some of them) to have some continued contact with their father (or offender).

Although it was usual for all family members to be seen during the evaluation period very young siblings were often not included in ongoing therapeutic sessions. In some families much older siblings showed marked resistance to attending and their mothers were not always either able to persuade them to attend or convinced that it was appropriate to do so. In several families girls of around 11 were most reluctant to include sisters aged around 8 on the grounds that they were afraid that the sister concerned would "gossip" about the abuse and so bring more distress to the abused child. This anxiety was shared by the mothers on what seemed to be realistic grounds. Nonetheless it was important to work towards some inclusion of younger sisters in order to identify any help she might need in dealing with the sudden loss of a father or stepfather.

Similarly it must be emphasised that those working on the team did not perceive the family system and any dysfunction in it as causing the abuse; the responsibility for that was seen as lying clearly with the individual abuser. But it was felt essential to try and identify any elements in the family system which might either have contributed to the vulnerability of a particular child or children or which might maintain an abusive context. Again it should be stressed that this was not always felt to be the case; part of the purpose of the evaluation and treatment was to try to identify families which did not have inherently dysfunctional patterns. It was also recognised that even if there had not been a dysfunction in the relationship between mother and victim prior to the abuse and its disclosure the disclosure, itself might make the relationship more vulnerable because of the feelings of guilt, rivalry and responsibility, however misplaced.

The twelve family treatment aims have already been detailed. We saw the work as falling into two stages, *disclosure and treatment*, although there were undoubtedly overlaps between them.

Work in Initial Crisis

At this time the family was in crisis and the primary aims of work were to ensure appropriate protection for the victim(s) and any other children. This frequently meant helping them with Court proceedings, both civil and criminal, and being involved actively in such proceedings ourselves. Secondly, we had to try and prevent the family closing up and so refusing or becoming unavailable for treatment. This was most likely to occur either if they denied that abuse had occurred or if they minimised its seriousness to such a degree that they refused professional help, insisting that there was no risk of future abuse and no harmful sequelae to the child or family from the previous abuse.

Offender Attitudes

Offenders invariably denied the possibility of their abusing again at this early crisis stage. Since most of them at this point had not been through the criminal process and as none of them had received any treatment this is hardly surprising. Some admitted only some of the offences and at this stage it was not possible to be clear whether denial of the abuses carrying more severe penalties was made out of protective self-interest or from a need to minimise and deny even to themselves.

Case example: Girl of 12 abused by her biological father

Amy's[8] father was one such man. Although her mother had always believed her disclosures she had also made it clear that she wished for eventual reunion with her husband. Indeed she had, unusually in our experience, insisted on having a time immediately after the disclosure when Amy left home to go to relatives and she could confront and discuss with her husband. During this period she had heard from him

---

[8] All names and some details of cases have been changed to prevent identification.

something of his more pervasive fantasising about this daughter. Consequently in family work it was possible for her to recount these to underline the serious risk. In addition the therapist asked the Probation Officer to detail all the offences specified and helped Amy clarify and reiterate her own experiences. Amy's mother then understood and accepted the need for her husband to live away from home for a significant period while he received treatment and also appreciated the possible risk to the other children in the family. It was then possible to support her through the subsequent wardship and criminal proceedings.

This work does not necessarily end when the initial crisis is over. Amy's father subsequently continued his denial of some of Amy's allegations and she was pressurised into retracting them rather than being able, with her parents' help to clarify any confusions which might have arisen from her own experience of the abuse and her father's technical knowledge of sexual activity. A focus of work during the treatment stage was an attempt to clarify this and support Amy.

In the same family the father who had gone into a treatment programme for adult offenders later became extremely negative to all professionals and refused to continue treatment. His wife was more concerned about this despite her continued wish for him to remain in the family, and the children closing ranks in support of their father. It was necessary for the agencies with statutory responsibility to return to the civil court to express the renewed concern.

This example illustrates the need for continued work on offender responsibility and appropriate child protection throughout the whole treatment period.

Denying Attitudes of Mothers

A significant number of mothers did deny that their child had been abused and remained totally allied to the offender. In these cases there was no possibility of family work and all that could be offered was therapeutic work with the victim(s), either in groups and/or with their substitute carers and social workers. The child/children thus had to face the double crisis of having disclosed abuse, of being rejected and of being moved to alternative homes, with all the attendant uncertainties about long-term placement

Children who remained uncertain about their long-term future were likely to be preoccupied with that and so were often unavailable emotionally for treatment on the effects of the abuse.

Case example: Girl of 12 abused by an uncle

Claudia had been in care at the time of her abuse by an uncle; her mother suffered from a chronic psychiatric illness and her father had died many years earlier. Claudia presented an apparently mature facade which was her attempt to respond to the failure to everyone to provide a stable home. The abuser had been more 'caring' to her than most other people in her life and indeed she denied that she was 'abused' by him although she admitted that sexual activities from others earlier in her life had been abusive. During the treatment state it became clear that she was both very vulnerable to further abuse and at times very infantile in her behaviour. Her dependency needs could be recognised by the professionals but Claudia herself had to deny them and fight attempts to help because of the real lack of stable carers in her life. Ultimately she was helped to accept placement in a therapeutic community.

Similarly families were unlikely to have the emotional energy to enter into the treatment stage until all the practical and legal arrangements about child protection were dealt with. Offenders could not truly engage themselves until criminal proceedings were over.

Treatment Stage

We would suggest that there were three different family contexts where treatment was offered.

1. Families where there was no change of membership because the abuser had left it for ever prior to disclosure.
2. Families which hoped to reintegrate the offender or the victim if she had been moved out.
3. Families which were 'new' where there was no intention to re-unite with the offender.

In all these situations we continued to work on the various conflicted feelings, open communications and difficulties in relationships.

Working with guilt and rivalry in mother-child relationships

We found that guilt was one of the most prevailing feelings among both mothers and victims. Much of the treatment centred on helping families modify this, since much was unreasonably attributed. Offenders of course contributed to this, often actively, since the more they could engender it the more they shifted responsibility from themselves and avoided having anger directed at them.

In families where the mother wished the offender to return and the victim did not there was always a painful tension because of this conflict of need and wish. This usually made both mother and the child/young person feel guilty and contributed to continuing or increased problems in open communication between them. It also increased the child's feelings of being let down and her anger with the mother; yet frequently because of the child's dependency on the mother was well as their emotional bond it was extremely difficult for these feelings to be shown openly. On the mother's side, even if she was totally protective, there was anger that her child's needs could prohibit her meeting her own emotional needs. Mothers also had to cope with feelings of jealousy that their daughters had been 'another woman' for the partners even when they themselves completely rejected the partner and were supportive and protective of the child. Sexual jealousy and rivalry were more likely to be felt by mothers of adolescent daughters than in relation to younger children.

Although we tried hard to work with the mother/daughter dyad in most cases it was extremely difficult to achieve significant change in their relationship where these feelings predominated. It was particularly so where there were longer standing inhibitions in open and verbal communication. It would seem that the mother's own unmet needs prevented her attending to her daughter enough.

Case example: Girl of 12 abused by her stepfather

Teresa had disclosed abuse by her stepfather by writing her mother a note; her mother believed her, was able to protect but did for a long time hope for ultimate reunion. Despite many attempts to help them talk and share over routine activities and feelings rather than very emotive areas the mother was never able to do so. Teresa continued to leave notes when she felt able to communicate something important and her mother would then use third parties such as professionals to address these issues with her daughter. Only after the mother definitely gave up any wish to be reunited with the abuser did the relationship between Teresa and her mother become rather more relaxed. However, we could have no confidence that in any future situation where there was a significant conflict between the mother's interests or views and Teresa's they would be able to communicate any more freely or directly.

Even in situations where there never had been any question of the offender returning to the family and no wish on anyone's part that he should do so maternal guilt, however unfounded or exaggerated, could seriously inhibit closeness with the victim.

Case example: Girl of 14 abused by stepfather

Daphne aged 14 was abused by an ex-cohabitee of her mothers. The mother had ended her relationship with him some time before Daphne disclosed sexual abuse, on the basis of his physical abuse of her and punitiveness to the children, Daphne and John (16). Although the mother was very distressed on learning of the abuse and wanted her daughter to have treatment she was not able to encourage her to do anything that she found difficult. It was also clear that the mother felt very guilty at having had a partner who turned out to be an abuser. This family was not in the sample offered group work and Daphne refused family work after the initial assessment. It seemed that it was too painful for either mother or Daphne to share much when they had acknowledged the existence of some guilt (mother's) and anger (Daphne's). After the follow-up they were offered a review appointment a year later to which they came. Daphne revealed that she was having quite severe post-traumatic flashbacks, nightmares and phobias. With encouragement from her mother she was able, albeit erratically at first, to accept some individual work and both came to a group about six months later. Daphne did extremely well and nine months later was symptom free and performing far better at school and socially. Significantly her mother was able to bring her but not communicate much spontaneously around Daphne's symptoms until her daughter began to improve. The more Daphne improved the more her mother was able to relate spontaneously to her around painful areas. It seemed possible that this mother's guilt had made her feel so inadequate as a parent that she became frozen in a way she probably had not been before and only as her daughter got better could she feel some unburdening. Whether she would have been helped by more individual time at the outset, or whether she had to experience her child genuinely overcoming the negative effects of the abuse can not be known.

In general we found that although painful and conflicted feelings between family members were often expressed openly we were much less often confident that there was any resolution of these feelings. This was especially marked in conflicted feelings between mothers and victims. There was more often some resolution of sibling conflicts. When the abuser was categorically out of the family there could be more clarity; although in such situations there was also often far less conflict. The feelings were more likely to be universally negative and very angry.

Children showed rather less strongly negative feelings towards sibling abusers although they often had marked anxiety initially. Their parents, on the other hand, often had very considerable difficulty and often disagreements between themselves as to how to deal with the abusing boy, when or whether to allow him back home and so on. When the abusing boy had himself been abused earlier this created even greater distress and disagreement.

Work on Sexual Feelings, Beliefs and Activities

An area of work which is not clearly distinguished in the original treatment aims, but one which was always attended to was that of keeping open work on sexual feelings, beliefs and activities. It probably fits best in Aim 5, that of developing relationships sufficiently open to work on painful issues. Certainly many victims showed marked problems of sexualisation or of major anxieties and inhibitions about touching of any kind. In family settings we would usually work with the mother and child together in order to try and ensure that the mother was actively enabled to help her child overcome confusions and learn appropriate boundaries for sexual behaviour. Many mothers did find this very difficult. Both social norms and/or their own personal upbringing did not facilitate open discussion of the normality of masturbation and childhood erotic pleasure. We found that books and drawings were very helpful tools in facilitating such work; and could be given to mothers to take home and continue work with the child. Even for older children and young adolescents such publications can be helpful.

The individual religious and cultural views of families in this area had to be understood and respected. While these may have impinged at various points in the treatment they were always important to identify over this issue.

Unless there were young children close in age when the work could largely be perceived as educative we did not see this work as appropriate to share in a family context. To do so would have been to break an ordinary boundary between personal privacy and family business.

The greatest difficulties arose when an abused child demonstrated not only overtly sexualised behaviour with other children but also involved younger children; inevitably this always raised the anxious question about whether the child was in turn abusing. Understandably the anger and punitiveness of mothers was likely to be greatest in such circumstances; although some mothers were paralysed by their own anxiety, a response which was likely to be equally unhelpful to the child.

Recognising Individual Needs

Identifying any particular difficulties and clarifying whether any problems shown were results of the abuse or consequent family upheaval, or were behaviours which might have occurred anyway in that child's development was always a potential task. Most time went to trying to help mothers deal with difficult behaviours shown by the victim, such as wetting, tantrums, sleeping problems. At times we did consider such behaviours in non-abused siblings but on the whole only if the mother or another professional actively brought it to our notice.

Loss and Bereavement

Given that in many families there was no proposal to reunite with the offender and that many had become single parent families without wishing to do so, therapeutic attention was paid to issues of loss and bereavement. In that area it was important to try and involve all the family members since all children experienced the change. Non-abused siblings sometimes felt the loss very differently from the victims and there needed to be acceptance of differences.

Mothers needed support often in these situations and in some families there was some active involvement of extended family members who might be supportive in a session. More often mothers were encouraged and helped to share openly either with extended family or friends in order to gain some social support. Since research findings (Finkelhor, 1984) indicate that social isolation is more likely to have a positive correlation with abuse than most other factors this seemed a particularly important thing to do.

Case example: Girl of 14 abused by stepfather

Jane's mother, who had not dared to tell her sister-in-law for fear of rejection, discovered to her amazement that she gained not only a sympathetic ear but also an occasional babysitter. Encouraging ordinary social contact was a very important way of helping the family feel positive about itself and its future. It was often essential to help mothers do this since their children often felt guilty at going out to enjoy themselves if the mother had no outlets or pleasures in her life.

Although we described families as falling into one of three groups we did not find that the work required was necessarily significantly different. At times it was simplified when there was no question of reunion with the abused and no conflicted feelings about that, but as the example of Daphne illustrates this was not always the case. Obviously, if the offender was going to have continued contact more work had to be done with and including him.

The following case example illustrates most facets of the family treatment work.

Case example: girl of 15 abused by her father

Susan had been abused by her father for some years. He had at one time made an advance to her elder sister Elizabeth (17) which she had repulsed and then 'forgotten'. The girls' mother had died when they were quite young and their father had then remarried. There were three boys of this marriage. Susan and her stepmother had always had a very conflicted relationship which everyone agreed was associated with Susan's resentment at having another person come into the family and be especially close to her father. Although it was generally acknowledged that the father was primarily responsible for the abuse, Susan was partially scapegoated for wishing to be so close to her father. Elizabeth, guilty at not having disclosed the father's attempt on her blamed, her sister for not having been able to say "no". The stepmother also blamed herself for her failure to establish a closer relationship with Susan although it was very clear that she had tried very hard and had genuinely wanted to do so. At the disclosure stage of work the main task was to help the stepmother understand the potential risks to the younger children and support her husband's living away and only having supervised access. The work on modifying the confused attributions of responsibility and scapegoating continued throughout the treatment stage.

Much of the work concentrated on trying to help the family (who were seen usually without the offender) appreciate that Susan's wish to be close to her father was natural, as was her reluctance to disclose the abuse; that Elizabeth's non-communication of what had genuinely seemed to her a very minor approach which had never been repeated was also understandable and in no way made her responsible for her sister's prolonged abuse and that the stepmother had really cared for her stepdaughters and tried to develop close relationships with them. In reducing these various feelings of guilt and the resultant scapegoating the family became more appropriately angry with the offender and much less ready to protect him or make excuses (he had himself been a victim and came from an emotionally uncaring family). At the same time the stepmother tried hard to help him maintain a regular and positive contact with his sons and for a considerable period maintained her wish to be reunited some day. The relationship between the sisters improved and although it remained volatile there was more closeness and positive contact between them than there had been for a long time. Susan became noticeably closer to her stepmother, although again this was punctuated by episodes of conflict.

Some work was done with the whole sibship and mother on the loss they had experienced. This was not only the father's absence from the household but also his increasingly noticeable failure to meet his sons' needs. He either failed to keep access arrangements, or else turned up unexpectedly demanding attention or ignored the children when he was in the home. The oldest boy, Richard (12), became quite openly angry with him. He also went through a rather depressed period and his mother was concerned that he might feel negatively identified with his 'failure' of a father. She asked for a specific session for him. Richard was not willing to be seen on his own but was able to tell his mother and the workers that he had had an upsetting experience of being bullied by a teacher. When his mother pursued this matter she learnt that other boys had also complained. Richard became more cheerful and sociable again. It was clarified that his distress and temporary difficulty had really been related to an extra-familial event; the only apparent link was that the teacher may have reminded him of his bullying and emotionally cold father.

In this family work most sessions were held with the stepmother and the two daughters plus Richard on occasions. However, all four were offered some individual time and all except Richard accepted. The father also had individual time in addition to two sessions with his wife, who was by then clearly telling him of the changes she needed to see him make before she could contemplate reunion. There were also two sessions with the father, stepmother, and girls and one with the whole family. Community professionals of course attended all sessions.

2. Group Treatment for Non-abusing Parents and Victims

In group treatment the primary problems being addressed are not the same as in family treatment although group experiences may assist the individual to address certain issues in the family. Groupwork is predominantly focused on problems arising from the experience of victimisation. These are:

1. Low self-esteem

This is almost universally found. Sgroi (1982) described it as the 'damaged goods syndrome'. Finkelhor (1987) describes this as stigmatisation in his description of traumagenic dynamics. Victims seldom believe on an emotional level that they are not intrinsically 'bad', unique, some kind of freak. Mothers who have been victimised share this response. Even those victims who have a high self-esteem based on their value as a sexual commodity have no other positive valuation of themselves.

2. Confusion of feelings

As discussed in the previous section both victims and mothers experience guilt even when objectively they may be seen to have no responsibility either for the abuse or for the consequent disruption of family life. Both may have very ambivalent feelings towards the abuser especially if he was a caring relative or friend. Even in the warmest relationships between mother and daughter there are likely to be moments of painful and ambivalent rivalry, jealousy and anger.

Both may experience overwhelming rage. This rage and the fantasies which accompany it may of its own accord engender guilt or impotence.

Self-hate and depression are common. Many victims are so afraid of the well of painful emotions and so inhibited by that and by the secrecy enjoined on them during the abuse that they dissociate and present as passive and empty.

3. Lack of confidence

The secrecy surrounding sexual abuse does not necessarily end with its disclosure. Families and victims often continue to maintain a high degree of secrecy in relation to others, including extended family. Fear of further stigmatisation can mitigate against sharing even with friends. The ordinary norms of privacy around sexual activity in our society reinforces a taboo of silence; families fear that any discussion will be met with hostile or cold responses.

4. Sexual difficulties and distorted beliefs

Many victims suffer from their premature exposure to erotic awareness either by becoming overtly sexualised and so vulnerable to further abuse or by being so disgusted and horrified by the abuse that they can hardly bear to acknowledge the existence of their own genital or anal areas. It should be noted that forced sexual experience does not educate either in physiology or in norms of sexual behaviour. Indeed the abusiveness of the act is likely to inhibit the capacity to learn.

5. Poor sense of self

Children who have been abused have had their wishes overruled more or less violently. Helplessness reduces awareness of individual needs and wishes since such knowledge might be either dangerous or unbearably painful.

Victims may suffer a range of additional problems such as soiling, eating or sleeping difficulties, or post-traumatic symptoms like flashbacks, phobias and nightmares. In our experience most of these require more individual treatment although groupwork may identify other sufferers.

Mothers have to contend with additional problems. Managing and caring for a child who has been abused without over-compensating and failing to establish ordinary controls is difficult. They are often forced to do this having become unwillingly a single parent family.

There may practical problems arising from this, predominantly financial but also child-care based. Social isolation is common and a greater or lesser degree of rejection by extended family may occur.

Aims and Focus of Groupwork

1. Normalisation

Meeting with others in a like situation can help to reduce the sense of stigmatisation and isolation.

2. Increasing self-confidence and self-esteem
   a) improving self-assertion and open communication.
   b) modifying distorted beliefs and feelings.
   c) validating any feelings; work on painful feelings to decrease risks of self/other-destructiveness.
   d) facilitating members care for each other.

3. Work on sexual constructs

This includes straightforward education, correction of distortions and work on actual/prospective relationships.

4. Encouragement of individuation

Identifying differences as well as similarities and helping the group tolerate and appreciate these.

Practical Considerations

These issues need to be addressed prior to starting a group.

1. Resources for group work

Rooms, workers, reception facilities, supervision, consultation all need to be agreed and organised.

2. Membership of groups

We found that more than 6 children or adolescents could present problems both of control and more importantly of providing adequate emotional space for each member. Even a steady three could constitute a viable group. Adult groups could be somewhat larger.

With children under 8 we had boys and girls together; this is possible with 8–10 year olds but there may be advantages in this age group also being single sex. Over 10/11 the groups were always single sex.

It is very important to have at least two of any gender if the group was mixed. Similarly it is important to have two members from any racial group represented (therapists are also members of racial groups). It is necessary to consider any factors which may prevent a client from being an effective group member, eg, florid psychiatric disturbance, major learning disability.

We have found difficulties in mixing mothers where the abuser was a family member with those where abuse was extra-familial in the longer groups because the former have so many painful personal issues.

3. Co-therapy

The gender of workers must be considered. We would strongly advocate that two men never work together. We suggest that male workers do not have a role with mothers' groups. Groups containing children of both sexes may be helped by having a male as well as a female worker. On the whole we advocated women therapists for girls over 11 although older adolescents may benefit from a mixed pair. Groups of boy victims required one male therapist. Co-therapy is essential to provide more adult attention, more flexibility, better control and sharing for workers.

4. Duration

Groups for victims over age 10 lasted approximately four months. There were parallel groups for mothers and substitute carers. Groups for children between 8-10 lasted around 6-10 weeks and those for young children 6 weeks. There were parallel groups for the carers (mothers, foster parents and social workers combined).

The duration of the small childrens' groups was an hour; the groups for older victims lasted an hour and a quarter. All groups were held weekly.

5. Links with professional network

These were actively maintained throughout. They included negotiating who took responsibility for work outside the group and for organising/facilitating any practical arrangements eg, transport.

Group Activities

Children and young people, especially when deprived and confused often find it very difficult to talk. Young children developmentally require activity and cannot be expected to sit for an hour, let alone talk. Adults also learn and gain understanding in a variety of ways. Consequently, even if a group membership is articulate, we use a range of activities such as writing, drawing, role play, sculpting and watching videos. We have often found that even the most verbal members may share and confide more painful material in non-verbal ways.

Although the activities described below were often intended to address a specific issue experience has shown that a group might use it for other problems. The most successful activities may help deal with several issues. It should be emphasised that the primary purpose of any exercise was to help trigger sharing of personal problems, especially when this could not be done spontaneously. The way in which any activity was done and the details would vary according to the age and needs of members.

We found it helpful to give children between 8–14 notebooks in which they could write personal messages and in which therapists could respond to them.

The activities will be described in relation to the primary therapeutic focus intended for each.

Normalisation

1. Sharing personal experiences

This was not only the abuse, but also other painful events (eg, removal from home, illness, pregnancy, etc). This happened throughout group. Included positive experiences as well as traumas. Most often done verbally in an ongoing way but writing painful experiences often helped as it allowed some distancing. All ages, including adults.

2. Videos

There are a number of professionally made videos which are very helpful in showing how others have felt/coped. Some are suitable for young children, others for adolescents and some for adults. All ages including adults.

3. Agony letters

Finding letters in teenage magazines and discussing/answering. Alternatively members may write their own. This can identify commonly shared concerns, around looks, relationships, etc. It also gives the therapists opportunity to introduce issues which may not have arisen yet in group life. Ages 11 and upwards.

Increasing self-confidence and self-esteem

1. News

At the start of each group members brought a piece of good and bad news. This both allowed for discussion of important events in members' lives and helped them get to know each other better. Listening and sharing were encouraged. It seemed equally important for members to report very minor and mundane events since many victims find it very difficult to identify any small pleasures in their lives. Suitable for all age groups.

2. Paying a compliment

A good way of ending a group on a positive note, and possibly defusing some of the painful angry feelings aroused. Therapists could model by introducing compliments either about looks or about interactions if this was not happening spontaneously. Therapists took part in both these activities, a message that good and bad news and pleasure in being noticed and valued are universal experiences and needs. Suitable for children under 13.

3. Body image

This exercise involves members filling in, in any way they choose, a life size image of themselves. The intention is for them to create a picture of their feelings and even 10 year olds understand this task. While many victims find this very difficult and may finish with an almost empty body in our experience all have been able to indicate something; often the depth of their pain or despair with they had not been able to do using words and may not have indicated in their manner of interaction. Helpful both in informing therapists of needs and in building up trust and sharing among the group. All ages above 10; could be used with adults.

4. Name tags

This may be most useful for children as a way of helping them get to know each other. We have used it for adults at a later stage as a game where members indicate what they do know. Labels with descriptions (for younger children, picture faces) are given by members to each other and also to themselves to see if the feelings attributed by self and others match or not. Specific painful feelings commonly felt by victims such as dirty, damaged, should be included as well as a whole range of emotions. Members often chose to write additional labels. Ages 8 upwards.

Self-assertion and Self-protection

Self-assertion needs to be encouraged in two main ways. Firstly, members must have their anger validated and be given opportunities to practise standing up for themselves. They also need to be able to identify risk situations. Secondly, it is essential that they develop some trust that they will be listened to and some ability in confiding.

1. Expressing anger towards the abuser

The anger felt about the abuse and towards the abuser can be quite incapacitating. This can be so for non-abusive parents as well as victims. It is important to help members realise that it is a valid emotion and not one they should be ashamed of. Drawing pictures of the abuser, sharing fantasies of what they would like to do or say, using cushions, etc., as a punch ball, writing letters and rehearsing what they may actually say if they are going to see him can all be helpful. The latter has been particularly so for mothers who are going to meet a partner in prison. All age groups.

2. Expressing anger to others

Many victims have markedly poor peer relationships. Role plays of confrontations with peers and adults are helpful. It is particularly important to do these both in the context where the member wants to stay friendly with the proponent as well as in situations where they do not.

Assertion in relation to stranger adults is often difficult for mothers as well as children and young people and role plays of such situations can be helpful. It allows for discussion of the risks of behaving in an unreasonably aggressive ways because all the frustrated fury felt towards the abuser may be projected on to anyone who aggravates. All age groups.

3. Trigger pictures

There are a number of books showing a variety of interactions and situations which are useful in starting discussions about possible risks, normal contacts and ways of coping if worried. Suitable for children 10 and under.

4. Role plays

Role plays of how to deal with strangers can be helpful; for example someone who tries to offer a lift, someone who touched inappropriately in a train, or someone asking the way. As with the trigger pictures contexts which are not necessarily abusive can be included to allow for discussion about how to distinguish the inappropriate from the normal, how to behave when in doubt, etc. Ages 8–14.

5. Screaming

Social taboos against making oneself a focus of attention are very strong. It is helpful to give members the opportunity of using their voices as loudly as possible and discussing the way in which shouting, screaming, standing up, etc., are all assertive and may alarm possible assailants. Ages 8–14.

6. Confiding worries and problems

Rehearsing the sharing of worries is a core task in all victim groups. Older adolescents and adults may do this spontaneously; if not then some kind of role play may help.

Given that most victims have kept a painful secret for a considerable time and that many meet with some rejection when they do tell this is a very difficult exercise. Children may experience some catharsis if they are instructed to role play a bad listener before they role play a good one. They also show themselves much more able to be positively helpful to the 'actors' than when acting themselves. Ages 5–15.

7. Storybuilding

The group tells a story about abuse taking it in turns and dividing into subgroups to provide a variety of endings. Gives opportunity to confirm the rightness of disclosure and to discuss what should happen to abuser as well as to victims. Also chance to hear about wider family responses. Ages 8–14.

Sexual Constructs

1. Education
Straightforward education, using books, diagrams, drawings and discussion. Normality of genital pleasure can be confirmed, as well as discussion of other pleasurable touches. Information about procreation for all except the youngest age groups. Contraception discussed from 11 onwards, more in adolescence. Health issues around sexuality for adolescents and adults. Discussion of the variety of sexual choices which people make and validation of those choices. All ages, differentially.

2. Sexual relationships
Need to emphasise the importance of choice and identify vulnerability to reabuse. Discussion of issues such as clothes and sexual posture and relate to messages which may be given, whether intended or not. For young adolescents role playing a relationship with a friend who wants to touch more than player does can be helpful. Adolescents and adults will spend far more time discussing this area.

3. Pregnancy
This is likely to be most discussed by adolescents, some of whom may already have had children or terminations. Important to try and establish whether those who are expressing a determination never to have children are in fact expressing a horror of any sexual relationship.

4. Games about sexuality
Board games and quizzes can be fun in this area as well as introducing further information. Suitable for adults as well as youngsters.

5. Abusing behaviour
Adults, adolescents and even children from 11 up raise the question 'why?'. Some discussion of why and education about the grooming behaviour of many abusers, as well as of treatment possibilities and uncertainties of success. Important to include discussion of abusive behaviour to adults or peer partners as well as to children.

Individuation

Throughout the life of the group it is important to allow for differences especially when one group member holds a view which is opposite to that held by all the others, eg, a mother who wishes to reunite with the abuser, or a child who has no anger.

1. Future biographies

Everyone is asked to write a brief account of how they think they would like their lives to be in say, 10 years, and to be realistic rather than bring dreams. Share a hope and asking the group to discuss the qualities that person has which will help them achieve their wish. Suitable for 11 upwards.

2. Parenting

Need for discussion of the qualities which make for 'good' and 'bad' parenting. Members can then discuss in relation to their own parents if they wish; what they would do the same and what different. For older adolescents and adults the demands which may appear particularly burdensome and pleasurable can be shared.

3. Sculpting

This could be helpful for mothers as a method of portraying their family, either of birth or procreation in order to get a new light on the relationships and interactions.

Specimen Programmes for Groups

4–8 years. 6 sessions

1. Meeting strangers safely. Good, bad and "yucky" touches.
2. Private parts of body; identify different touches.
3. Recap of touching. Discussion of secrets. Rehearsal of standing up to peers and asserting oneself.
4. Rehearsal of saying 'go away' and discussion of risky situations.
5. Telling worries; role play.
6. Recap of all work done; identification of who you would tell a worry to in future.

Picture cards, dolls, drawing materials and plasticine all may be helpful for these age groups.

8–10 years. 8 sessions

1. Introductions, who you live with, trust games.
2. Draw the perpetrator, say his relationship to you; discuss and enact what you would like to do to him/her.
3. Draw yourself before and after disclosure. Discuss feelings. List unpleasant experiences.
4. Bring photographs, etc., of people, places and pets you like to share pleasant experiences. Talk of hopes.
5. Nice and not nice touches, private parts and body feelings.
6. Saying no. Practice standing up for yourself with other children and perhaps adults. Screaming, possible use of trigger pictures to discuss risky situations.
7. Role play confiding worries with a child or grown up.
8. Using cards or mime share ideas about what makes a good parent and a bad one. Identify who you would tell a worry to in future. Say goodbye.

This programme can easily be extended to last 10 weeks or more. Additional time can be spent on sessions 6 and 7 and other topics particularly pertinent to a given group included.

Programme for 11–15 year olds. 16 sessions.

It is desirable to divide this age group if possible into two bands one from 11–13 and the other from 14–15. The activities may be done differently according to age. They may also be used with older adolescents if wished.

In the first session any group rules—eg, no damage to be inflicted on people or furnishings—spelt out. Also confidentiality and its limits clarified, namely that confidentiality would only be broken if the leaders heard something which showed that a member was at some risk.

1. Introductions, who you live with, shared interests, trust games, sharing of abuse if the groups seems ready.
2. How hard was it to tell? Keeping secrets and why. Who knows now and is trusted.
3. Body image exercise.
4. Sharing range of bad and upsetting experiences.
5. What are the effects of abuse; the biggest worry now. Looking at the range of difficulties and problems being experienced.
6. Drawing your abuser. What would you say to him now? Rehearsing that if possible in role play, writing, etc. Discussing what should have happened to him.
7. Self-assertion; dealing with strangers, risky or difficult situations. Screaming.

8. Knowledge of body function and sexuality; an educative session although can be done in various ways.
9. Game; eg, quiz, grapevine, etc.
10. Sharing worries; rehearsal of telling worry to someone, discussion of how to deal with hearing problems.
11. Growing up and relationships. How to deal with problems. Discussion of interests which may be shared by both boys and girls, other than sex.
12. Agony letters.
13. Planning for the future; future biographies.
14. Qualities which make for good mothers/fathers and reverse. Discussion of parenting and anxieties.
15. Watching video of survivors, etc.
16. Endings, future needs, recap dealing with any issue which requires more time.

In a 4 month group it is likely that there will be a natural holiday break at some point. We have found it best to have this about half way through if possible.

As described in the introduction, all groups start with "news" each week.

Groups for Mothers and non-abusive parents

Although the groups for mothers do not have as structured a programme as those for victims, there are a number of crucial issues which always arise and have to be addressed. As mentioned above, a range of activities may be helpfully employed such as sculpting, role play and rehearsal, drawing and quiz games. These can both facilitate discussion in particularly painful areas and may enhance learning and understanding as well as adding enjoyment.

The issues which commonly are central are:
1. Mother's (or non-abusive parents') relationship with the abuser.
2. Guilt.
3. Anger.
4. Pain and loss.
5. Relationship with victim, including rivalry as well as love.
6. Coping with difficult behaviours, and fears in victim.
7. Victim's behaviour before and since disclosure.
8. Sexualised behaviour in victim.
9. The victim's friends, victim's fears about growing up, her body image, developing sexual relationships.
10. Relationship with the natural father of victim if he is not the abuser.

11. Telling the extended family and/or friends.
12. Tackling social isolation.
13. Practical problems of being a single parent, and/or arising out of the abuse, eg, finance, change of housing, etc.
14. Non-abusive parents' own history, including any abuse ever experienced.
15. Partner choices; sexual anxieties; marital relationship.
16. Why did it happen?
17. Grooming by offenders.
18. Help for offenders or what?

Groups for Carers

These are groups run for substitute carers, foster parents, residential workers and social workers who bring children to their groups. Although the agenda will vary with the membership and particular concerns of any group there are again a number of issues which regularly arise.

These groups are also used to let the adults know in general terms the activity/topic planned for the concurrent girls group.

1. Abuse; range and variations.
2. Why and how it occurs; examining models explaining offender behaviour, grooming, etc.
3. What maintains secrecy prior to disclosure. Pressure to retract.
4. Identifying signs of continued secrecy. Balancing victim's right to privacy with concern that secrecy or dysfunctional inhibition is occurring.
5. Coping with difficult behaviour.
6. Coping with fears, phobias, etc.
7. Coping with suicidal or self-damaging behaviour.
8. Sexualised behaviour; how to help with children who are either eroticised or phobically anxious.
9. Physical contact between carer and victim; possible modifications of family life style which may be required in order to minimise vulnerability and confusion.
10. Consequences of sexual abuse.
11. Talking to families and within families.
12. Protecting the victims.
13. Legal aspects of child sexual abuse and possible implications for victims and carers.
14. Personal impact on the carer and his/her own family and network; how to look after ourselves.

References

ACHENBACH, T.M. & EDELBROCK, C.S. (1983) *Manual for the Child Behavior Checklist and revised Child Behavior Profile.* University of Vermont, Department of Psychiatry, Burlington, VT.

ADAMS-TUCKER, C (1982) Proximate effects of sexual abuse in childhood: report on 28 children. American Journal of Psychiatry 139, 1252–1256.

American Academy of Child & Adolescent Psychiatry (1988) Guidelines for the clinical evaluation of child and adolescent abuse. Position Statement of the AACAP Journal of the American Academy of Child & Adolescent Psychiatry.

American Association for Protecting Children (1986) *Highlights of Official Child Neglect & Abuse Reporting, 1984.* Denver, CO: American Humane Association.

BAKER, A.W. & DUNCAN, S.P. (1985) Child sexual abuse: a study of prevalence in Great Britain. Child Abuse & Neglect, **9**: 457–467.

BARRNETT, R.J, DOCHERTY, J.P, & FROMMELT, G.E. (1991) Review of child psychotherapy research since 1963. J American Academy Child Adolesc Psychiatry 30:1–14.

BEITCHMAN, J.H., ZUCKER, K.J., HOOD, J.E., DA COSTA, G. & AKMAN, D. (1991) Review of the short-term effects of child sexual abuse. Child Abuse & Neglect, 15: 537–556.

BENE, E. & ANTHONY, J. (1978) Manual for the Children's Version if the Family Relations Test—revised. Windsor, UK: NFER Publishing Company.

BENTOVIM, A. (1991) Clinical work with families in which sexual abuse has occurred. In CR Hollin & K. Howells (eds), *Clinical approaches to sexual offenders and their victims.* London, UK: John Wiley & Sons.

BENTOVIM, A, & BOSTON, P. (1988) Sexual Abuse—basic issues—characteristics of children & families. In A.Bentovim, A.Elton, J.Hildebrand, M.Tranter, E.Vizard (eds), *Child sexual abuse within the family.* London, UK: Wright.

BENTOVIM, A, VAN ELBERG, A, & BOSTON, P (1988) The results of treatment. In A.Bentovim, A.Elton, J.Hildebrand, M.Tranter, E.Vizard (eds), *Child sexual abuse within the family.* London, UK: Wright.

BENTOVIM A, ELTON A, HILDEBRAND J, TRANTER M, & VIZARD E. (1988) *Child sexual abuse within the family*. London, UK: Wright.

BERLINER, L. (1991) Clinical work with sexually abused children. In CR. Hollin & K. Howell (eds), *Clinical approaches to sex offenders and their victims*, Chichester, UK: John Wiley & Sons Ltd.

BERLINER, L. & BARBIERI, M.K. (1984) The testimony of the child victim of sexual assault. J Social Issues 40: 125–137.

BERLINER, L. & CONTE, J. (1995) The effects of disclosure and interventions on sexually abused children. Child Abuse & Neglect, 19: 371–384.

BRIERE, J. (1987) The long-term clinical correlates of childhood sexual victimization. Invited Paper: New York Academy of Sciences Conference: Human sexual aggression: current perspectives.

BRIERE, J. & RUNTZ M (1988) Symptomatology associated with childhood sexual victimization in a non-clinical adult sample. Child Abuse & Neglect, 12: 51–59.

BROWN, G. (1989) Life events and measurement. In GW. Brown & TO. Harris (eds), *Life Events and Illness*. London, UK: Unwin Hyman.

BURGESS, A.W., HOLMSTROM, L.L., & MCCAUSLAND, M.P. (1978). In: A.W. Burgess, A.N. Groth, L.L. Homstrom & S.M. Sgroi (eds), *Counseling young victims and their families*. Lexington, M.A.: Lexington Books.

COLLINGS, S.J. (1995) The long-term effects of contact and noncontact forms of child sexual abuse in a sample of university men. Child Abuse & Neglect, 19: 1–6.

Committee on Sexual Offenses against Children & Youth. (1986) *Sexual offenses against children*. Ottawa, Canada: Canadian Publishing Centre.

CONTE, J.R & SCHUERMAN, J.R. (1987) Factors associated with an increased impact of child sexual abuse. Child Abuse & Neglect, 11: 201–211.

COURTOIS, C. & WATTS, D. (1982) Counselling adult women who had experienced incest in childhood or adolescence. Personnel & Guidance Journal, 275–279.

CREIGHTON, S. (1992) *Child abuse trends in England & Wales 1988–1990*. London: National Society for the Prevention of Cruelty to Children (p64, Figure 7).

CREIGHTON S & NOYES, P. (1989) *Child abuse trends in England & Wales 1983–1987*. London: NSPCC.

DEBLINGER, E., MCLEER, S.V. & HENRY, D. (1990) Cognitive behavioral treatment for sexually abused children suffering post-traumatic stress: preliminary findings. J American Academy Child Adolescent Psychiatry, 29: 747–752.

DE JONG, A.R., HERVADA, A.R. & EMMETT, G.A. (1983) Epidemiologic variations in childhood sexual abuse. Child Abuse & Neglect, 7: 155–162.

DE LUCA, R.V., BOYES, D.A., GRAYSTON, A.D. & ROMANO, E. (1995) Sexual abuse: effects of group therapy on pre-adolescent girls. Child Abuse Review, 4: 263–277.

Department of Health (1990) Children and young persons on child protection registers (year ending March 1990). London: Department of Health.

DEVINE, R.A. (1980) The sexually abused child in the emergency room. In: Jones, BM, Jenstrom, LL. & MacFarlane, K. (ed), Sexual abuse of children: selected readings. DHSS Publication No. OHDS 78–30161. Washington, DC: Government Printing Office.

DIETZ, C.A. & CRAFT, J.L. (1980) Family dynamics of incest: a new perspective. Social Casework: the Journal of Contemporary Social Work, 61: 602–609.

DUBOWITZ, H. (1990) Costs and effectiveness of interventions in child maltreatment. Child Abuse & Neglect, 14: 177–186.

ECKENRODE, J., MUNSCH, J., POWERS, J. & DORIS, J. (1988) The nature and substantiation of official sexual abuse reports. Child Abuse & Neglect, 12: 311–319.

EVERILL, J. & WALLER, G. (1995) Disclosure of sexual abuse and psychological adjustment in female undergraduates. Child Abuse & Neglect, 19: 93–100.

FALLER, (1987) Women who sexually abuse children. Victims & Violence, 2: 263–276.

FINKELHOR, D. (1979) *Sexually victimised children*. New York: Free Press.

FINKELHOR, D. (1984) *Child sexual abuse: new theory and research*. New York: The Free Press.

FINKELHOR, D. (1987) The trauma of sexual abuse: two models. J. Interpersonal Violence, 2: 348–366.

FINKELHOR, D. & BERLINER, L. (1995) Research on the treatment of sexually abused children: a review and recommendations. Journal of the American Academy of Child & Adolescent Psychiatry, **34**: 1–16.

FRIEDRICH, W.N. (1988) Behavior problems in sexually abused children: an adaptational perspective. In GE.Wyatt & GJ.Powell (eds), *Lasting effects of child sexual abuse*. Newbury Park, CA: Sage.

FRIEDRICH, W.N., LUECK, W.J., BEILKE, R.L. & PLACE, V. (1992) Psychotherapy outcome of sexually abused boys. J Interpersonal Violence, 7: 396–409.

FRIEDRICH, W.N., URQUIZA, A.J. & BEILKE, R. (1986) Behaviour problems in sexually abused young children. J. of Pediatric Psychology, 11: 47–57.

FRITZ, G., STOLL, K. & WAGNER, N. (1981) A comparison of males and females who were sexually molested as children. J. Sex & Marital Therapy, 7: 54–59.

FROMUTH, M.E. (1983) The long-term psychological impact of childhood sexual abuse. PhD dissertation, Auburn University, Australia. Quoted by D. Finkelhor (1979) op.cit.

FROMUTH, M.E. (1986) The relationship of child sexual abuse with later psychological and sexual adjustment in a sample of college men. Child Abuse & Neglect, 10: 5–15.

FURNISS, T. (1983) Family process in the treatment of intra-familial child sexual abuse. J. of Family Therapy, 4: 263–279.

GALE, S., THOMSON, R., MORAN, T. & SACK, W. (1988) Sexual abuse in young children: its clinical presentation and characteristic patterns. Child Abuse & Neglect, 12: 163–170.

GIARETTO, H. (1981) A comprehensive child sexual abuse program. In P.B.Mrazek & C.H.Kempe (eds), *Sexually abused children and their families.* London, UK: Pergamon.

GLASER, D. (1991) Treatment issues in child sexual abuse. British J. of Psychiatry, 159: 769–782.

GOLD, E.R. (1986) Long-term effects of sexual victimization in childhood: an attributional approach. J. Consulting & Clinical Psychology, 54: 471–475.

GOLDBERG D (1978) *Manual of the General Health Questionnaire.* Windsor, UK: NFER Nelson.

GOLDSTEIN, J., Freud A. & Solnit, A.J. (1979) *Before the best interests of the child.* New York: Free Press.

GOMES-SCHWARTZ B, HOROWITZ M & CARDARELLI AP, (1990) *Child Sexual Abuse: the initial effects.* Newbury Park, CA: Sage.

GOODYER, I. (1990) Family relationships, life events and child psychopathology. J. Child Psychology & Psychiatry, 31: 161–192.

GREENWALD, E., LEITENBERG, H., CADO, S. & TARRAN, M.J. (1990) Childhood sexual abuse: longterm effects on psychological and sexual functioning in a non-clinical and non-student sample of adult women. Child Abuse & Neglect, 14: 503–513.

GROTH AN (1978) Sexual trauma in the life histories of rapists and child molesters. Victimology, 4: 6–10.

HALL, R.C.W., TICE, L., BERESFORD, T.P., WOOLEY, B. & HALL, A.K. (1989) Sexual abuse in patients with anorexia nervosa and bulimia. Psychosomatics, 30: 73–79.

HARTER S (1985) *The self-perception profile for children.* Denver, CO: University of Denver.

HARTER S (1987) *The self-perception profile for adolescents.* Denver, CO: University of Denver.

HARTER S & PIKE R (1982) The pictorial scale of perceived competence and social acceptance for young children. *Child Development,* 55: 1969–1982.

HARTMAN, C.R. & BURGESS, A.W. (1989) Sexual abuse of children: causes and consequences. In D. Cicchetti & V. Carlson (eds), *Child Maltreatment: theory and research on the causes and consequences of child abuse & neglect.* Cambridge, UK: Cambridge University Press.

HOBBS C (1990) Child Sexual Abuse—paediatric aspects. In The Consequences of Child Sexual Abuse: Association for Child Psychology and Psychiatry, Occasional Papers No 3.

HOBBS C. & WYNNE, J. (1990) The sexually abused battered child. Archives of Disease in Childhood, 65: 423–427.

HOOPER, C-A. (1992). *Mothers surviving child sexual abuse.* London & New York: Tavistock/Routledge.

HYDE C, BENTOVIM A, MONCK E (1995) Some clinical implications of a treatment outcome study of sexually abused children. Child Abuse & Neglect 19: 1387–1397.

JOHNSON, B.K., & KENKEL, M.B. (1991) Stress, coping and adjustment in female adolescent incest victims. Child Abuse & Neglect 15: 293–305.

JONES, D.P.H. (1988) Individual psychotherapy for the sexually abused child. Child Abuse & Neglect, 10: 377–385.

KELLY, L., REGAN, L. & BURTON, S. (1995) *An exploratory study of the prevalence of sexual abuse in a sample of 16–21 year olds.* London, UK: HMSO

KITCHUR, M. & BELL, R. (1989) Group psychotherapy with pre-adolescent sexual abuse victims: literature review and description of an inner-city group. International Journal of Group Psychotherapy, 39: 285–310.

KOLKO, O.J, MOSNER, J.T, WELDY, S.R. (1988) Behavioural/emotional indicators of sexual abuse in child psychiatric in-patients: a controlled study comparison with physical abuse. Child Abuse & Neglect, 12: 529–541.

KOVACS, M. & BECK, A.T. (1977) An empirical approach towards a definition of childhood depression. In J.G. Schulterbrandt & A. Raskin (ed), *Depression in children: diagnosis, treatment and conceptual models.* New York: Raven.

LOVE, A.J. (1989) Process and outcome: evaluation of the sexual abuse treatment project. Early Child Development & Care 42: 113–125.

MACFARLANE, K. (1978) Sexual abuse of children. In: JR.Chapman & M. Gates (eds), *The Victimization of Women.* Newbury Park, CA: Sage.

MCGUIRE, I.S. & WAGNER, N.N. (1978) Sexual dysfunction in women who were molested as children: response patterns and suggestions for treatment. Journal of Sex & Marital Therapy, 4: 11–15.

MCLEER, S.V., DEBLINGER, E., HENRY, D. & ORVASCHEL, H. (1992) Sexually abused children at high risk for post-traumatic stress disorder. J. American Academy Child & Adolescent Psychiatry, 31: 875–879.

MACVICAR, K. (1979) Psychotherapeutic issues in the treatment of sexually abused girls. J. American Academy of Child Psychiatry 19: 342–353.

MADDOCK, J.W., LARSON, P.R. & LALLY, C.F. (1991) An evaluation protocol for incest family functioning. In: MQ.Patton (ed), *Family Sexual Abuse.* Newbury Park, CA: Sage.

MADONNA, P.G., VAN SCOYK S. & JONES, D.P.H. (1991) Family interactions within incest and non-incest families. American Journal of Psychiatry, 148: 46–49.

MANNARINO, A.P. & COHEN, J.A. (1986) A clinical-demographic study of sexually abused children. Child Abuse & Neglect, 10: 17–23.

MANNARINO, A.P. & COHEN, J.A. (1987) Psychological symptoms of sexually abused children. Paper presented at the Third National conference of Family Violence Research, U. of New Hampshire, July 1987.

MANNARINO, A.P, COHEN, J.A, SMITH, J.A, & MOORE.-MOTILY, S. (1991) Six and twelve-month follow-up of sexually abused girls. *J Interpersonal Violence* 6:494–511.

MARKOWE, H.L.J. (1988) The frequency of childhood sexual abuse in the UK. Health Trends, 20: 2–6.

MEISELMAN, K.C. (1978) Incest: a psychological study of causes and effects with treatment recommendations. San Francisco, CA: Jossey-Bass.

MESSER, B. & HARTER, S. (1986) *The adult self-perception profile.* Denver, CO: University of Denver.

MORRISON, J. (1989) Childhood sexual histories of women with Somatization disorder. American J. Psychiatry, 146: 239–241.

Mrazek, P.J., Bentovim, A. & Lynch, M. (1983) Sexual abuse of children in the United Kingdom. Child Abuse & Neglect, 7: 147–153.

Mullen, P.E., Romans-Clarkson, S.E., Walton, V.A. & Herbison, G.P. (1988) Impact of sexual abuse on women's mental health. Lancet, 1: 841–845.

NCH (National Children's Homes) (1992) The Report of the Committee of Enquiry into Children and Young People who Sexually Abuse Other Children. Introduction (p. v–vi). London: NCH.

Office of Population Censuses and Surveys (OPCS) (1980) Classification of Occupations. London: HMSO.

Peters, S.D. (1988) Child sexual abuse and later psychological problems. In GE. Wyatt & GJ. Powell (eds), *Lasting effects of child sexual abuse*. Newbury Park, CA: Sage.

Peters, S.D., Wyatt, G.E. & Finkelhor, D. (1986). Prevalence. In D. Finkelhor (ed), *A Sourcebook on Sexual Abuse*. Newbury Park, CA: Sage.

Peters, DK. & Range, LM. (1996) Childhood sexual abuse and current suicidality in college women and men. Child Abuse & Neglect, 19: 335–341.

Research Team, The (1990) *Child sexual abuse in Northern Ireland*. Antrim, NI: Greystone Books.

Resick, P.A., Calhoun, K.S., Atkeson, B.N. & Ellis, E.M. (1981) Social adjustment in victims of sexual assault. J. Consulting & Clinical Psychology, 49: 705–712.

Rimza M.E., Berg, R.A. & Locke, C. (1988) Sexual abuse: Somatic and emotional reactions. Child Abuse & Neglect, 12: 201–208.

Rogers, C. & Terry, T. (1984) Clinical intervention with boy victims of sexual abuse. In I. Stuart & J. Greer (eds), *Victims of sexual aggression*. New York: Van Nostrand & Reinhold.

Rosenberg, M. (1965) *Society and adolescent self-image*. Princeton University Press.

Russell, D. (1984) *Rape, incest and sexual exploitation*. Newbury Park, CA: Sage.

Rutter M (1967) A children's behaviour checklist for completion by teachers: preliminary findings. J. Child Psychology & Psychiatry, 8: 1–11.

Rutter, M. (1983) Some issues and some questions. In N. Garmezy & M. Rutter (ed), *Stress, coping and development in children*. New York: McGraw Hill.

RUTTER, M (1985) Resilience in the face of adversity: protective factors and resistance to psychiatric disorder. British J. of Psychiatry, 147: 598–611.

SALZINGER S, KAPLAN S, PELCOVITZ D, SAMIT C, KRIEGER R (1984). Parent and teacher assessment of children's behaviour in child maltreating families. J. American Academy Child & Adolescent Psychiatry, 23: 458–464.

SAUZIER, M., SALT, P. & CALHOUN, R. (1990) The effects of sexual abuse. In: B. Gomes-Schwartz, JM. Horowitz & AP. Cardarelli, *Child Sexual Abuse: the initial effects.* Newbury Park, CA: Sage.

SEDNEY, M.A. & BROOKS, B. (1984) Factors associated with a history of childhood sexual experience in a nonclinical female population. J. American Academy of Child Psychiatry, 23: 215–218.

SGROI, S.M. (1982) Handbook of Clinical Intervention in Child Sexual Abuse. Lexington, M.A.: Lexington Books.

SILBERT, M.H. & PINES, A.M. (1981) Sexual abuse as an antecedent to prostitution. Child Abuse & Neglect, 5: 407–411.

SIRLES, E.A. & SMITH, J.A. (1990). Behavioral profiles of pre-school, latency and teenage female incest victims. Paper given at the Eighth International Congress of Child Abuse & Neglect, Hamburg, September, 1990.

SMITH, M. & JENKINS, J. (1991) The effects of marital disharmony on prepubertal children. J. Abnormal Child Psychology, 19: 625–639.

SMITH, M. & GROCKE, M. (1995) *Normal Family Sexuality and Sexual Knowledge in Children.* Royal College of Psychiatrists/ Gorkill Press.

STEIN, J.A., GOLDING, J.M., SIEGAL, J.M., BURNAM, M.A. & SORENSON, S.B. (1988) Long-term psychological sequelae of child sexual abuse: The Los Angeles Epidemiologic Catchment Area Study. In GE. Wyatt & GJ. Powell (ed), *Lasting effects of child sexual abuse.* Newbury Park, CA: Sage.

SUDMAN, S. & BRADBURN, N. (1973) Effects of time and memory factors on response in surveys. J. American Statistical Association, 68: 344.

SUMMIT, R. & KRYSO, J. (1978) Sexual abuse of children: a clinical spectrum. American Journal of Orthopsychiatry, 48: 237–251.

TAITZ, L.S., KING, J.M., NICHOLSON, J. & KESSEL, M. (1987) Unemployment and child abuse. British Medical Journal, 294: 1074–1076.

TERR LC (1991) Childhood traumas: an outline and an over view. American J Psychiatry, 148: 10–20.

TONG, L., OATES, K. & MCDOWELL, M. (1987) Personality development following sexual abuse. Child Abuse & Neglect, 11: 371–383.

WALKER, E.A., KATON, W.J., HANSOM, J., HARROP-GRIFFITHS, J., HOLM, L., JONES, M.L., HICKOK, L. & JEMELKA, R.P. (1992) Medical and psychiatric symptoms in women with childhood sexual abuse. Psychosomatic Medicine, 54: 658–664.

WALLER, G. (1991) Sexual abuse as a factor in eating disorders. British J. of Psychiatry, 159: 664–671.

WATTAM, C. (1992) *Making a case in child protection.* London, UK: Longman.

WEISS, E.H.& BERG, R.F. (1982) Child victims of sexual assault: impact of court procedures. J. American Academy Child Psychiatry, 21: 513–518.

WHITE, S., HALPIN, B., STROM, G. & SANTELLI, G. (1988) Behavioral comparisons of young sexually abused, neglected and non-referred children. J. Clinical Child Psychology, 17: 53–61.

WOODWORTH, D.L. (1991) Evaluation of a multiple-family incest program. M.G.Patton (ed): *Family sexual abuse.* Newbury Park, CA: Sage

WOZENCRAFT, T. WAGNER, W. & PELLEGRINI, A. (1991) Depression and suicidal ideation in sexually abused children. Child Abuse & Neglect, 15: 505–511.